D0917112

THE ALKALOIDS

Chemistry and Pharmacology

Volume XXII

THE ALKALOIDS

Chemistry and Pharmacology

Edited by
Arnold Brossi
National Institutes of Health
Bethesda, Maryland

Founding Editor
R. H. F. Manske

VOLUME XXII

1983

ACADEMIC PRESS
A Subsidiary of Harcourt Brace Jovanovich, Publishers

New York ● London
Paris ● San Diego ● San Francisco ● São Paulo ● Sydney ● Tokyo ● Toronto

ACADEMIC PRESS, INC.
111 Fifth Avenue, New York, New York 10003

United Kingdom Edition published by
ACADEMIC PRESS, INC. (LONDON) LTD.
24/28 Oval Road, London NW1 7DX

LIBRARY OF CONGRESS CATALOG CARD NUMBER: 50-5522

ISBN 0-12-469522-1

PRINTED IN THE UNITED STATES OF AMERICA

83 84 85 86   9 8 7 6 5 4 3 2 1

195938

CONTENTS

Chapter 1. Ipecac Alkaloids and β-Carboline Congeners

TOZO FUJII AND MASASHI OHBA

Chapter 2. Elucidation of Structural Formula, Configuration, and Conformation of Alkaloids by X-Ray Diffraction

ISABELLA L. KARLE

vi

CONTENTS

LIST OF CONTRIBUTORS

Numbers in parentheses indicate the pages on which the authors' contributions begin.

H. C. BEYERMAN (281), Laboratory of Organic Chemistry, Delft University of Technology, 2628 BL Delft, The Netherlands

TOZO FUJII (1), Faculty of Pharmaceutical Sciences, Kanazawa University, Kanazawa 920, Japan

ARMIN GUGGISBERG (85), Organisch-chemisches Institut der Universität Zürich, CH-8057 Zürich, Switzerland

MANFRED HESSE (85), Organisch-chemisches Institut der Universität Zürich, CH-8057 Zürich, Switzerland

ISABELLA L. KARLE (51), Laboratory for the Structure of Matter, Naval Research Laboratory, Washington, D.C. 20375

L. MAAT (281), Laboratory of Organic Chemistry, Delft University of Technology, 2628 BL Delft, The Netherlands

TAKEAKI NAITO (189), Kobe Women's College of Pharmacy, Kobe, Japan

ICHIYA NINOMIYA (189), Kobe Women's College of Pharmacy, Kobe, Japan

MASASHI OHBA(1), Faculty of Pharmaceutical Sciences, Kanazawa University, Kanazawa 920, Japan

PREFACE

Sophisticated physical techniques have had a major impact on alkaloid research, particularly on the determination of novel structures. Progress in this area will be reviewed periodically, and the chapter on "Elucidation of Structural Formula, Configuration, and Conformation of Alkaloids by X-Ray Diffraction" is the first of a series of such reviews. Attention is given to chemical methodology, found to be most valuable in alkaloid synthesis, presented here in the chapter on "Application of Enamide Photocyclization in Alkaloid Synthesis." Of a more traditional nature is the presentation of alkaloid classes, which have not been reviewed for many years and where considerable new material has accumulated. Hence, two chapters, namely, "The Imidazole Alkaloids," last reviewed in Vol. III (1953) and "Ipecac Alkaloids and β-Carboline Congeners," reviewed in Vol. XIII (1971), are included.

An interesting group of alkaloid substances, not previously discussed in these volumes, are the polyamine alkaloids; they are reviewed in a chapter entitled "Putrescine, Spermidine, Spermine, and Related Polyamine Alkaloids." Recently available books on alkaloids of a general nature, primarily intended for students of natural products, are not going to replace a treatise that continues to be detailed, complete, and up-to-date.

Arnold Brossi

CONTENTS OF PREVIOUS VOLUMES

Contents of Volume IV (1954)

edited by R. H. F. Manske and H. L. Holmes

Contents of Volume V (1955)

edited by R. H. F. Manske

Contents of Volume VI (1960)

edited by R. H. F. Manske

Contents of Volume IX (1967)

edited by R. H. F. Manske and H. L. Holmes

Contents of Volume X (1967)

edited by R. H. F. Manske and H. L. Holmes

Contents of Volume XI (1968)

edited by R. H. F. Manske and H. L. Holmes

Contents of Volume XII (1970)

edited by R. H. F. Manske and H. L. Holmes

Contents of Volume XIII (1971)

edited by R. H. F. Manske and H. L. Holmes

Contents of Volume XIV (1973)

edited by R. H. F. Manske and H. L. Holmes

Contents of Volume XV (1975)

edited by R. H. F. Manske and H. L. Holmes

Contents of Volume XVI (1977)

edited by R. H. F. Manske and H. L. Holmes

—— Chapter 1 ——

IPECAC ALKALOIDS AND β-CARBOLINE CONGENERS

Tozo Fujii and Masashi Ohba

Faculty of Pharmaceutical Sciences,
Kanazawa University, Kanazawa, Japan

I. Introduction

There have been three earlier chapters on the ipecac alkaloids in this series (*1–3*), the latest having appeared in Volume XIII in 1971 (*3*) with coverage of the analogs and β-carboline congeners isolated from the Indian plant *Alangium lamarckii* Thwaites. Since that time special phases of the subject have been treated in an unclassified form in Volume XIV, page 508 and Volume XVI, pages 513 and 531. A few reviews are also pertinent to the discussion before 1972 (*4–6*), and a supplementary one has covered the period from 1972 to mid-1977 (*7*). The significant recent advances in structure establishment, stereochemistry, and synthesis of the *Alangium* alkaloids, structurally related to the ipecac alkaloids, emphasize the need for the present chapter, which supplements the previous one (*3*) by updating the literature through mid-1982.

THE ALKALOIDS, VOL. XXII
Copyright © 1983 by Academic Press, Inc.
All rights of reproduction in any form reserved.
ISBN 0-12-469522-1

TOZO FUJII AND MASASHI OHBA

By 1971 synthetic routes to the main groups of the ipecac alkaloids and the β-carboline congeners had been well established, but during the last decade many studies were directed toward synthesis, featuring the use of new approaches and synthetic strategies that might be adaptable to the synthesis of structurally analogous bases such as the *Corynanthe*-type indoloquinolizidine alkaloids. Considerable progress has also been made in understanding the biosynthesis of the ipecac and *Alangium* alkaloids and in the isolation of about ten either new or previously known bases from *A. lamarckii* that do not fall within the general category of the ipecac alkaloids and β-carboline congeners but will be referred to briefly later. The sections

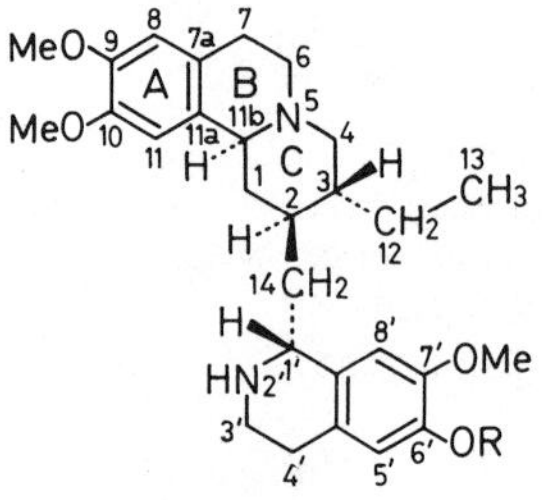

1 Emetine*,[†] R = Me

2 Cephaeline*,[†] R = H

3 Isocephaeline[†,‖] R = H
(1′α-H instead of 1′β-H)

4 Alangamide[†] R = H
MeNH—CO— for 2′-H
(possibly an artifact)

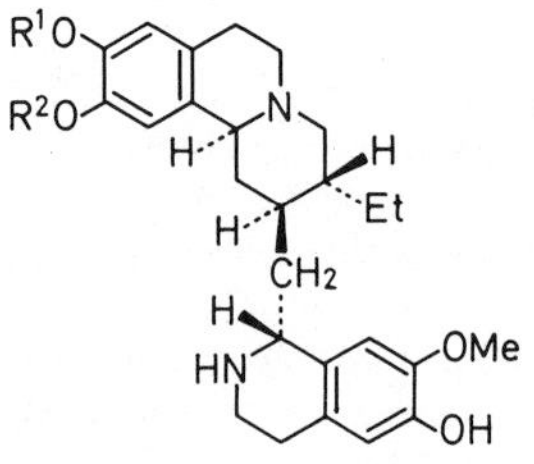

5 Demethylcephaeline[†,‡]
R[1] or R[2] = Me, R[2] or
R[1] = H

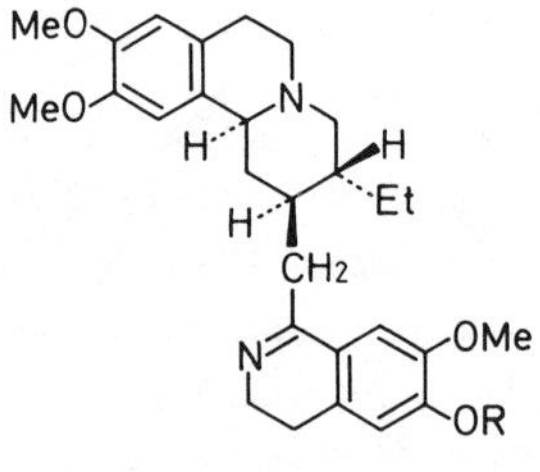

6 Psychotrine*,[†] R = H

7 O-Methylpsychotrine*
R = Me

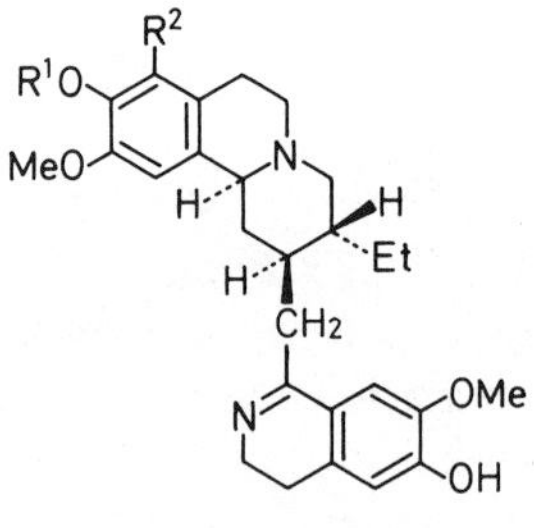

8 Demethylpsychotrine[†,§]
R[1] = R[2] = H

9 Alangicine[†,§] R[1] = Me,
R[2] = OH

---

[1] For the next few pages: * Ipecac alkaloids, † *Alangium* alkaloids, ‡ New structure, § Revised or established structure, ‖ Newly isolated structure.

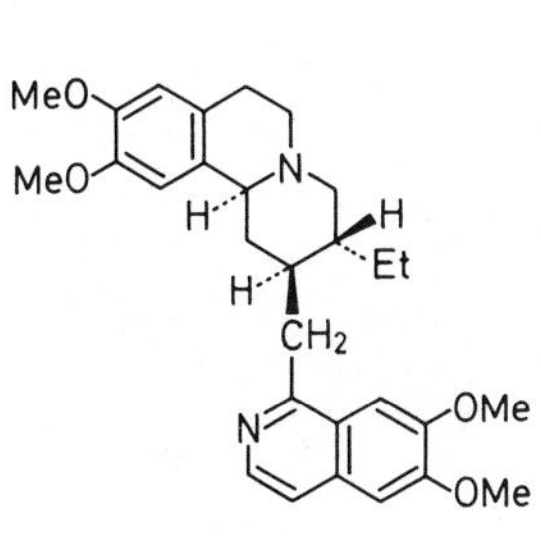

10 Emetamine*

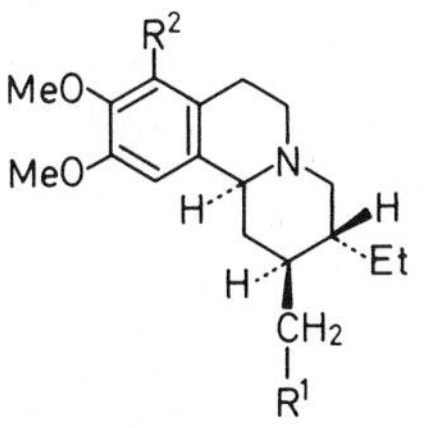

11 Protoemetine* R¹ = CHO,
R² = H

12 Protoemetinol† R¹ = CH₂OH,
R² = H

13 Ankorine†,§ R¹ = CH₂OH,
R² = OH

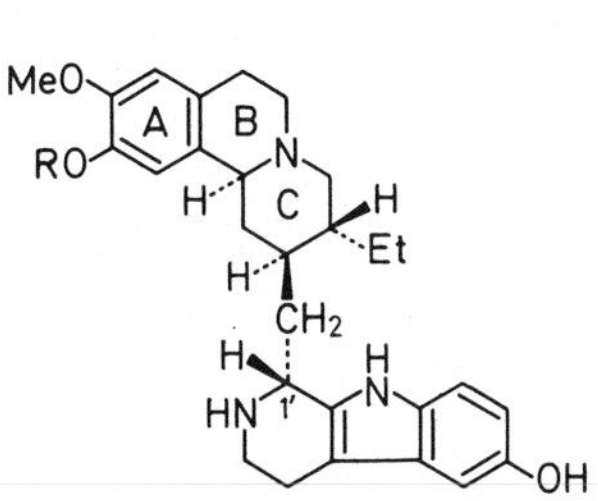

14 Tubulosine† R = Me
15 Isotubulosine†
(1′α-H instead of 1′β-H)
16 Demethyltubulosine†,§
R = H

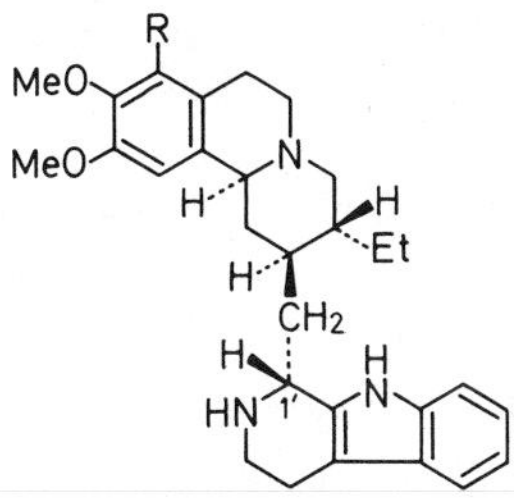

17 Deoxytubulosine†
R = H
18 Alangimarckine†,§
R = OH

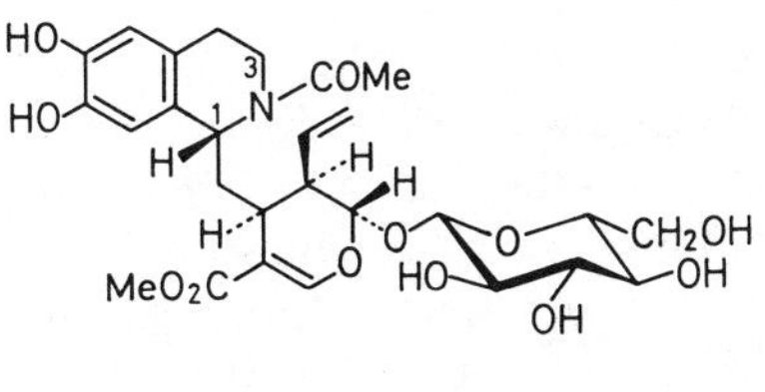

19 Ipecoside*,§

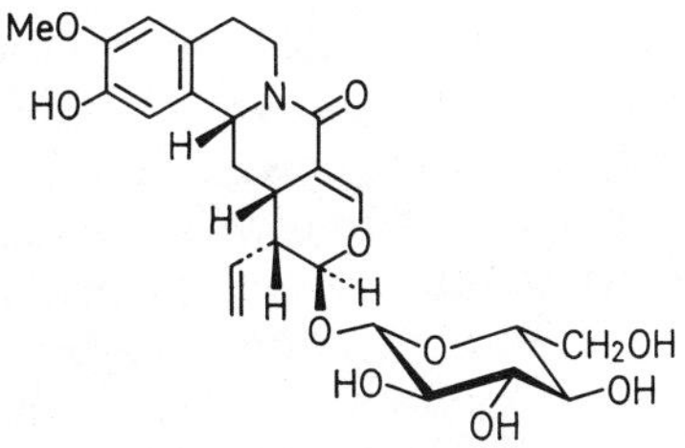

20 Alangiside†,‡

on related compounds, analytical methods, and biological activity are not intended to be exhaustive; however, they should serve as guides to current general aspects of these areas for the subject during the period under review.

Unless otherwise noted, the structural formulas of optically active com-

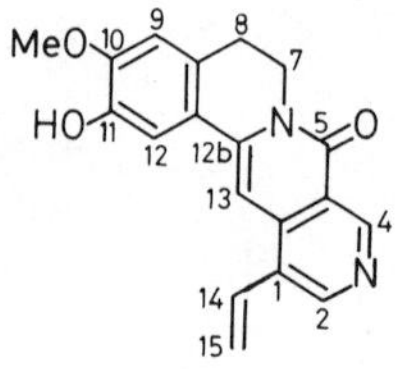

21 Alangimarine[†, ‡]

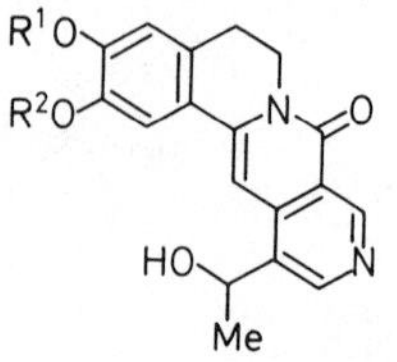

22 Alamarine
(racemic)[†, ‡]
R^1 = Me, R^2 = H

23 Isoalamarine[†, ‡]
R^1 = H, R^2 = Me

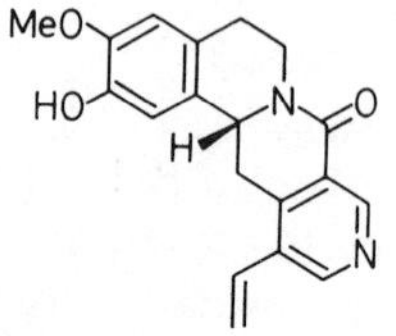

24 Alangimaridine[†, ‡]

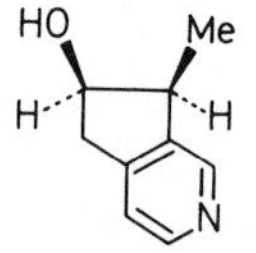

25 Venoterpine[†, ||]

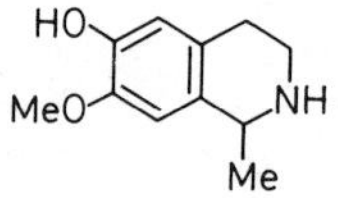

26 (±)-Salsoline[†, ||]

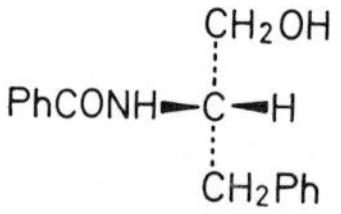

27 N-Benzoyl-L-
phenylalaninol[†, ||]

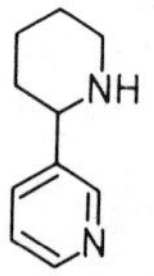

28 (±)-Anabasine[†, ||]

pounds in this chapter represent their absolute configuration, and the numbering system employed for the benzoquinolizidine alkaloids is identical with that used previously (3). The structures of all of the 7 ipecac alkaloids and 24 *Alangium* alkaloids reported so far are shown in formulas **1–28**, where the ipecac alkaloids are marked with an asterisk and the *Alangium* alkaloids with a dagger. In addition, the new structures, the revised or established structures, and the known but newly isolated structures encountered since the last review (3) are marked with a double dagger, section, and parallel, respectively. A group of bisindole alkaloids from other plant sources contains the *Corynanthe*–tubulosine hybrid structure analogous to the ipecac–β-carboline structure from *A. lamarckii,* but this group has been reviewed separately by Cordell and Saxton in Chapter 1 of Volume XX in this treatise.

II. Occurrence

At the time when the ipecac alkaloids were last reviewed in Volume XIII of this treatise (3), seven alkaloids, namely, emetine (1), cephaeline (2), psychotrine (6), O-methylpsychotrine (7), emetamine (10), protoemetine (11), and ipecoside (19) had been isolated from ipecacuanha plants (family Rubiaceae). The occurrence of the first 3 alkaloids had been duplicated in *Alangium lamarckii* Thwaites (Alangiaceae), from which 10 more bases closely related to the ipecac alkaloids had been isolated. They were alangamide (4) (possibly an artifact), demethylpsychotrine (8), alangicine (9), protoemetinol (dihydroprotoemetine) (12) (not isolated in pure form), ankorine (13), tubulosine (14), isotubulosine (15), demethyltubulosine (16), deoxytubulosine (17), and alangimarckine (18).

Since that time the same plant has been found to contain 10 other either new or known alkaloids, among which isocephaeline (3) [$C_{28}H_{38}N_2O_4$; mp 108–116°C; $[\alpha]_D^{20}$ −69.8° (CHCl$_3$)] (8), alangimarine (21) ($C_{19}H_{16}N_2O_3$; mp 247°C) (9), alamarine (22) [$C_{19}H_{18}N_2O_4$; mp 288°C; $[\alpha]_D \pm 0°$] (9), alangimaridine (24) [$C_{19}H_{18}N_2O_3$; mp 278°C; $[\alpha]_D$ +429° (CHCl$_3$)] (9), venoterpine (25) (8), and (±)-salsoline (26) (8) have been isolated from the seeds; demethylcephaeline (5) [$C_{27}H_{36}N_2O_4$; mp 147–149°C; $[\alpha]_D$ −53.5° (CHCl$_3$)] (10) and isoalamarine (23) ($C_{19}H_{18}N_2O_4$) (9) from the stem bark; alangiside (20) [$C_{25}H_{31}NO_{10}$; mp 164–187°C dec.; $[\alpha]_D^{26}$ −105° (MeOH)] from the roots, leaves, or fruit (11,12); N-benzoyl-L-phenylalaninol (27) from the leaves (13). Of these ten alkaloids, isocephaeline (3) has long been known (1,14) but not encountered before in nature. The monoterpene alkaloid venoterpine (25) (15), simple isoquinoline alkaloid (±)-salsoline (26) (16), and N-benzoyl-L-phenylalaninol (27) (17) are also known compounds, but have not been isolated previously from *A. lamarckii*.

Cephaeline (2), psychotrine (6), ankorine (13), venoterpine (25), and (±)-anabasine (28) have also been found in some other species of *Alangium* (18,19) and the alkaloid, $C_{28}H_{35}N_3O_3$ [mp 200°C; $[\alpha]_D^{20}$ −40° (pyridine)], in the bark of *A. vitiense* (20). It is interesting to note that emetine (1) has been isolated from *Hedera helix* L. (family Araliaceae) (21,22).

III. Chemistry and Synthesis

A. PROTOEMETINE, PROTOEMETINOL, AND ANKORINE

In the early investigations discussed in Chapter 3 of Volume XIII (3), the ipecac alkaloid protoemetine (11) was synthesized (23,24) both in its racemic and natural (−) form by diisobutylaluminum hydride reduction of the

($\pm$)- and ($-$)-tricyclic ester **29**, the key intermediates in many of earlier syntheses (*3–7*) of ($\pm$)- and ($-$)-emetine (**1**). The compound corresponding to the *Alangium* alkaloid protoemetinol (**12**) (the absolute configuration shown is as yet tentative) was prepared (*25*) from ($-$)-protoemetine (**11**) by catalytic hydrogenation or NaBH$_4$ reduction. In view of these conversions, later new syntheses of ($\pm$)- and ($-$)-**29** constitute formal syntheses of the two alkaloids, and they are summarized in Section III,B.

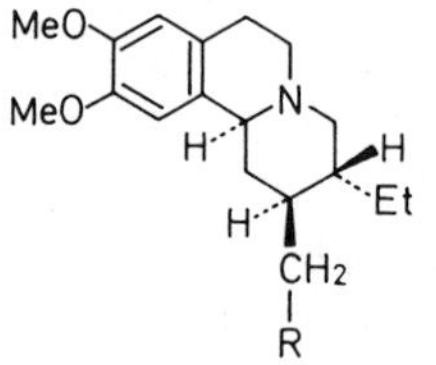

29 R = CO$_2$Et
30 R = CO$_2$Me
31 R = CN

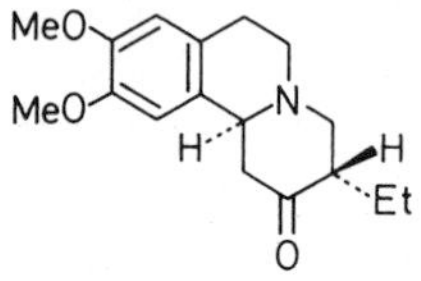

32

A similar synthetic approach to ($\pm$)- and ($-$)-protoemetine (**11**) was reported by Szántay *et al.* (*26,27*) who prepared the ($\pm$)- and ($-$)-tricyclic nitrile **31** from the known ($\pm$)- and ($-$)-aminoketone **32** (*3*) by condensation with dialkyl cyanomethylphosphonate followed by hydrogenation. Treatment of ($\pm$)- and ($-$)-**31** with diisobutylaluminum hydride then produced ($\pm$)- and ($-$)-**11**. In different approaches, a few research groups also utilized some intermediates generated in their own emetine syntheses. Brown *et al.* (*28*) obtained ($-$)-protoemetine (**11**) from the (+)-piperidone **33** by Bischler–Napieralski cyclization followed by NaBH$_4$ reduction and hydrolysis of the hydrazone moiety with aqueous Cu(OAc)$_2$. Kametani *et al.* (*29–31*) synthesized ($\pm$)-**11** from the ($\pm$)-lactam ester **34** through ($\pm$)-**35** and ($\pm$)-**36** and Takano *et al.* (*32,33*) synthesized ($\pm$)-**11** from the ($\pm$)-lactam **40** through the ($\pm$)-aminodithiane **41**. ($\pm$)-Protoemetinol (**12**) was prepared by

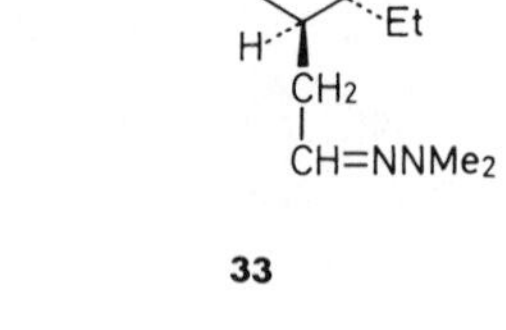

33

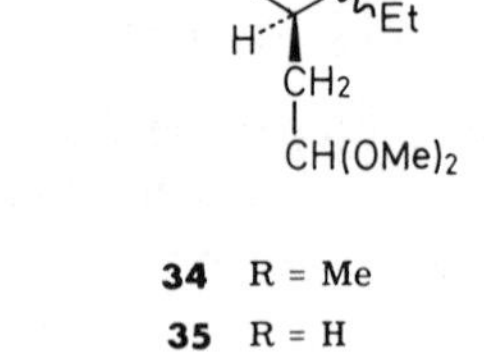

34 R = Me
35 R = H

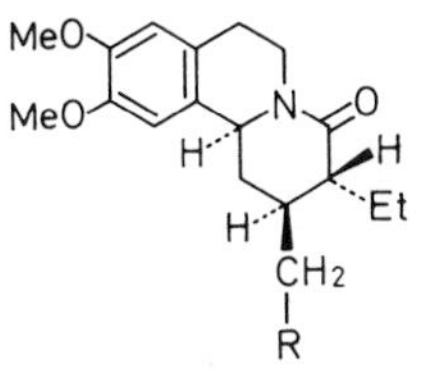

36 R = CH(OMe)$_2$	**40** Z = O
37 R = CHO	**41** Z = H$_2$
38 R = CO$_2$H	
39 R = CO$_2$Me	

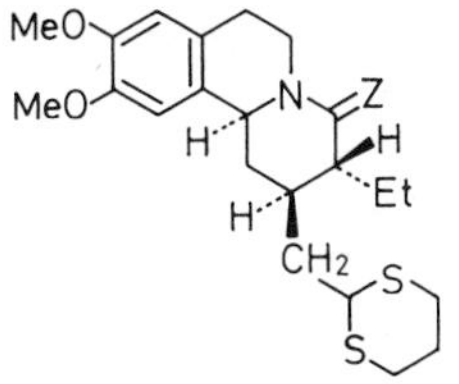

LiAlH$_4$ reduction of any one of the (±)-lactam ester **39** (*34*), (±)-lactam acid **38** (*32,33,35*), (±)-lactam aldehyde **37** (*29,30*), or (±)-tricyclic ester **29** (*36*) or **30** (*33,37*).

The structure and stereochemistry of the *Alangium* alkaloid (−)-ankorine have been established as **13** by synthesis. Szántay *et al.* (*38,39*) synthesized all four possible racemic stereoisomers of the tentatively proposed gross structure (**42**) (*40*) for ankorine, starting from the (±)-aminoketone **43** by a route that paralleled their previous protoemetine synthesis (*23*), proceeding through (±)-**32** and (±)-**29**, and by a route that paralleled the earlier (±)-*O*-methylpsychotrine synthesis (*41*) involving the condensation of (±)-**32** with ethyl cyanoacetate. As none of these four stereoisomers matched natural ankorine, they proposed a revised structure bearing a hydroxyl group at C-8 instead of C-11, and suggested the normal-type relative stereochemistry as in **13** for this alkaloid after considering the results of their mass spectral study (*38*). Fujii *et al.* (*42*) confirmed the correctness of the newly proposed structure by synthesizing the racemic target molecule **13**. Treatment of the (±)-lactim ether **46**, obtained from the (±)-lactam ester **45** by a method given in the literature (*43,44*), with 2-benzyloxy-3,4-dimethoxyphenacyl bromide yielded the (±)-lactam ketone **47**. Reduction of (±)-**47** with NaBH$_4$, followed by catalytic hydrogenolysis and benzylation

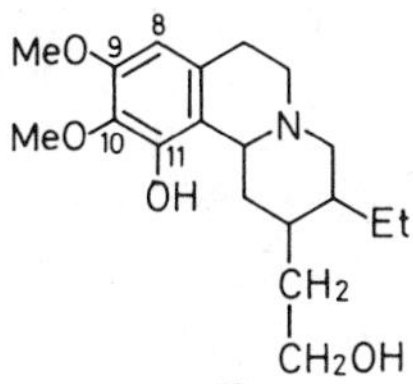

42	**43** R^1 = OCH$_2$Ph, R^2 = H
	44 R^1 = H, R^2 = OCH$_2$Ph

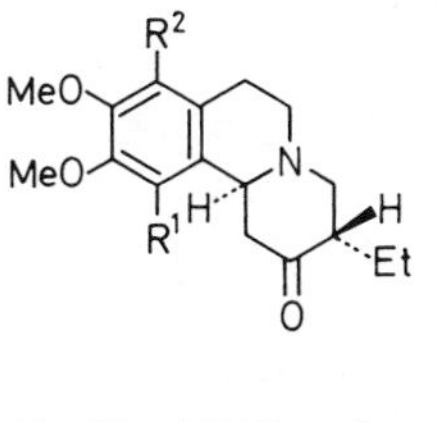

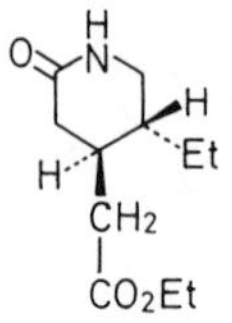

45 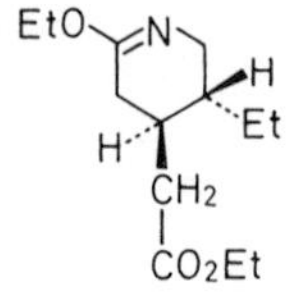   46 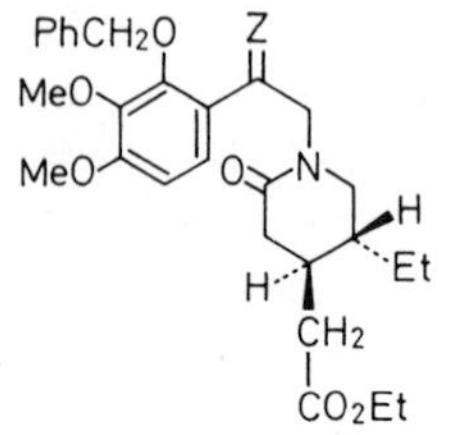   47 Z = O
                                            48 Z = H$_2$

with benzyl bromide, gave the (±)-lactam **48**, which was then converted to the (±)-tricyclic ester **49** by Bischler–Napieralski cyclization and subsequent catalytic hydrogenation. Reduction of (±)-**49** with LiAlH$_4$ and removal of the benzyl group afforded (±)-**13**, found to match ankorine. This was also confirmed by Szántay et al. (39) who independently prepared (±)-**13** by repeating the above synthetic route to the normal-type isomer of (±)-**42**, but with the isomeric (±)-aminoketone **44** in place of (±)-**43**.

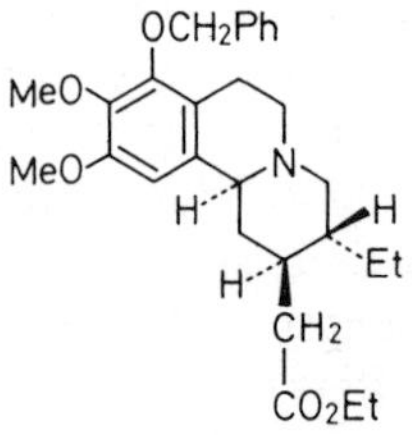

49 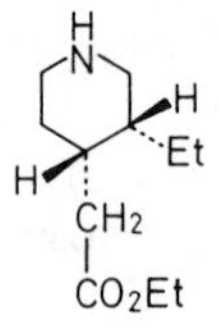   50 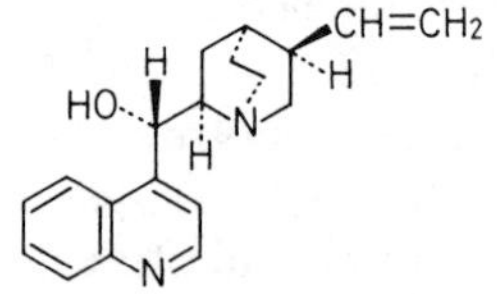   51

The solution of the absolute stereochemistry of ankorine came with the synthesis of the chiral target molecule **13**, eventually found to correspond to the (−) form by Fujii and co-workers (45,46). The starting material for this synthesis was (+)-cincholoipon ethyl ester (**50**) available from commercial (+)-cinchonine (**51**), one of the major Cinchona alkaloids, in 50% overall yield according to the classical degradation procedure (47,48). Condensation of (+)-**50** with 2-benzyloxy-3,4-dimethoxyphenacyl bromide in the presence of K$_2$CO$_3$ furnished the (−)-aminoketone **52**, which was then reduced with NaBH$_4$ to give a mixture of the diastereomeric alcohols **53**. On oxidation with Hg(OAc)$_2$–EDTA, the mixture **53** produced the 6-piperidone **54** in 57% yield and an oily substance, presumed to be a mixture of the 2-piperidones **55**, in 19% yield. Catalytic hydrogenolysis of the 6-piperidone **54** followed by alkaline hydrolysis yielded the (−)-cis-lactam acid **56**, which

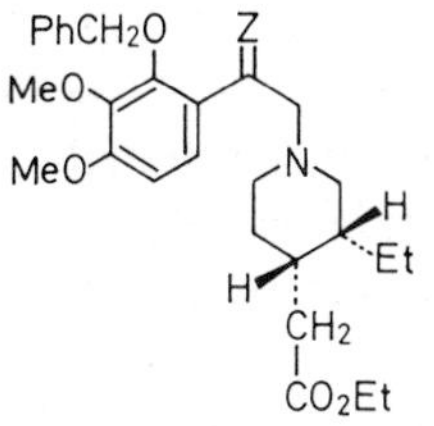

52 Z = O
53 Z = H,OH

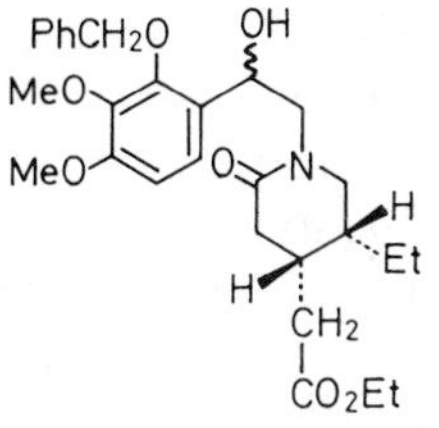

54

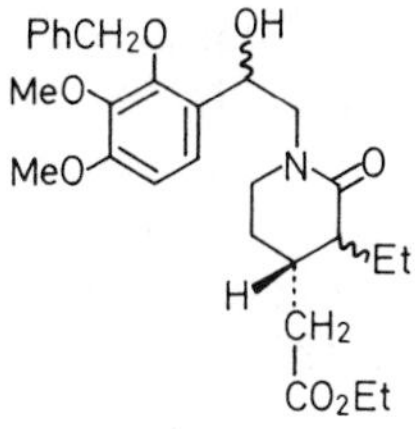

55

was heated neat at 180°C for 90 min to afford a 67 : 33 mixture of the trans and the cis isomers. The (+)-*trans*-lactam acid **58**, isolated from the equilibrated mixture, was then transformed into the (+)-lactam **48** by esterification and subsequent O-benzylation. Conversion of (+)-**48** to (−)-**13** through the (−)-tricyclic ester **49** was effected as in the case of the racemic series described earlier, and the synthetic (−)-**13** was identical with natural (−)-ankorine. Thus, structure **13** is a complete expression for ankorine; that is to say, ankorine is the 8-hydroxy congener of dihydroprotoemetine (protoemetinol) (**12**).

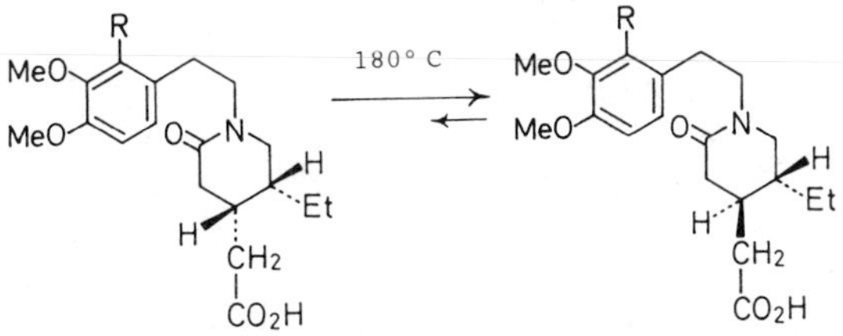

56 R = OH
57 R = H

58 R = OH
59 R = H

B. EMETINE, *O*-METHYLPSYCHOTRINE, AND EMETAMINE

Already reviewed in several places (*1–7,49,49a*), numerous attempts to synthesize emetine (**1**) (*23,25,34,36,37,41,50–73*) and about a score of successful syntheses of this alkaloid (*23,25,34,36,37,51,52,61–73*) during the three decades ending in 1977, culminated in the development of an elegant and commercially applicable version by Openshaw and Whittaker (*72,73*). Notwithstanding this fact, the emetine molecule **1** has still been a favorite target of synthetic organic chemists for evaluating their own methodologies of transformation of organic materials and for checking the stereochemical outcome of new synthetic routes adaptable to analogous but new alkaloids of unestablished stereochemistry.

A formal synthesis of emetine (**1**) was achieved by Fujii *et al.* (*42,74*) via the synthesis of the (±)-*trans*-lactam ester **60** from the (±)-lactim ether **46**. The steps involved were condensation of (±)-**46** with 3,4-dimethoxyphenacyl bromide and NaBH$_4$ reduction of the resulting (±)-lactam ketone **61**, followed by catalytic hydrogenolysis. The (±)-ester **60** has already been shown (*36*) to lead to the (±)-tricyclic ester **29**, the known emetine precursor. The same workers (*75*) synthetically incorporated the (+)-ethyl cincholoiponate skeleton **50** in the (−)-tricyclic ester **29** (*25,72*), the key intermediate for the earlier synthesis of *O*-methylpsychotrine (**7**) (*72*), emetine (**1**) (*72*), psychotrine (**6**) (*76*), protoemetine (**11**) (*23*), and tubulosine alkaloids (**14,17**) (*77*), by a series of conversions [(+)-**50** → (+)-**62** → **63** → **64** → (−)-**65** → (−)-**57** → (+)-**59** → (+)-**60** → (−)-**29**] similar to that adopted for the synthesis of (−)-ankorine (**13**) (Section III,A).

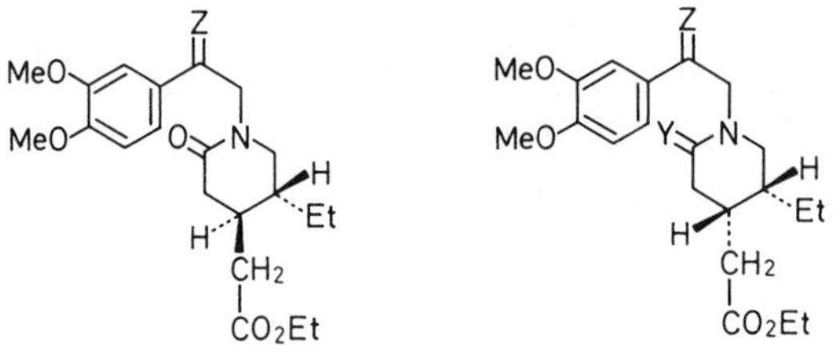

60 Z = H$_2$
61 Z = O

62 Y = H$_2$, Z = O
63 Y = H$_2$, Z = H, OH
64 Y = O, Z = H, OH
65 Y = O, Z = H$_2$

Some modifications of the previous processes (*23,36,72*) were also reported (*78,79*) for the preparation of the (±)- and (−)-ethyl esters **29** and the corresponding (±)-methyl ester **30**, another emetine precursor (*34,37,61,72*). The latter (±)-precursor **30** was alternatively synthesized by Takano *et al.* (*80*) who utilized cleavage of an α-diketone monothioketal intermediate. Birch reduction of the starting (±)-tetrahydroprotoberberine

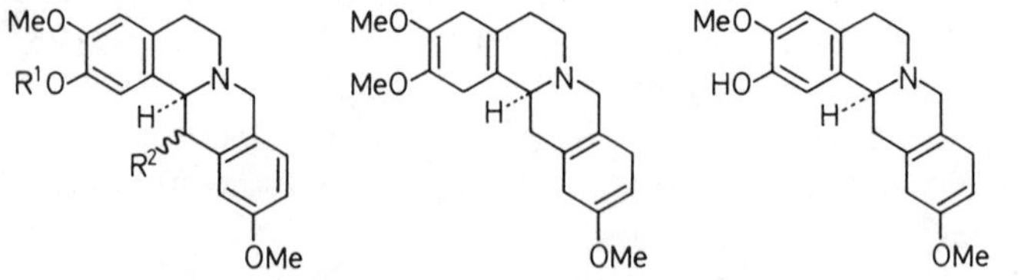

66 R^1 = Me, R^2 = H or CN
67 R^1 = CH$_2$Ph, R^2 = H or CN

68

69

66 and treatment of the resulting enol ether **68** with *N*-chlorosuccinimide gave the (±)-tetracyclic ketone **70**, which was also obtained from the (±)-tetrahydroprotoberberine **67** via the enol ether **69**. The (±)-unsaturated ketone **70** underwent a stereoselective catalytic hydrogenation to yield the (±)-saturated ketone **71**. The (±)-α-diketone monothioketal **72**, derived from (±)-**71** by treatment with pyrrolidine followed by trimethylene dithiotosylate, was then converted to the desired (±)-tricyclic ester **30** by alkaline cleavage and subsequent esterification with diazomethane and desulfurization with Raney nickel of the resulting (±)-thioacetal ester **73**.

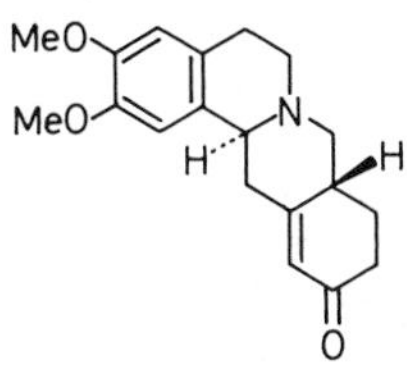

70

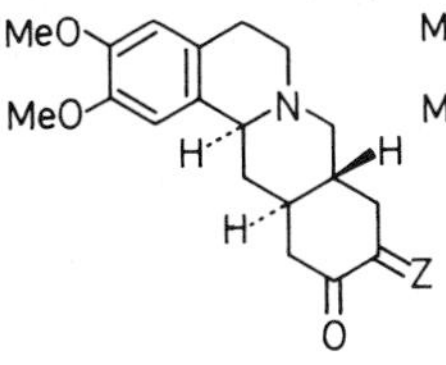

71 Z = H$_2$

72 Z = —S—(CH$_2$)$_3$—S—

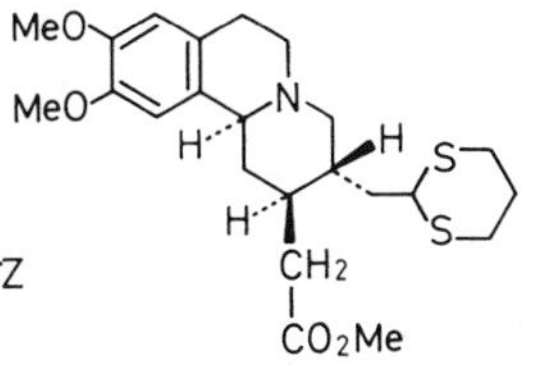

73

The (±)-tricyclic lactam ester **39**, yet another emetine precursor (*34*), was also prepared by Takano's group (*32,33,35*) from (±)-norcamphor (**74**). Baeyer–Villiger oxidation of (±)-**74** and subsequent ethylation yielded the (±)-lactone **75**, which was then condensed with 3,4-dimethoxyphenethylamine and oxidized. The resulting (±)-ketone **76** was cleaved through the (±)-thioketal **77** to afford the (±)-amido acid **78**. Exposure of (±)-**78** to MeI

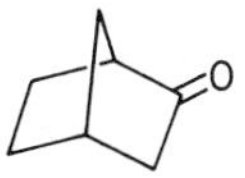

74

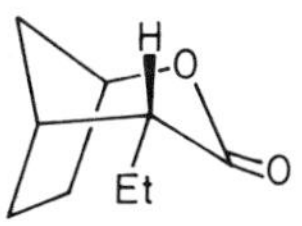

75

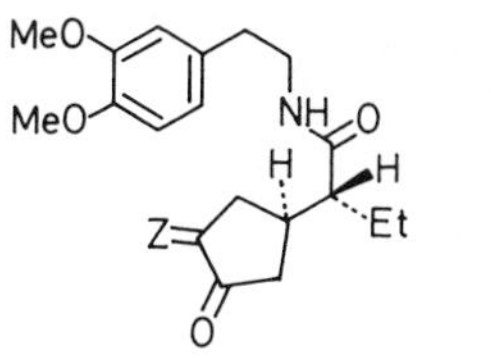

76 Z = H$_2$

77 Z = —S—(CH$_2$)$_3$—S—

produced a mixture of the epimeric lactam acids (±)-**38** and (±)-**79**, and the former was esterified to the emetine precursor (±)-**39** and reduced to (±)-protoemetinol (**12**) as described in Section III,A. On the other hand, thermal cyclization of (±)-**78** followed by NaBH$_4$ reduction and subsequent treatment with *p*-toluenesulfonic acid provided a mixture of the isomeric lactams (±)-**80**, (±)-**81**, and (±)-**82**. The first of these three lactams was

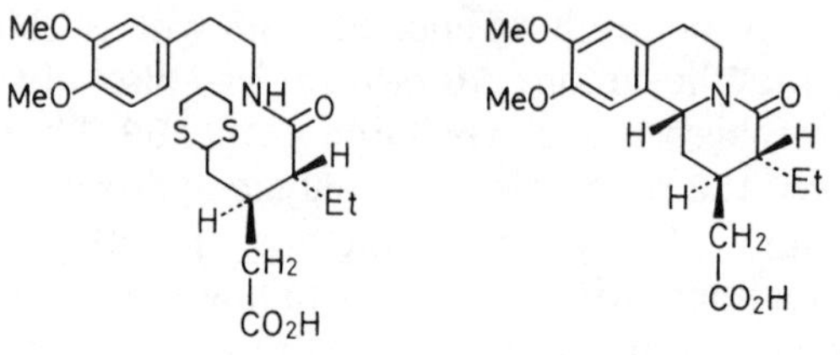

78

79

found to lead to (±)-**38**, and the other two isomerized to (±)-**80** on treatment with a Lewis acid.

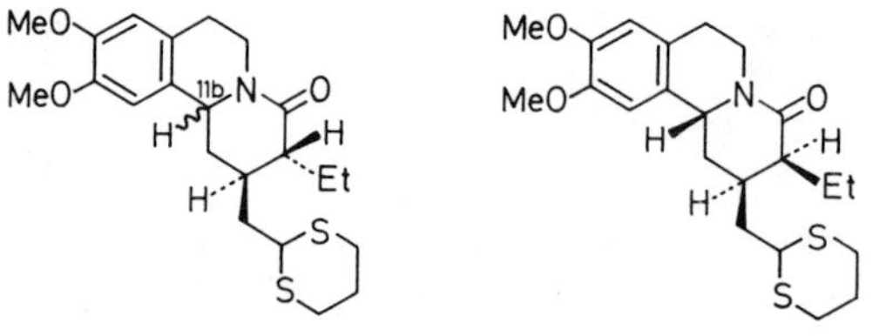

80 11bα-H
81 11bβ-H

82

Kametani *et al.* (*29,30*) have reported another stereoselective synthesis of (±)-emetine (**1**). Condensation of the dihydroisoquinoline **83** with the unsaturated malonate **84** gave the tricyclic lactam **85** which furnished

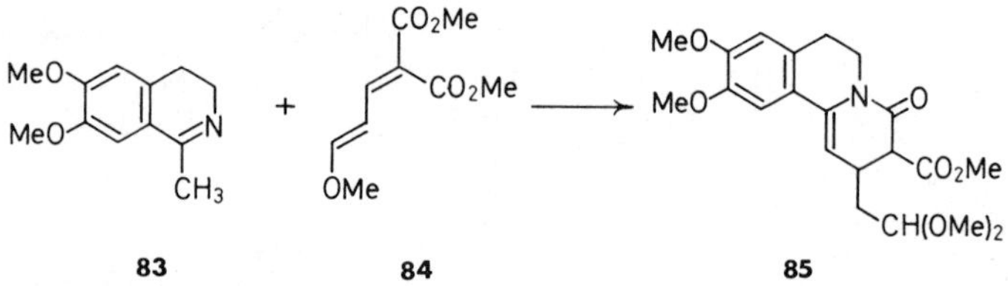

83   **84**   **85**

(±)-**34** on ethylation followed by catalytic reduction. Alkaline hydrolysis and decarboxylation followed by acid hydrolysis converted (±)-**34** to (±)-**37**, a dihydroprotoemetine precursor (see Section III,A), through (±)-**36**, a protoemetine precursor (Section III,A). Pictet–Spengler cyclization of (±)-**36** or (±)-**37** with 3-hydroxy-4-methoxyphenethylamine afforded a 7:3 mixture of the cephaeline derivative (±)-**86** and its 1′-epimer. Methylation of the major isomer with diazomethane, followed by LiAlH₄ reduction, provided (±)-emetine (**1**).

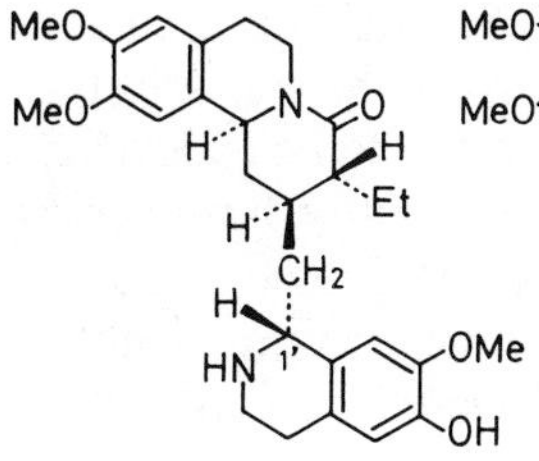

86 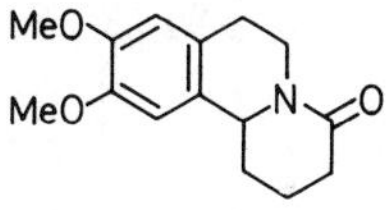   **87** 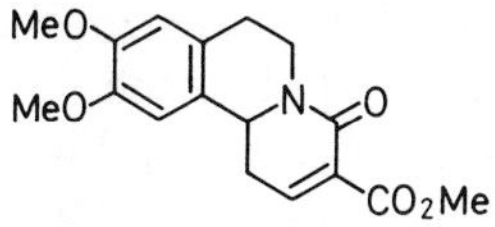   **88**

Another synthesis of emetine by Kametani's group (*81*) utilized the main mass spectral fragments of this alkaloid for generation of important synthons. Michael addition of the dihydroisoquinoline **83** to the (±)-unsaturated lactam ester **88**, synthesized from the (±)-tricyclic lactam **87** in several steps, produced the (±)-adduct **89** as a single diastereomer. Ethylation of (±)-**89** at C-3 and subsequent processes involving removal of the methoxy-carbonyl group and reduction of the lactam carbonyl group gave the (±)-base **90**, the 11b-epimer of *O*-methylpsychotrine (**7**). Dehydrogenation of (±)-**90** with Hg(OAc)$_2$ and NaBH$_4$ reduction of the resulting hexahydrobenzoquinolizinium salt yielded (±)-emetine (**1**) and its 1′-epimer, (±)-isoemetine.

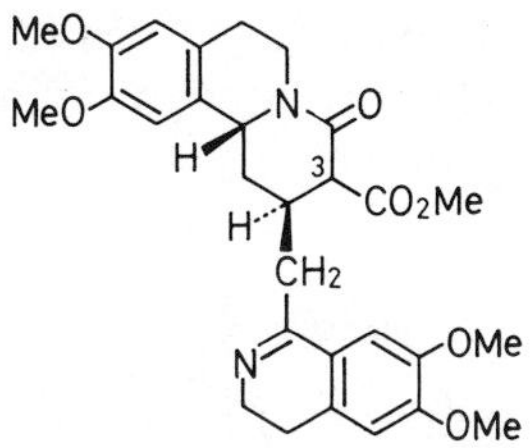

89 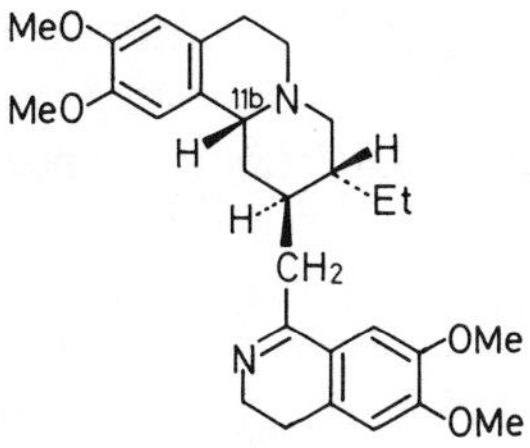   **90**

Brown *et al.* (*28*) have reported a stereoconservative synthesis of (−)-cephaeline (**2**) from secologanin via (+)-**33** and (−)-protoemetine (**11**) (see Section III,A and D). It represents a formal synthesis of (−)-emetine (**1**) in view of the earlier conversion (*23*) of (−)-**2** to (−)-**1**.

The (±)- and (−)-aminoketones **32** were also the key intermediates used in earlier emetine syntheses (*23,41,63,64,72,73*), and a few new synthetic routes to them have been reported (*82–84*).

(+)-*O*-Methylpsychotrine has been assigned structure **7** in which the double bond in the dihydroisoquinoline moiety is endocyclic (*3–5*), and the formation of the *N*-benzoyl derivative **91** (*85,86*) was interpreted in terms of a tautomeric shift of the double bond (*87*). Nonetheless, Schuij *et al.* (*88*) claimed the double bond in *O*-methylpsychotrine to be exocyclic, as in **92**, on the basis of their mass spectral study. More recently, however, Fujii *et al.* (*89*) presented ¹H- and ¹³C-NMR and UV spectroscopic evidence that this alkaloid has the genuine 3,4-dihydroisoquinoline structure **7**, not the exocyclic double-bond structure **92**, in the free-base form as well as in the protonated form. A minor modification of the previous procedure (*36*) for resolution of (±)-*O*-methylpsychotrine (**7**) has been reported (*90*).

91 R = COPh
92 R = H

Emetamine (**10**) and *O*-methylpsychotrine (**7**) were among many other fragmentation and/or oxidation products obtained by photochemical or thermal decomposition of emetine (**1**) (*91*). Battersby *et al.* (*92*) synthesized (±)-**10** from their emetine precursor (±)-**29** (*36*) through cyclization of the corresponding β-3,4-trimethoxyphenethylamide.

C. Psychotrine, Demethylpsychotrine, and Alangicine

(+)-Psychotrine (**6**) has already been synthesized by condensation of the (−)-tricyclic ester **29** with 3-benzyloxy-4-methoxyphenethylamine, followed by Bischler – Napieralski cyclization and debenzylation (*76*). Thus, the syntheses of (−)-**29** reviewed last in Section III,B constitute formal syntheses of this alkaloid.

In the early work (*93*) summarized in Chapter 3 of Volume XIII (*3*), two alternative structures **8** and **93** were considered for (+)-demethylpsychotrine on the basis of its conversion to *O*-methylpsychotrine (**7**) and its formation from psychotrine (**6**) by acid hydrolysis and from the mass spectral evidence. The structure of (+)-demethylpsychotrine has now been established as **8** as a result of the syntheses of (±)-9-demethylpsychotrine (**8**), (±)-10-demethylpsychotrine (**93**), and (+)-9-demethylpsychotrine (**8**) by Fujii *et al.* (*94,95*). The syntheses of (±)-**8** and (±)-**93** (*94*) consisted of initial condensations of the (±)-lactim ether **46** with 3-benzyloxy-4-methoxyphen-acyl bromide and with 4-benzyloxy-3-methoxyphenacyl bromide and sub-sequent steps similar to those of their emetine and ankorine syntheses (*42,74*) starting with (±)-**46**. The synthesis of (+)-**8** starting from (+)-**50** and 3-benzyloxy-4-methoxyphenacyl bromide was accomplished (*95*) by the "cincholoipon-incorporating method" as adopted for their syntheses of (−)-ankorine (**13**) (*45,46*) and (−)-emetine (**1**) (*75*).

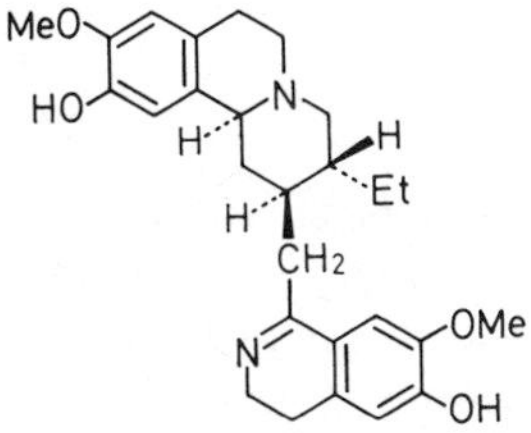

93

The originally proposed gross structure **94** for (+)-alangicine (*93*) has now been revised to the complete expression **9** as a result of the syntheses of (±)-**9** and (+)-**9** by Fujii *et al.* (*96,97*). Alkaline hydrolysis of the (±)-tricyclic ester **49**, the racemic ankorine precursor (*42*), furnished the (±)-amino acid **95**

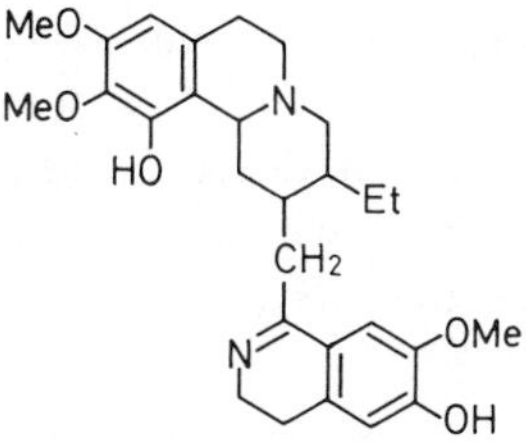

94 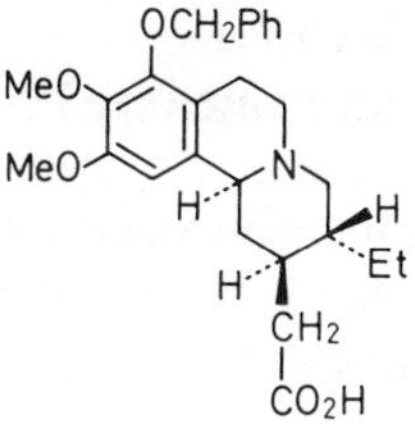   **95**

that was condensed with 3-benzyloxy-4-methoxyphenethylamine by the diethyl phosphorocyanidate method (*98*) to give the (±)-amide **96**. Bischler–Napieralski cyclization of (±)-**96** and debenzylation of the result-ing (±)-base **97** produced (±)-**9** (*96*), which matched natural alangicine. A parallel series of conversions starting with the (−)-tricyclic ester **49**, the (−)-ankorine precursor (*45,46*), and proceeding through (−)-**95**, (−)-**96**, and **97**, yielded (+)-**9** (*97*), which was identical with (+)-alangicine.

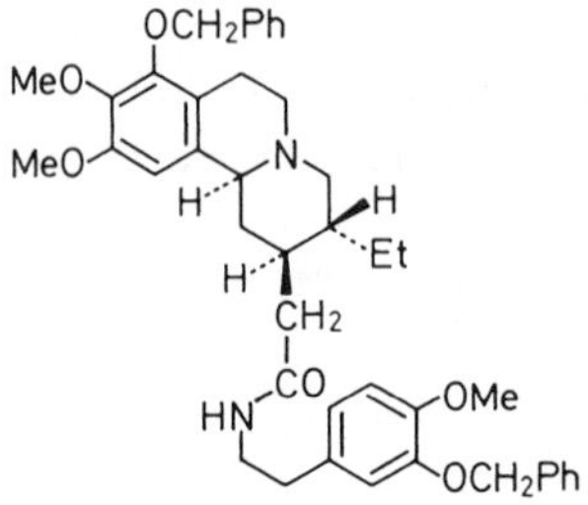

96 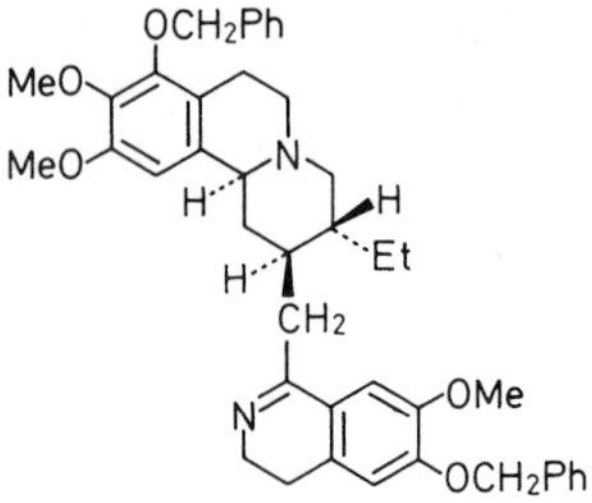   **97**

No direct evidence has been presented for the position of the double bond in the dihydroisoquinoline moiety of psychotrine, demethylpsychotrine, and alangicine. However, Fujii *et al.* (*89*) considered the double bond to be endocyclic, as shown in formulas **6**, **8**, and **9**, analogous to that established for *O*-methylpsychotrine (**7**).

D. CEPHAELINE, ISOCEPHAELINE, AND DEMETHYLCEPHAELINE

The previous synthesis of cephaeline (**2**) and isocephaeline (**3**) in both their racemic and (−) forms by Szántay *et al.* (*23,24*) included Pictet–Spengler condensation of (±)- and (−)-protoemetine (**11**) with 3-hydroxy-4-methoxyphenethylamine as the final step. Brown *et al.* (*28*) repeated this step with (−)-**11**, derived from the (+)-piperidone **33** (see Section III,A), obtaining this piperidone from secologanin (**98**), the biogenetic precursor for cephaeline and emetine, by a stereoconservative, multistep synthesis. The precursor **98** was converted to the (−)-ester **99** by acetylation, oxidation, methylation, and catalytic reduction. On deacetylation and enzymatic hydrolysis, (−)-**99** gave the aglycone **100**, which was then treated with 3,4-dimethoxyphenethylamine and NaBH$_3$CN to afford a mixture of two

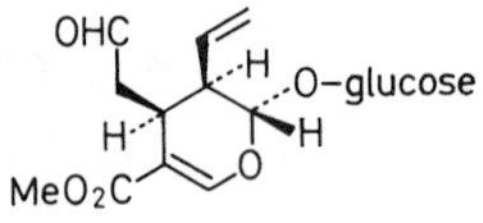

98 Secologanin 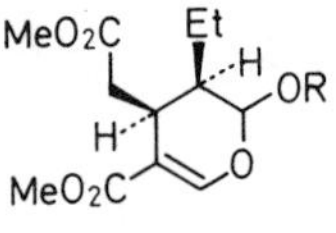   **99**   R = Glc(OAc)$_4$
                                       100   R = H

diastereomers **101** and (+)-**102**. Selective hydrolysis and decarboxylation of the major product (+)-**102** yielded the carbinolamine **103**. On treatment with *N,N*-dimethylhydrazine, **103** underwent ring opening and lactam formation to produce the desired (+)-piperidone **33**.

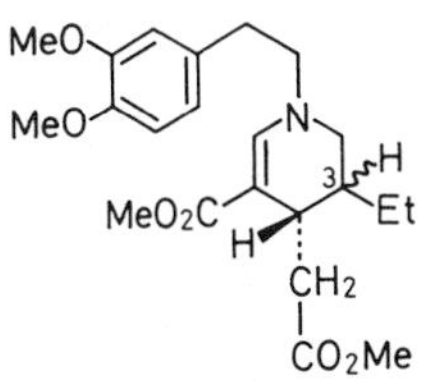

101 3α-H
102 3β-H

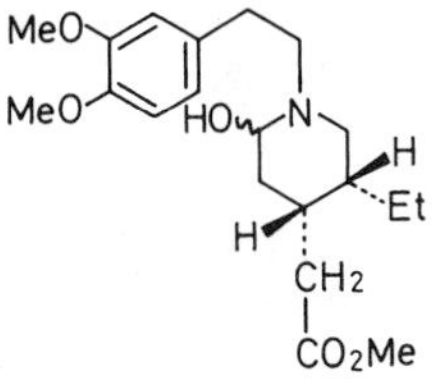

103

Although known synthetically for a long time (*1,14*), (−)-isocephaeline (**3**) was not encountered in nature until quite recently. In 1980, Pakrashi's group (*8*) isolated this (−) base from the seeds of *Alangium lamarckii* Thw. and confirmed its identity with authentic isocephaeline prepared by NaBH₄ reduction of psychotrine (**6**).

The problem of the structure of (−)-demethylcephaeline has been essentially the same as that of (+)-demethylpsychotrine (**8**) and (−)-demethyltubulosine (**16**). Based on chemical correlation with cephaeline (**2**) and emetine (**1**) and on spectral evidence, Pakrashi and Achari (*10*) have assigned the alternative structure **5** (**104** or **105**) to (−)-demethylcephaeline. Fujii *et al.* (*99,100*) synthesized both (−)-9-demethylcephaeline (**104**) and (−)-10-demethylcephaeline (**105**) from (+)-**50** and appropriate phenacyl bromides by applying their cincholoipon-incorporating method that worked well for the syntheses of (−)-ankorine (**13**) (*45,46*), (−)-emetine (**1**) (*75*), (+)-alangicine (**9**) (*97*), and (+)-demethylpsychotrine (**8**) (*95*). However, lack of a sufficient amount of natural (−)-demethylcephaeline for a minute and direct comparison precluded identification of either (−)-**104** or (−)-**105** with this alkaloid, thus leaving its chemistry incomplete.

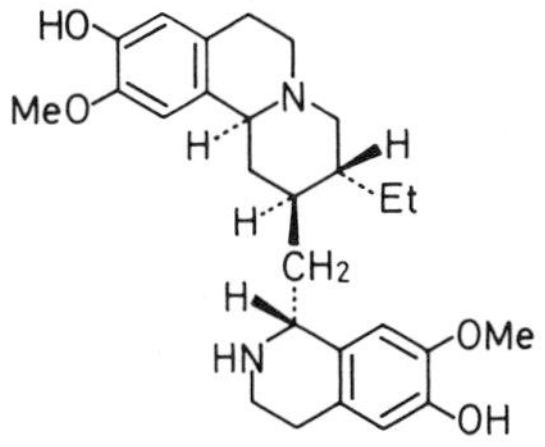

104

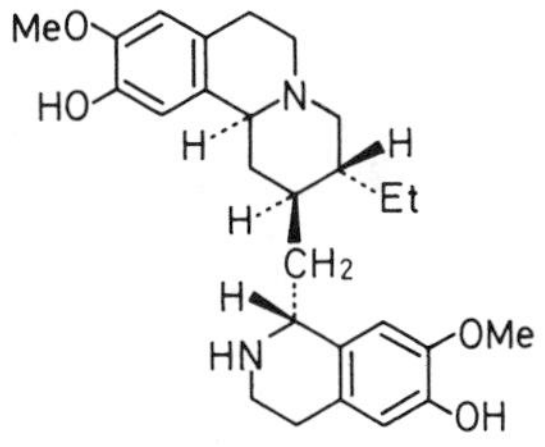

105

E. TUBULOSINE, ISOTUBULOSINE, AND DEMETHYLTUBULOSINE

In addition to the previous racemic and chiral syntheses (*77,101–103*) of tubulosine (**14**) and isotubulosine (**15**), Kametani *et al.* (*104*) prepared (±)-**14** and (±)-**15** from their emetine precursor (±)-**37** by Pictet–

Spengler reaction with serotonin (106), followed by reduction with NaAlH$_2$(OCH$_2$CH$_2$OMe)$_2$ in pyridine.

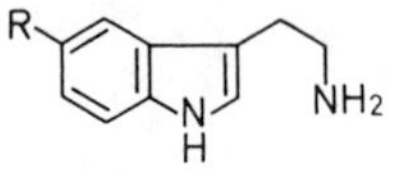

106 R = OH
107 R = OCH$_2$Ph
108 R = H

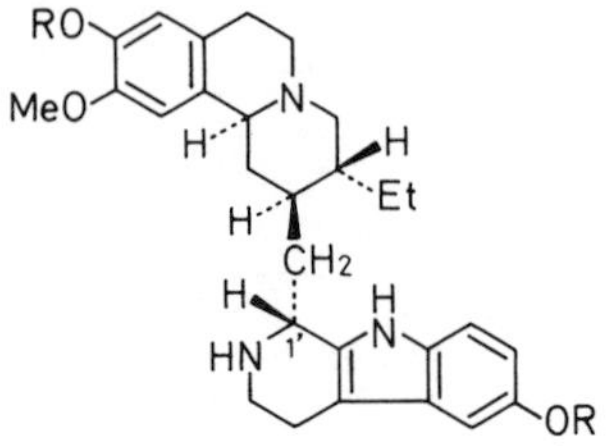

109 R = H
110 R = CH$_2$Ph

A definite decision regarding the choice between two alternative structures, 109 and 16, being considered (105) for (−)-demethyltubulosine (93,105) was made on the basis of synthesis of both structures (±)-109 and (±)-16 by Fujii's group (106,107). Condensation of the (±)-tricyclic amino acid 111, the (±)-9-demethylpsychotrine precursor (94), with 5-benzyloxy-tryptamine (107) by the diethyl phosphorocyanidate method (98) and Bischler–Napieralski cyclization of the resulting amide provided the (±)-di-hydro-β-carboline 113, which on reduction gave (±)-O,O-dibenzyl-9-de-methyltubulosine (110) as well as its 1′-epimer. Debenzylation of (±)-110 furnished the desired (±)-9-demethyltubulosine (109), but it did not match natural demethyltubulosine (106). On the other hand, (±)-10-demethyltu-bulosine (16) obtained from the (±)-tricyclic amino acid 112, a precursor of (±)-10-demethylpsychotrine (94), by a parallel synthesis through (±)-114

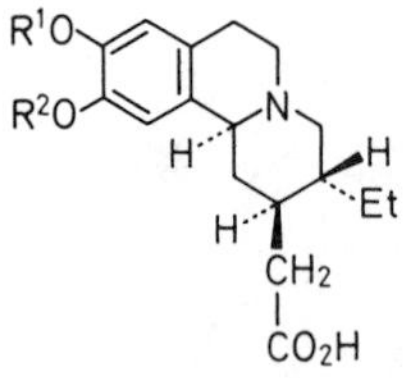

111 R^1 = CH$_2$Ph, R^2 = Me
112 R^1 = Me, R^2 = CH$_2$Ph

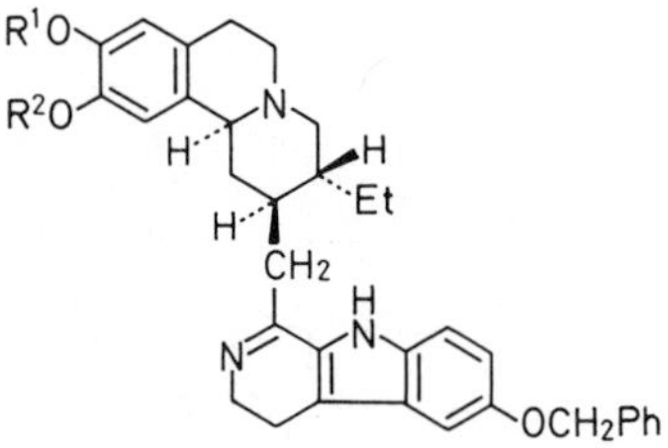

113 R^1 = CH$_2$Ph, R^2 = Me
114 R^1 = Me, R^2 = CH$_2$Ph

and (±)-115 was found to match the natural alkaloid (107). This fact and the previous chemical correlation (105) with tubulosine (14) have thus established the structure and absolute configuration of (−)-demethyltubulosine as **16**. It is interesting that the positions of the methoxyl and the hydroxyl groups in ring A of this base are just the reverse of those of (+)-demethylpsychotrine (**8**).

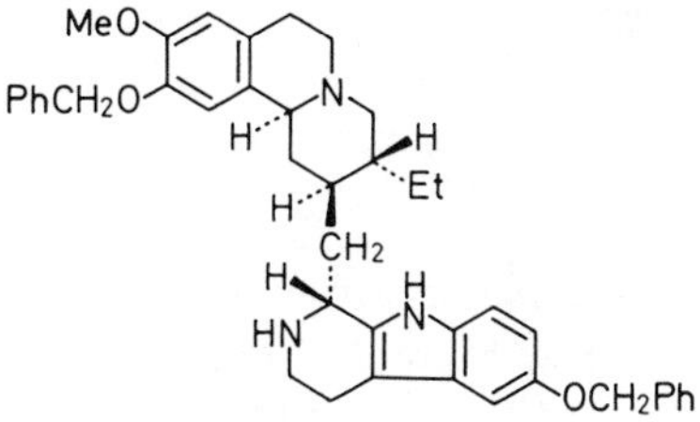

115

F. Deoxytubulosine and Alangimarckine

Recent new synthetic approaches to deoxytubulosine (**17**) have multiplied the number of its known syntheses (*77,92,108*). Kametani *et al.* (*31*) carried out Pictet–Spengler reaction of (±)-protoemetine (**11**) with tryptamine (**108**) and obtained (±)-deoxytubulosine (**17**) and its 1′-epimer, but in poor yields. However, replacement of (±)-**11** by the protoemetinol and emetine precursor (±)-**37** in this reaction gave a 4:1 mixture of (±)-4-oxo-deoxytubulosine (**116**) and its 1′-epimer in high yield, and reduction of (±)-**116** with NaAlH₂(OCH₂CH₂OMe)₂ in pyridine yielded (±)-**17** (*104*).

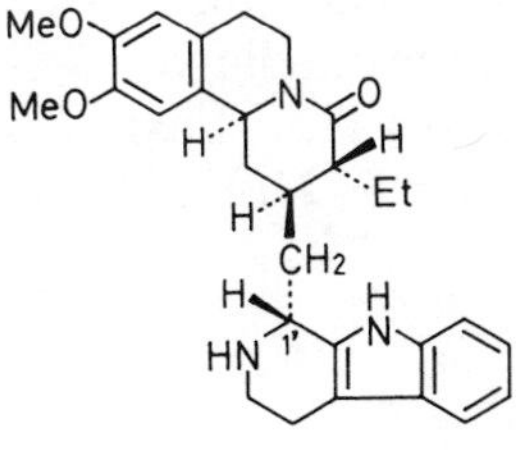

116

Brown *et al.* (*28*) obtained (−)-deoxytubulosine (**17**) by a similar condensation of tryptamine (**108**) with (−)-protoemetine (**11**) derived from secologanin (**98**) and alternatively, from the (+)-piperidone **33** through the corresponding aldehyde and the (+)-tetrahydro-β-carboline **117**.

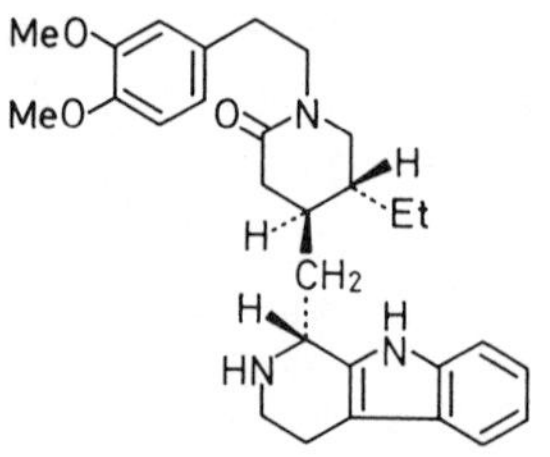

117

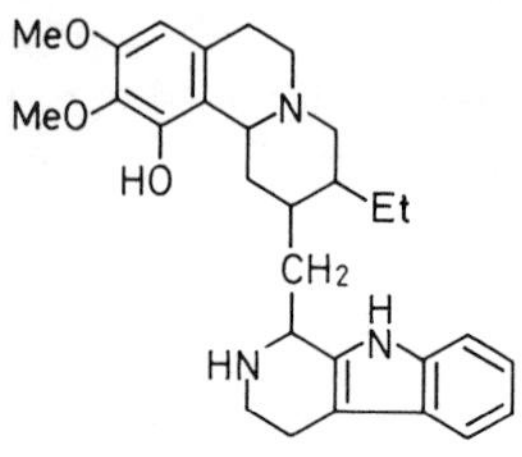

118

The originally proposed planar structure (**118**) for (−)-alangimarckine (*40*) has now been revised by Fujii *et al.* (*109,110*) who prepared the target molecules, (±)-**18** and (−)-**18**, selected for synthesis. Condensation of the (±)-tricyclic amino acid **95** (*96*), available as a key intermediate for the

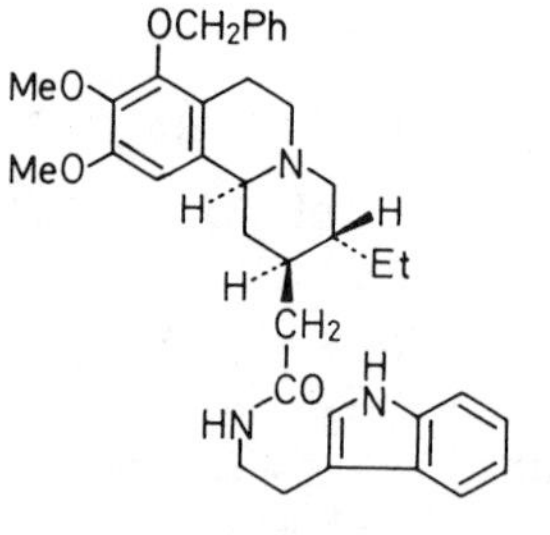

119

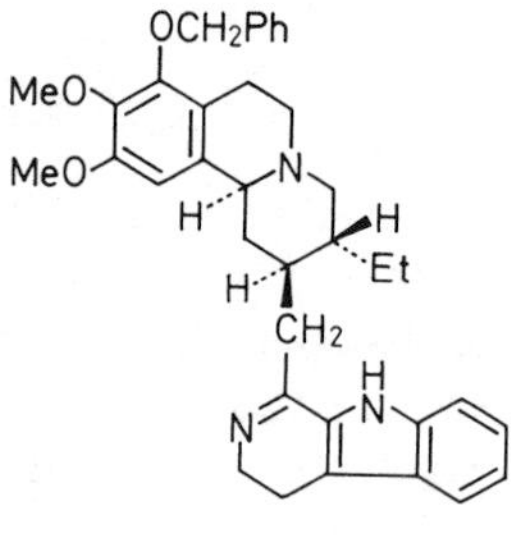

120

synthesis of (±)-alangicine (**9**), with tryptamine (**108**) by the diethyl phosphorocyanidate method (*98*) yielded the (±)-tryptamide **119** that was then cyclized with POCl₃ to give the (±)-dihydro-β-carboline **120**. Reduction of (±)-**120** with NaBH₄ afforded a 1 : 3.3 mixture of the (±)-tetrahydro-β-carboline **121** and its 1′-epimer (±)-**122**. Hydrogenolytic debenzylation of the

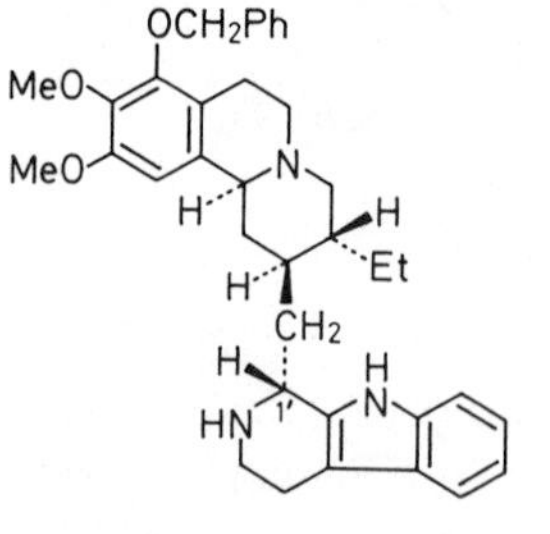

121

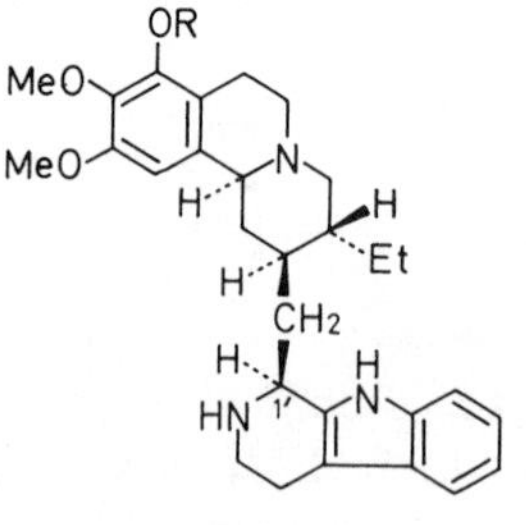

122 R = CH₂Ph
123 R = H

two bases furnished (±)-**18** and its 1′-epimer (±)-**123**, respectively, and the former was identical, apart from its racemic nature, with (−)-alangimarckine (*109*). A parallel transformation of the (−)-tricyclic amino acid **95** (*97*) produced the desired (−)-base **18,** identical with the natural material, as well as its 1′-epimer (−)-**123** (*110*). Thus, (−)-alangimarckine has the structure and absolute stereochemistry represented in formula **18.**

It has been reported (*111*) that permanganate oxidation of deoxytubulosine (**17**) yielded *m*-hemipinic acid (**124**), whereas selenium dehydrogenation and zinc dust distillation of **17** and alangimarckine (**18**) produced harman (**125**).

124   125

G. Ipecoside and Alangiside

When the last review on the ipecac alkaloids was written (for Volume XIII in this treatise), the chemistry of (−)-ipecoside (**19**) (*112–116*), the progenitor of a new series of monoterpenoidal alkaloid glycosides, was not well established, and the existence of (−)-alangiside (**20**) was not even known (*3,4*). The latter substance was then isolated from *A. lamarckii* and studied intensively (*11,12*). The main features of the two alkaloidal glycosides have now been elucidated and summarized in a few places (*5,7,117*). Since this subsection is intended not to duplicate material, the reader is referred to Kapil and Brown (*117*) (Volume XVII, Chapter 5, in this treatise) for an authoritative summary of the chemistry of (−)-ipecoside (**19**) and (−)-alangiside (**20**). Notable is the absolute configuration of both alkaloids at C-1 of the tetrahydroisoquinoline moiety, which is opposite to that of emetine (**1**) and its congeners at C-11b, the corresponding position.

H. Related Alkaloids of *Alangium lamarckii* Thwaites

Pakrashi's group (*9*) has isolated alangimarine (**21**), (±)-alamarine (**22**), and (+)-alangimaridine (**24**) from the seeds of *Alangium lamarckii* Thw. and characterized them as novel benzopyridoquinolizine bases on the basis of spectral evidence and interrelation with the congeners. The absolute configuration of (+)-alangimaridine (**24**) at C-12b was deduced from a comparison of its specific rotation with those of the closely related tetrahydroprotoberberines, such as (+)-tetrahydropalmatine (**126**). The presence of isoalamarine (**23**) in the stem bark of the same plant was also indicated. The

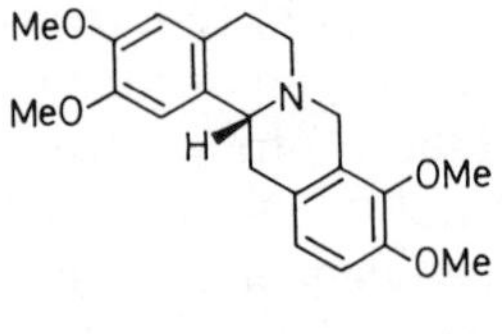

126

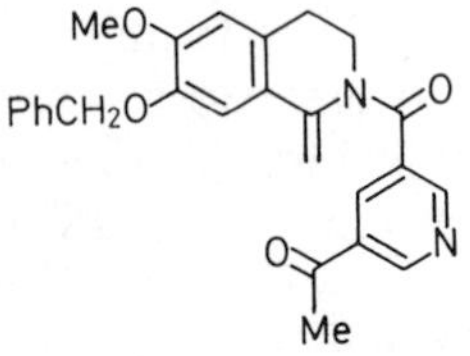

127

structure of (±)-alamarine (**22**), a typical member of this new group, has now been confirmed by Ninomiya and co-workers (*118*), who synthesized this racemic alkaloid from the enamide **127** via a route involving thermal or photochemical cyclization of **127** and debenzylation of the resulting tetracycle **128**, followed by NaBH$_4$ reduction.

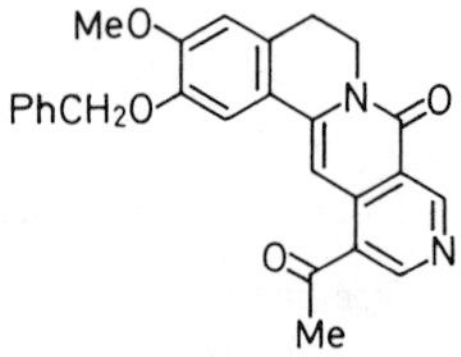

128

IV. Related Compounds

The compelling interest in the chemistry and biological activities of the ipecac alkaloids and their congeners and the attempts to synthesize related compounds have greatly multiplied the number of known benzo[*a*]quinolizine derivatives.

The structure of the (+)-rubremetinium cation (**129**) (*119–123*), the mild oxidation product from emetine (**1**) (*1,121*) or *O*-methylpsychotrine (**7**) (*1,86*), and those of its hydrogenation products, (−)-A(or α)-dihydrorubremetine (**131**) (*120,121,124–126*), (+)-B(or β)-dihydrorubremetine (**132**) (*120,121,124–126*), and (+)-tetrahydrodehydroemetine (**133**) (*121,124, 126*), have been inferred from the early studies summarized in Chapter 3 of Volume XIII and elsewhere (*4,5*). In early synthetic studies of (±)-emetine (**1**) and (±)-*O*-methylpsychotrine (**7**), (±)-rubremetinium bromide (**129·Br⁻**) (*41,54,55,57*) and iodide (**129·I⁻**) (*51,52*) were prepared by dehydrogenation of synthetic alkaloids of mixed or unknown stereochemistry in order to compare with the natural materials. On the basis of structure

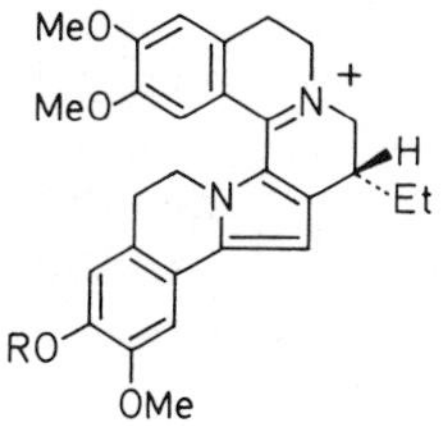

129 R = Me
130 R = H

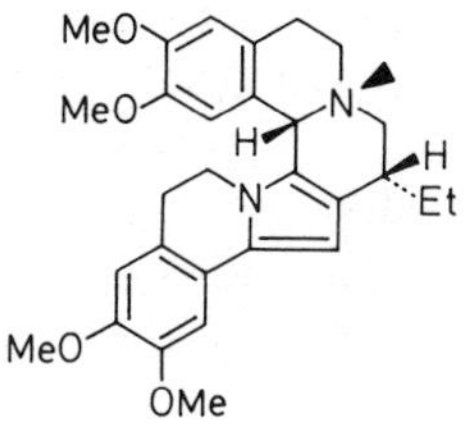

131

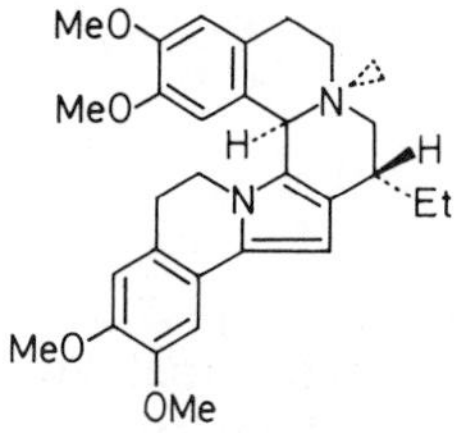

132

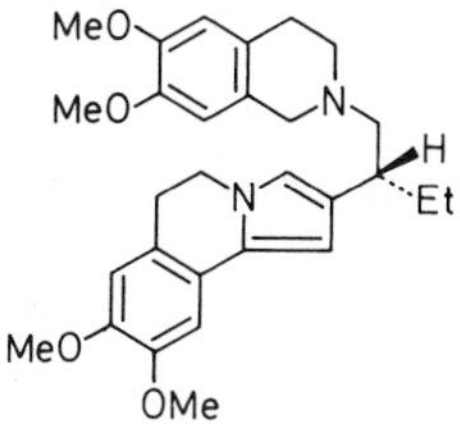

133

129 for rubremetine, structure **134** follows for rubremetamine (*127*) which is a similar oxidation product of emetamine (**10**) (*86,127,128*) and is hydrogenated to give isomeric α- and β-dihydrorubremetamines analogous to the above dihydrorubremetines. Kovar *et al.* (*129*) reported the preparation of rubrocephaelinium chloride (**130·Cl⁻**) and didehydrocephaelinium acetate (**135**) from cephaeline (**2**) by mercuric acetate oxidation.

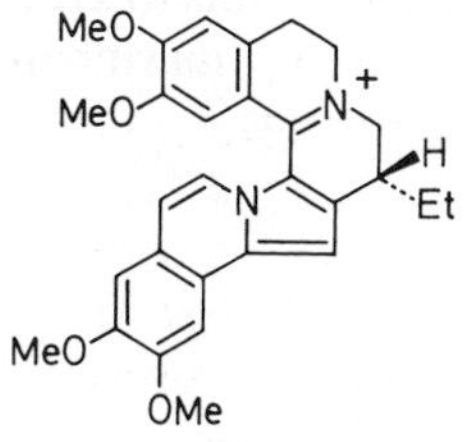

134

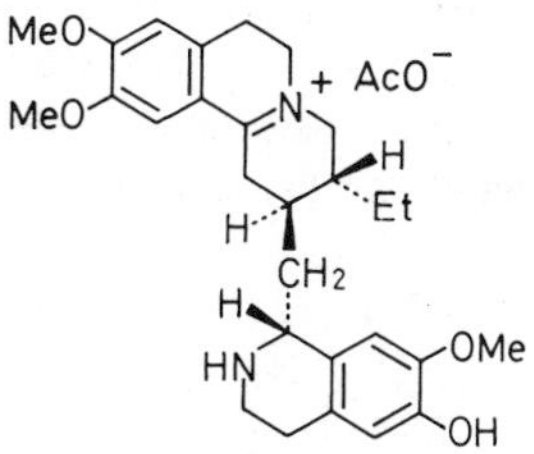

135

One of the highlights covered in the last review on the ipecac alkaloids (*3*) was the discovery of the active amebicide (±)-2,3-dehydroemetine (**136**) by Brossi and co-workers (*63*), and its synthesis (*63,130–132*) and amebicidal activity have also been summarized (*3*), together with those of the 1′-epimer [(±)-2,3-dehydroisoemetine] and the pair of (−)-2,3-dehydroemetine (**136**) (*132,133*) and its (+)-enantiomer (*133*). In addition, new synthetic approaches to 2,3-dehydroemetine (**136**) (*134,135*) and its modified structures such as the tetraethoxy analog (*135,136*), $N^{2'}$-substituted derivatives (*134,137*), and didehydro derivative **137** (*138*) have been reported.

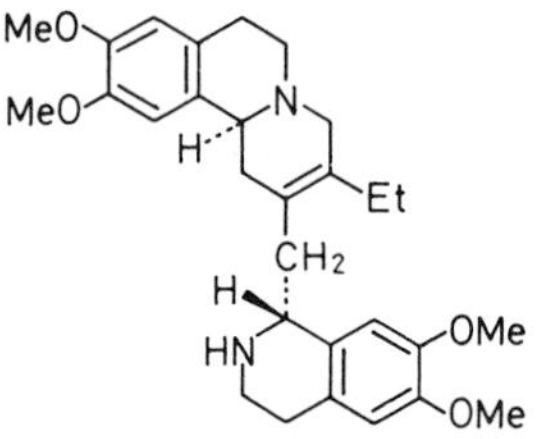

136 2,3-Dehydroemetine 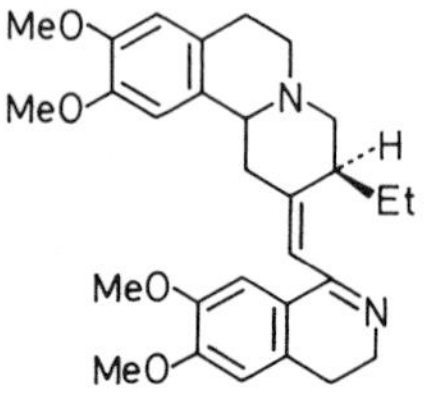   **137**

Of the two types of seco emetines covered in the previous review (*3*), the benzazecine **138** was synthesized by an alternative route involving mono-perphthalic acid oxidation of $N^{2'}$-benzyloxycarbonylemetine and cleavage of the resulting N^{5}-oxide with Li in liquid NH_3 containing 1-methoxy-2-propanol (*139*). Many other compounds possessing modified emetine structures have been synthesized: stereoisomers of (±)-2,3-*cis*-emetine (*52,61,65,140,141*); $N^{2'}$-substituted emetines (*137*); $N^{2'}$-aminoacylemetines (*142–144*); the tetraethoxy analog of emetine (*145*); 3-azaemetine (*146*); (−)-13-noremetine (*147*); (±)-12,13-bisnoremetine (*147–153*); (±)-bisnor-rubremetinium bromide (*122,149,152,153*); (±)-1-methyl- and (±)-1-ethyl-12,13,-bisnoremetines (*154*); compound **139** (*155,156*); compounds **140** (*50*) and **141** (*156,157*) [corresponding to the incorrect structures proposed by Brindley and Pyman (*85*) in 1927 for emetine and emetamine, respectively]; (±)-14-hydroxyemetamine (*158*); and four diastereomers of optically active "pseudotubulosine" (**142**) (regarded as the *Corynanthe*–emetine hybrid structure) (*159,160*).

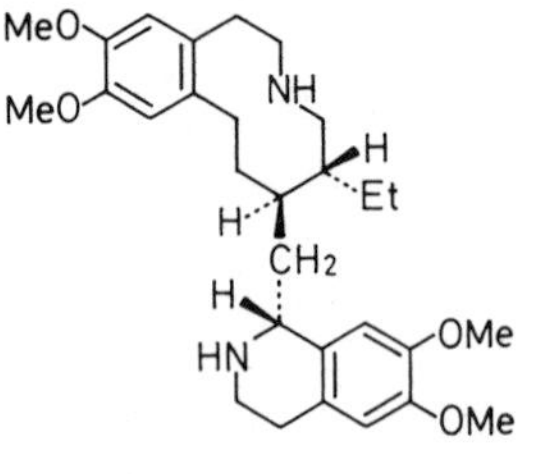

138

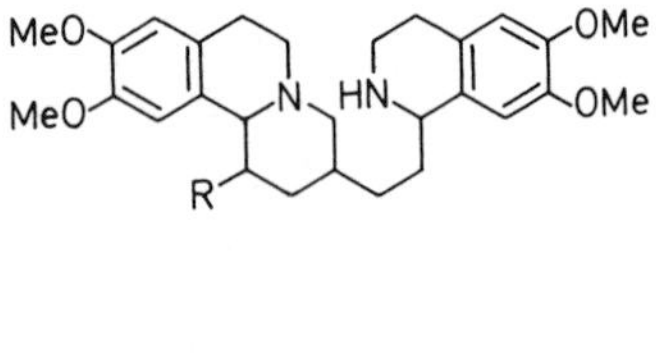

139 R = H
140 R = Me

Many other synthetic benzo[a]quinolizine derivatives have been available, and the reader is referred to Popp and Watts (49) and to Saraf (49a) for recent reviews on these compounds.

141

142

V. Analytical Methods

The clinical usefulness of emetine (**1**) and its various biological properties (*1,3,5,7*) (Section VII) are responsible for the development of many methods suitable for both rapid and accurate analysis of this base and related compounds. Table I lists these methods and the literature contained in *Chemical Abstracts,* Volumes 66 (1967)–96 (1982), with the exception of a few references.

Apart from their general application to the structure establishment of the new alkaloids discussed in Section III, spectroscopic means have played a still more important role in the elucidation of stereochemistry. For example, mass spectrometry (*3–5,278*) was useful in the cases of ankorine (**13**) and the stereoisomers of **42** (*38,39*); ¹H- and/or ¹³C-NMR spectroscopy, for the stereochemical assignment at C-1' of alangimarckine (**18**) (*109*), (−)-9-demethylcephaeline (**104**) (*99*), (−)-10-demethylcephaeline (**105**) (*99*), (±)-9-demethyltubulosine (**109**) (*106*), and (±)-10-demethyltubulosine (**16**) (*107*) [by comparison with the ¹H-NMR spectra of a pair of tubulosine (**14**) and isotubulosine (**15**) (*279*), and/or the ¹³C-NMR spectra of the pair emetine (**1**) (*7,280*) and isoemetine (*109*)], and in the cases of methylalangiside tetraacetate and methylisoalangiside tetraacetate (*281*); CD spectroscopy, in the cases of alangicine (**9**) (*97*) and alangimarckine (**18**) (*110*) [by comparison with the CD spectra of psychotrine (**6**) (*97*), emetine (**1**) (*110*), and isoemetine (*110*)], and four diastereomers of the optically active *Corynanthe*-emetine hybrid structure (**142**) (*160*). In some cases (*38,39,99,106,107,109,279*), TLC mobility has also served as a criterion on which stereochemical assignment is possible for a set of diastereomers. Previously, ORD spectroscopy contributed effectively to the solution of the

TABLE I

ANALYTICAL METHODS FOR IDENTIFICATION AND DETERMINATION OF IPECAC
ALKALOIDS AND RELATED COMPOUNDS

Method	Sample[a]	Reference
Available methods (review)	Emetine-HCl	*161*
Color reaction	Emetine	*162*
	Cephaeline	*162*
Salt formation	Emetine	*163–167*
Precipitation reaction	Emetine	*168*
	Emetine in C	*169*
	Emetine in D	*170*
	Ipecac alkaloids in B	*171*
Complexometry	Ipecac alkaloids in B	*172*
Extraction	Emetine	*168–171, 173–178*
	Emetine in B	*171, 179*
	Emetine in E	*180–184*
	Cephaeline	*177, 179*
	Psychotrine	*177, 179*
	O-Methylpsychotrine	*177, 179*
	Alkaloids in A	*177, 185, 186*
	Ipecac alkaloids in B	*171, 187, 188*
Dialysis	Emetine in D	*189*
Paper chromatography	Emetine	*182, 190, 191*
	Emetine in B	*179, 192*
	Cephaeline in B	*179, 192*
	Psychotrine in B	*179*
	O-Methylpsychotrine	*179*
	Ipecac alkaloids in B	*193, 194*
Thin-layer chromatography	Emetine	*168, 169, 190, 195–213*
	Emetine in A	*214–217*
	Emetine in B	*192, 215, 219–225*
	Emetine in D	*222, 226, 227*
	Cephaeline	*196, 202, 208, 228, 229*
	Cephaeline in A	*214–217, 230*
	Cephaeline in B	*192, 215, 219–222, 224*
	Cephaeline in D	*222, 226*
	Psychotrine in D	*226*
	O-Methylpsychotrine in D	*226*
	Emetamine in D	*226*
	Rubremetine in D	*226*
	Ipecac alkaloids	*177, 200, 231*
	Ipecac alkaloids in B	*171, 193, 194, 232*

TABLE I (Continued)

Method	Sample[a]	Reference
Liquid chromatography	Emetine in A	*233, 234*
	Emetine in B	*234, 235*
	Cephaeline in A	*233, 234*
	Cephaeline in B	*234, 235*
	Psychotrine in A	*233*
	Alkaloids in A	*236*
	Ipecac alkaloids in B	*231, 232*
High-performance liquid	Emetine	*237–239*
chromatography	Emetine in A	*218, 240*
	Emetine in B	*241*
	Emetine in C	*242*
	Cephaeline	*237*
	Cephaeline in A	*218, 240*
	Cephaeline in B	*241*
Thin-layer electrophoresis	Emetine	*243*
Titrimetry	Emetine	*244–250*
	Emetine in B	*251–254*
	Ipecac alkaloids in B	*255*
Spectrophotometry	Emetine	*169, 217, 256–259*
	Emetine in A	*259, 260*
	Emetine in B	*251, 259–262*
	Cephaeline	*217*
Colorimetry	Emetine	*259, 263–265*
	Emetine in A	*259, 266*
	Emetine in B	*259, 266–268*
	Cephaeline in A	*230, 236*
Fluorimetry	Emetine	*175, 216, 238, 239, 269, 270*
	Emetine in B	*271*
	Emetine in C	*242, 272*
	Cephaeline	*216, 270*
	2,3-Dehydroemetine in C	*272*
Atomic absorption spectrometry	Emetine	*272a*
Polarimetry	Emetine	*273*
	Emetine salts	*274*
Polarography	Emetine	*275*
	Emetine in D	*276, 277*

[a] Key: A, ipecac roots; B, pharmaceutical preparations; C, biological fluids and/or tissues; D, solution; E, buffered aqueous solution.

stereochemistry of emetine (**1**) (*37*) and four optically active 2,3-dehydro-
emetines (*133*). The ORD data have been available (*76*) for psychotrine (**6**)
and *O*-methylpsychotrine (**7**).

The ^{13}C-NMR spectral data of benzo[*a*]quinolizidine compounds have
been reported (*154,282,283*), and many examples from the isoquinoline
alkaloids have been reviewed by Hughes and MacLean (*284*) (Volume
XVIII, Chapter 3, in this treatise). As described in Section III, B, ^{13}C-NMR
spectroscopy, together with ^{1}H-NMR and UV spectroscopy, has also
played the leading role in the confirmation of the endocyclic double-bond
structure in the dihydroisoquinoline moiety of *O*-methylpsychotrine (**7**)
(*89*).

Finally, X-ray crystallographic analysis of *O,O*-dimethylipecoside (**143**),
together with the knowledge that hydrolysis of ipecoside (**19**) produces
D-glucose, has established its structure and absolute stereochemistry (*115*).
The position of the double bond in the (±)-unsaturated lactam acid **144**, a
synthetic intermediate for the emetine precursor (±)-**45**, has been inferred
from chemical and spectroscopic evidence (*285*) and confirmed by an X-ray
study (*286*).

143   **144**

VI. Biosynthesis

The close biogenetic relationship of the ipecac and monoterpenoid indole
alkaloids has already been reviewed in several places (*3,5,117,287*). In the
case of the ipecac alkaloids, tracer experiments (*288*) have shown that
labeled geraniol (**146**) and loganin (**148**) are incorporated by *Cephaelis
ipecacuanha* into radioactive cephaeline (**149**) and ipecoside (**150**). Thus,
the C_9–C_{10} units of the ipecac alkaloid and ipecoside, represented by
thickened lines in formulas **149** and **150**, are of monoterpenoid origin from
geraniol (**145**), and loganin (**147**) acts as a precursor for both alkaloids. The
incorporation of glycine into the C_9 unit of cephaeline (**2**) in *Cephaelis*
plants has also been reported (*289*).

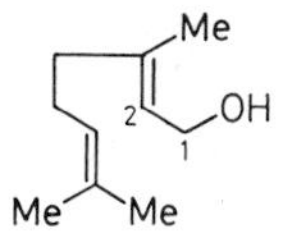

145 Geraniol
146 [2-¹⁴C]Geraniol

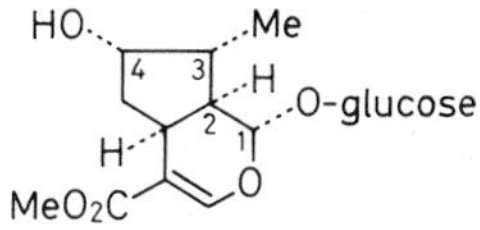

147 Loganin
148 [2-¹⁴C] Loganin

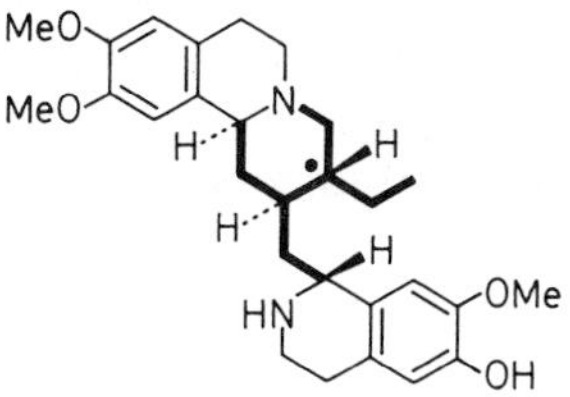

149

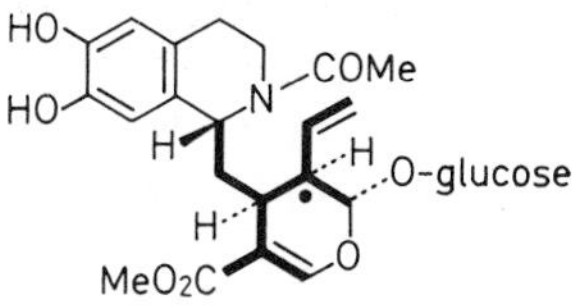

150

The suggested pathway (*290*) beyond loganin (**147**) involves loganin (**147**) → secologanin (**98**) → deacetylipecoside (**151**) (*116*) and its 1α-H isomer deacetylisoipecoside (**152**) (*116*) → ipecoside (**19**), cephaeline (**2**), and emetine (**1**). That the two isoquinoline moieties in cephaeline (**2**) are biogenetically derived from tyrosine and not from phenylalanine was proved by tracer experiments with [2-¹⁴C]tyrosine and [2-¹⁴C]phenylalanine in *C. ipecacuanha* (*291*). Support for the first and the second stages of the above pathway came from the established biosynthesis of secologanin (**98**) from loganin (**147**) in *Vinca rosea* and *Menyanthes trifloliata* plants (*116*), together with the reaction of secologanin (**98**) with dopamine to produce deacetylipecoside (**151**) and deacetylisoipecoside (**152**) (*116*).

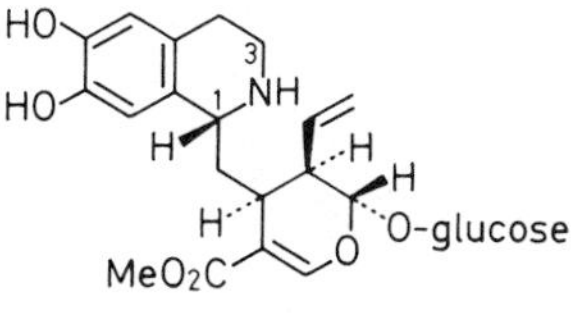

151 Deacetylipecoside

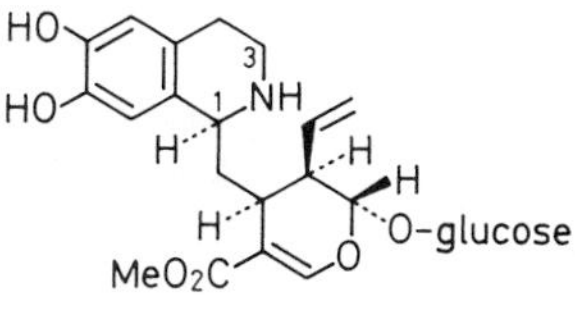

152 Deacetylisoipecoside

In sharp contrast to the earlier claim (*292*) that the 1β-H isomer deacetyli-pecoside (**151**) is the precursor for cephaeline (**2**) and emetine (**1**), it has now been well established (*293,294*) that the 1α-H isomer deacetylisoipecoside (**152**) is incorporated by both *C. ipecacuanha* (*293,294*) and *Alangium lamarckii* (*293*) into cephaeline (**2**) and emetine (**1**) with retention of configuration. The 1β-H isomer deacetylipecoside (**151**) was not incorporated into both alkaloids, but was exclusively and specifically metabolized to give ipecoside (**19**) in *Cephaelis* (*293,294*) and alangiside (**20**) in *Alangium* (*293*), both with retention of configuration. Thus, the known biosynthetic pathways for the ipecac, *Alangium,* and indole alkaloids (*293,294*) can be depicted as shown in Schemes 1 and 2.

It has been suggested (*293*) that the two N-acylated alkaloidal glycosides, ipecoside (**19**) and alangiside (**20**), are metabolic dead-end products from intermediates with wrong stereochemistry. Interestingly, however, the structures arrived at for (+)-alangimaridine (**24**), alangimarine (**21**), and alamarine (**22**), the novel benzopyridoquinolizine bases isolated recently from *Alangium lamarckii* (see Section III,H), also suggest (*9*) that they are

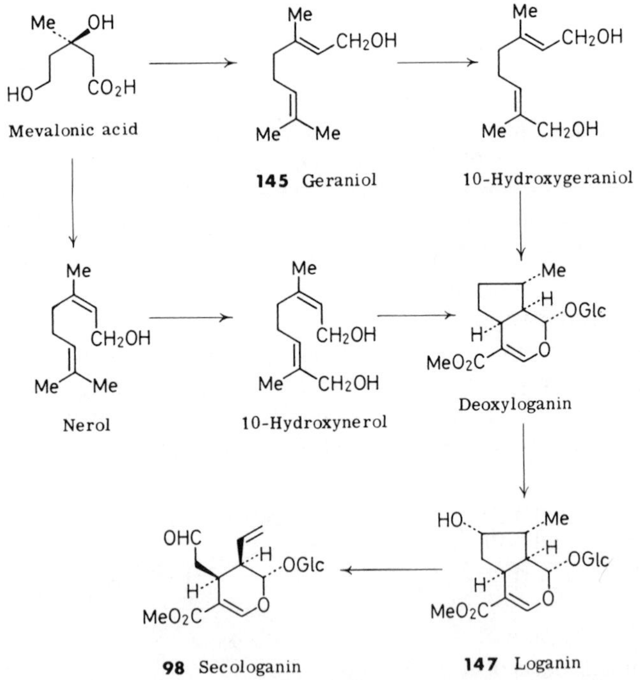

Mevalonic acid   **145** Geraniol   10-Hydroxygeraniol

Nerol   10-Hydroxynerol   Deoxyloganin

98 Secologanin   **147** Loganin

SCHEME 1. Biosynthesis of ipecac, *Alangium,* and indole alkaloids: C_9–C_{10} unit portion.

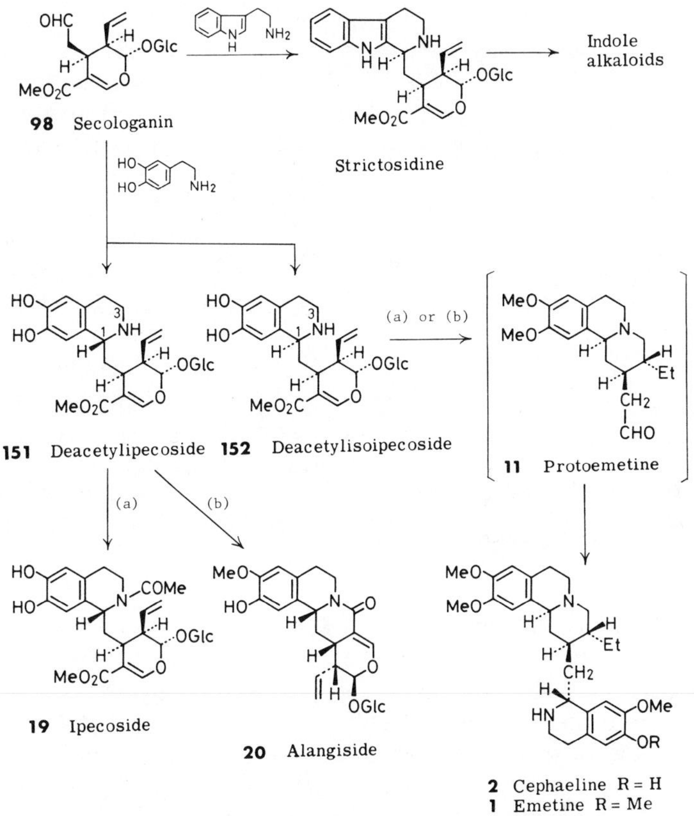

SCHEME 2. Biosynthesis of ipecac, *Alangium*, and indole alkaloids: nitrogenous portion. Pathway (a) in *Cephaelis ipecacuanha;* pathway (b) in *Alangium lamarckii.*

biogenetically derivable from alangiside (**20**) through hydrolysis, amination, and other unexceptional transformations.

VII. Biological Activity

Much of the early work on the pharmacological and biological properties of emetine (**1**) and both its natural and synthetic congeners has already been reviewed (*1,3,5,7,295–302*). Tables II and III supplement these reviews by summarizing the recent data from *Chemical Abstracts,* Volumes 72 (1970)–96 (1982), with the exception of some references.

The various parts of *Alangium lamarckii* Thwaites have been used in the indigenous Indian systems of medicine for a long time: They have been

TABLE II

PHARMACOLOGICAL AND BIOLOGICAL PROPERTIES OF EMETINE

Pharmacological and biological action	Reference
Amebicidal activity	*303–332a*
Other antiparasitic activities	*332a–335*
Antibacterial activity	*336–338*
Antiviral activity	*339–343*
Antitumor activity	*344–351*
No carcinogenicity	*352, 353*
Cardiovascular effect	*354–356*
Formation of emetine-resistant mutants	*357–367*
Neuromuscular effect	*368–374*
Inhibition of RNA, DNA, and protein synthesis	*314, 317, 325, 329, 340, 348, 349, 361, 375–424*
Toxicity	*347, 425–433*
Effects on other biochemical processes	*317, 339, 382, 411, 413, 415, 424, 432, 434–475a*

claimed useful as anthelmintic, alterative, purgative, emetic, diuretic, diaphoretic, antipyretic, alexipharmic, aphrodisiac, tonic, expectorant, alexe-
teric agents, and effective against leprosy, syphilis, dysentery, rheumatic
pain, lumbago, and diseases of the blood (*507–509*). An alkaloidal material
from the bark of this plant (*508*) and aqueous extracts of the roots of *A.
salvifolium* Wangerin (*510*) were reported to produce a drop in blood
pressure. Alkaloid extracts of *A. vitiense* have been reported to possess
oncostatic activity (*511*) and an alkaloid ($C_{28}H_{35}N_3O_3$) from the bark (see
Section II) to increase the survival time of mice infected with leukemia
L1210 or P388 (*20*). An amorphous alkaloidal mixture (AL 60) isolated
from the stem bark of *A. lamarckii* and eventually proved to be a mixture of
psychotrine (**6**), cephaeline (**2**), and demethylcephaeline (**5**) has been found
to exert dose-dependent biphasic action on blood pressure (*10*).

Addendum

Pakrashi's group (*512*) has isolated two new alkaloids from the seeds of
A. lamarckii and, on the basis of spectral and chemical evidence, inferred
them to be 9-demethylprotoemetinol (**153**) and 10-demethylprotoemetinol
(**154**). Fujii *et al.* (*513*) have confirmed the structure and stereochemistry of
the second alkaloid by synthesizing (±)-**154**, (−)-**154**, and their diacetates
from the (±)-ethyl ester (*94*) and the (−)-ethyl ester (*100*) of the tricyclic

TABLE III

PHARMACOLOGICAL AND BIOLOGICAL PROPERTIES OF EMETINE CONGENERS

Compound	Pharmacological and biological action	Reference
Cephaeline (2)	Antitumor activity	*346*
	RNA, DNA, and protein synthesis	*317, 422*
	Toxicity	*425*
	Effects on other biochemical processes	*446, 476*
2,3-Dehydroemetine (136)	Amebicidal activity	*303, 313, 323, 326, 328*
	Other antiparasitic activities	*333, 477–482*
	Antiviral activity	*339, 483*
	Antitumor activity	*344, 347*
	Cardiovascular effect	*484–488*
	Neuromuscular effect	*370, 371, 484, 489–493*
	RNA, DNA, and protein synthesis	*422, 494*
	Toxicity	*347*
	Effects on other biochemical processes	*438, 495–498*
Deoxytubulosine (17)	RNA, DNA, and protein synthesis	*317*
	Effects on other biochemical processes	*317*
2′-Substituted emetines	Amebicidal activity	*499*
	Toxicity	*499, 500*
Hydroxybutyl-2,3-dehydroemetine	Antitumor activity	*344*
Isoemetine	RNA, DNA, and protein synthesis	*377*
Aqueous extracts of ipecac roots	Toxicity	*501, 502*
O-Methylpsychotrine (7)	RNA, DNA, and protein synthesis	*377*
Tubulosine (14)	Amebicidal activity	*314*
	Antitumor activity	*346*
	RNA, DNA, and protein synthesis	*314, 317, 417, 422, 503–505*
	Toxicity	*425*
	Effects on other biochemical processes	*317, 460, 476, 503, 504, 506*

amino acid **112**. Although the isomeric (−)-amino alcohol **153** was similarly synthesized (*513*) from the (−)-ethyl ester (*95*) of the tricyclic amino acid **111**, a direct comparison with the other alkaloid inferred to be 9-demethyl-protoemetinol was not possible owing to paucity of the natural base.

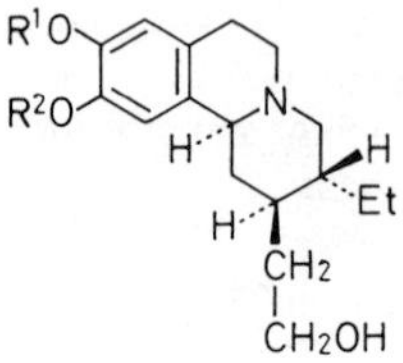

153 9-Demethylprotoemetinol
$R^1 = H, R^2 = Me$
154 10-Demethylprotoemetinol
$R^1 = Me, R^2 = H$

More recently, Pakrashi *et al.* (*514*) have isolated another new alkaloid named bharatamine [$C_{18}H_{19}NO_2$; mp 182–183°C; $[\alpha]_D \pm 0°$] from the seeds of the same plant and established its structure as **155** by an unequivocal synthesis. The structure is unique for its novel racemic tetrahydroprotober-berine skeleton unoxygenated in ring D.

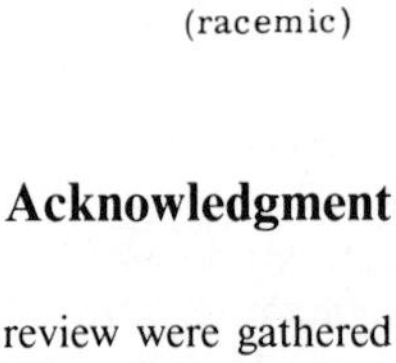

155 Bharatamine
(racemic)

Acknowledgment

A part of the references in this review were gathered by an on-line computer search of *Chemical Abstracts.* We thank Dr. Mitsuo Sasamoto for these references.

REFERENCES

1. M.-M. Janot, *in* "The Alkaloids" (R. H. F. Manske and H. L. Holmes, eds.), Vol. III, Chapter 24. Academic Press, New York, 1953.
2. R. H. F. Manske, *in* "The Alkaloids" (R. H. F. Manske, ed.), Vol. VII, Chapter 18. Academic Press, New York, 1960.

3. A. Brossi, S. Teitel, and G. V. Parry, *in* "The Alkaloids" (R. H. F. Manske, ed.), Vol. XIII, Chapter 3. Academic Press, New York, 1971.
4. H. T. Openshaw, *in* "Chemistry of the Alkaloids" (S. W. Pelletier, ed.), Chapter 4. Van Nostrand-Reinhold, Princeton, New Jersey, 1970.
5. M. Shamma, "The Isoquinoline Alkaloids," Chapter 23. Academic Press, New York, 1972.
6. T. Kametani, "The Chemistry of the Isoquinoline Alkaloids," Vol. 2, Chapter 18. Kinkodo, Sendai, Japan, 1974.
7. M. Shamma and J. L. Moniot, "Isoquinoline Alkaloids Research, 1972–1977," Chapter 27. Plenum, New York, 1978.
8. B. Achari, E. Ali, P. P. Ghosh Dastidar, R. R. Sinha, and S. C. Pakrashi, *Planta Med., Suppl.* 5 (1980).
9. S. C. Pakrashi, B. Achari, E. Ali, P. P. Ghosh Dastidar, and R. R. Sinha, *Tetrahedron Lett.* **21,** 2667 (1980).
10. S. C. Pakrashi and B. Achari, *Experientia* **26,** 933 (1970).
11. R. S. Kapil, A. Shoeb, S. P. Popli, A. R. Burnett, G. D. Knowles, and A. R. Battersby, *Chem. Commun.* 904 (1971).
12. A. Shoeb, K. Raj, R. S. Kapil, and S. P. Popli, *J. Chem. Soc., Perkin Trans. 1* 1245 (1975).
13. B. Achari, A. Pal, and S. C. Pakrashi, *Indian J. Chem.* **12,** 1218 (1974).
14. F. H. Carr and F. L. Pyman, *J. Chem. Soc.* **105,** 1591 (1914).
15. S. Baassou, H. Mehri, and M. Plat, *Phytochemistry* **17,** 1449 (1978).
16. E. Späth, A. Orechoff, and F. Kuffner, *Ber. Dtsch. Chem. Ges.* **67B,** 1214 (1934).
17. A. R. Battersby and R. S. Kapil, *Tetrahedron Lett.* 3529 (1965).
18. M.-J. Chen, L.-L. Hou, and H. Zhu, *Chih Wu Hsueh Pao* **22,** 257 (1980); *CA* **94,** 12818b (1981).
19. L. Hou, M. Chen, and H. Zhu, *Zhongcaoyao* **12,** 352 (1981); *CA* **96,** 139646r (1982).
20. E. Chenut, M. B. Hayat, H. P. Husson, C. Kan-San, G. Mathe, P. Potier, and T. Sevenet, *Fr. Demande* **2,393,004** (1978); *CA* **91,** 181442t (1979).
21. G. H. Mahran and S. H. Hilal, *Egypt. J. Pharm. Sci.* **13,** 321 (1972); *CA* **81,** 169679m (1974); G. H. Mahran, S. H. Hilal, and T. S. El-Alfy, *Planta Med.* **27,** 127 (1975).
22. R. H. F. Manske, *in* "The Alkaloids" (R. H. F. Manske, ed.), Vol. XVI, p. 531. Academic Press, New York, 1977.
23. C. Szántay, L. Töke, and P. Kolonits, *J. Org. Chem.* **31,** 1447 (1966).
24. C. Szántay, L. Töke, and P. Kolonits, *Magy. Kem. Foly.* **73,** 293 (1967); *CA* **68,** 3044g (1968).
25. A. R. Battersby, G. C. Davidson, and B. J. T. Harper, *Chem. Ind. (London)* 983 (1957); A. R. Battersby and B. J. T. Harper, *J. Chem. Soc.* 1748 (1959).
26. C. Szántay and L. Töke, *Fr. Demande* **2,234,285** (1973); *CA* **83,** 97682y (1975).
27. C. Szántay, L. Töke, and G. Blaskó, *Acta Chim. Acad. Sci. Hung.* **95,** 81 (1977); *CA* **90,** 39088t (1979).
28. R. T. Brown, A. G. Lashford, and S. B. Pratt, *J. Chem. Soc., Chem. Commun.* 367 (1979).
29. T. Kametani, Y. Suzuki, H. Terasawa, M. Ihara, and K. Fukumoto, *Heterocycles* **8,** 119 (1977).
30. T. Kametani, Y. Suzuki, H. Terasawa, and M. Ihara, *J. Chem. Soc., Perkin Trans. 1* 1211 (1979).
31. T. Kametani, Y. Suzuki, and M. Ihara, *Heterocycles* **13,** 209 (1979).
32. S. Takano, Y. Takahashi, S. Hatakeyama, and K. Ogasawara, *Heterocycles* **12,** 765 (1979).
33. S. Takano, S. Hatakeyama, Y. Takahashi, and K. Ogasawara, *Heterocycles* **17,** 263 (1982).

36 TOZO FUJII AND MASASHI OHBA

34. A. W. Burgstahler and Z. J. Bithos, *J. Am. Chem. Soc.* **81**, 503 (1959); **82**, 5466 (1960).
35. S. Takano, S. Hatakeyama, and K. Ogasawara, *Tetrahedron Lett.* 2519 (1978).
36. A. R. Battersby and J. C. Turner, *Chem. Ind.* (*London*) 1324 (1958); *J. Chem. Soc.* 717 (1960).
37. E. E. van Tamelen, P. E. Aldrich, and J. B. Hester, Jr., *J. Am. Chem. Soc.* **79**, 4817 (1957); E. E. van Tamelen and J. B. Hester, Jr., *ibid.* **81**, 507 (1959); E. E. van Tamelen, P. E. Aldrich, and J. B. Hester, Jr., *ibid.* 6214.
38. C. Szántay, E. Szentirmay, and L. Szabó, *Tetrahedron Lett.* 3725 (1974).
39. C. Szántay, E. Szentirmay, L. Szabó, and J. Tamás, *Chem. Ber.* **109**, 2420 (1976).
40. A. R. Battersby, R. S. Kapil, D. S. Bhakuni, S. P. Popli, J. R. Merchant, and S. S. Salgar, *Tetrahedron Lett.* 4965 (1966).
41. A. R. Battersby and H. T. Openshaw, *Experientia* **6**, 378 (1950); A. R. Battersby, H. T. Openshaw, and H. C. S. Wood, *J. Chem. Soc.* 2463 (1953).
42. T. Fujii, S. Yoshifuji, and K. Yamada, *Tetrahedron Lett.* 1527 (1975); *Tetrahedron* **36**, 965 (1980).
43. M. Uskoković, C. Reese, H. L. Lee, G. Grethe, and J. Gutzwiller, *J. Am. Chem. Soc.* **93**, 5902 (1971).
44. T. Fujii and S. Yoshifuji, *Chem. Pharm. Bull.* **27**, 1486 (1979).
45. S. Yoshifuji and T. Fujii, *Tetrahedron Lett.* 1965 (1975).
46. T. Fujii and S. Yoshifuji, *J. Org. Chem.* **45**, 1889 (1980).
47. A. Kaufmann, E. Rothlin, and P. Brunschweiler, *Ber. Dtsch. Chem. Ges.* **49**, 2299 (1916).
48. V. Prelog and E. Zalán, *Helv. Chim. Acta* **27**, 535 (1944).
49. F. D. Popp and R. F. Watts, *Heterocycles* **6**, 1189 (1977).
49a. S.-U.-D. Saraf, *Heterocycles* **16**, 803 (1981).
50. S. Sugasawa and K. Kobayashi, *Proc. Jpn. Acad.* **24**(9), 17 (1948); *Yakugaku Zasshi* **69**, 88 (1949); *CA* **44**, 1514d (1950).
51. R. P. Evstigneeva, R. S. Livshits, L. I. Zakharkin, M. S. Baïnova, and N. A. Preobrazhenskiĭ, *Dokl. Akad. Nauk SSSR* **75**, 539 (1950); *CA* **45**, 7577c (1951).
52. R. P. Evstigneeva and N. A. Preobrazhensky, *Tetrahedron* **4**, 223 (1958).
53. S. Sugasawa and T. Fujii, *Proc. Jpn. Acad.* **30**, 877 (1954); *Pharm. Bull.* **3**, 47 (1955).
54. S. Sugasawa and Y. Ban, *Proc. Jpn. Acad.* **31**, 31 (1955); Y. Ban, *Pharm. Bull.* **3**, 53 (1955).
55. T. Fujii, *Chem. Pharm. Bull.* **6**, 591 (1958).
56. N. Itoh and S. Sugasawa, *J. Org. Chem.* **24**, 2042 (1959).
57. M. Pailer and G. Beier, *Monatsh. Chem.* **88**, 830 (1957).
58. J. A. Berson and T. Cohen, *J. Am. Chem. Soc.* **78**, 416 (1956); *J. Org. Chem.* **20**, 1461 (1955); J. A. Berson and J. S. Walia, *ibid.* **24**, 756 (1959).
59. F. D. Popp and W. E. McEwen, *J. Am. Chem. Soc.* **80**, 1181 (1958).
60. E. G. Podrebarac and W. E. McEwen, *J. Org. Chem.* **26**, 1386 (1961).
61. M. Barash and J. M. Osbond, *Chem. Ind.* (*London*) 490 (1958); J. M. Osbond, *ibid.* 257 (1959); M. Barash, J. M. Osbond, and J. C. Wickens, *J. Chem. Soc.* 3530 (1959).
62. A. Brossi, H. Lindlar, M. Walter, and O. Schnider, *Helv. Chim. Acta* **41**, 119 (1958).
63. A. Brossi, M. Baumann, L. H. Chopard-dit-Jean, J. Würsch, F. Schneider, and O. Schnider, *Helv. Chim. Acta* **42**, 772 (1959).
64. A. Brossi, M. Baumann, and O. Schnider, *Helv. Chim. Acta* **42**, 1515 (1959).
65. A. Grüssner, E. Jaeger, J. Hellerbach, and O. Schnider, *Helv. Chim. Acta* **42**, 2431 (1959).
66. E. E. van Tamelen, G. P. Schiemenz, and H. L. Arons, *Tetrahedron Lett.* 1005 (1963); E. E. van Tamelen, C. Placeway, G. P. Schiemenz, and I. G. Wright, *J. Am. Chem. Soc.* **91**, 7359 (1969).
67. D. Beke and C. Szántay, *Chem. Ber.* **95**, 2132 (1962).

68. J. A. Weisbach, J. L. Kirkpatrick, E. L. Anderson, K. Williams, B. Douglas, and H. Rapoport, *J. Am. Chem. Soc.* **87**, 4221 (1965).
69. F. Zymalkowski and A. W. Frahm, *Arch. Pharm. Ber. Dtsch. Pharm. Ges.* **297**, 219 (1964).
70. K. Lénárd and P. Bite, *Acta Chim. Acad. Sci. Hung.* **38**, 57 (1963); *CA* **60**, 14471b (1964); see also *CA* **59**, 15256b (1963).
71. J. H. Chapman, P. G. Holton, A. C. Ritchie, T. Walker, G. B. Webb, and K. D. E. Whiting, *J. Chem. Soc.* 2471 (1962); D. E. Clark, P. G. Holton, R. F. K. Meredith, A. C. Ritchie, T. Walker, and K. D. E. Whiting, *ibid.* 2479 (1962); D. E. Clark, R. F. K. Meredith, A. C. Ritchie, and T. Walker, *ibid.* 2490 (1962).
72. H. T. Openshaw and N. Whittaker, *J. Chem. Soc.* 1449, 1461 (1963).
73. N. Whittaker, *J. Chem. Soc. C* 85 (1969); H. T. Openshaw and N. Whittaker, *ibid.* 89 (1969).
74. T. Fujii and S. Yoshifuji, *Chem. Pharm. Bull.* **27**, 1486 (1979).
75. T. Fujii and S. Yoshifuji, *Tetrahedron Lett.* 731 (1975); *Tetrahedron* **36**, 1539 (1980).
76. S. Teitel and A. Brossi, *J. Am. Chem. Soc.* **88**, 4068 (1966).
77. H. T. Openshaw and N. Whittaker, *J. Chem. Soc. C* 91 (1969).
78. C. Szántay, I. Jelinek, I. Turcsan, and A. Brencsan, *Br. UK Pat. Appl. GB* **2,039,902** (1980); *CA* **96**, 6605e (1982).
79. S. Takano, M. Sato, and K. Ogasawara, *Heterocycles* **16**, 799 (1981).
80. S. Takano, M. Sasaki, H. Kanno, K. Shishido, and K. Ogasawara, *Heterocycles* **7**, 143 (1977); *J. Org. Chem.* **43**, 4169 (1978).
81. T. Kametani, S. A. Surgenor, and K. Fukumoto, *Heterocycles* **14**, 303 (1980); *J. Chem. Soc., Perkin Trans. 1* 920 (1981).
82. I. Ninomiya, T. Kiguchi, and Y. Tada, *Heterocycles* **6**, 1799 (1977).
83. M. Barczai-Beke, I. Jelinek, and C. Szántay, *Acta Chim. Acad. Sci. Hung.* **95**, 77 (1977); *CA* **89**, 59991q (1978).
84. T. Shono, M. Sasaki, K. Nagami, and H. Hamaguchi, *Tetrahedron Lett.* **23**, 97 (1982).
85. W. H. Brindley and F. L. Pyman, *J. Chem. Soc.* **130**, 1067 (1927).
86. P. Karrer, C. H. Eugster, and O. Rüttner, *Helv. Chim. Acta* **31**, 1219 (1948).
87. A. Brossi, J. Würsch, and O. Schnider, *Chimia* **12**, 114 (1958).
88. C. Schuij, G. M. J. Beijersbergen van Henegouwen, and K. W. Gerritsma, *J. Chem. Soc., Perkin Trans. 1* 970 (1979).
89. T. Fujii, M. Ohba, O. Yonemitsu, and Y. Ban, *Chem. Pharm. Bull.* **30**, 598 (1982).
90. S. Simo and I. Jelinek, *Hung. Teljes* **5002** (1972); *CA* **78**, 84629h (1973).
91. C. Schuijt, G. M. J. Beijersbergen van Henegouwen, and K. W. Gerritsma, *Pharm. Weekbl., Sci. Ed.* **1**, 186 (1979); *CA* **91**, 57253t (1979).
92. A. R. Battersby, G. C. Davidson, and J. C. Turner, *J. Chem. Soc.* 3899 (1961).
93. S. C. Pakrashi and E. Ali, *Tetrahedron Lett.* 2143 (1967).
94. T. Fujii, M. Ohba, S. C. Pakrashi, and E. Ali, *Heterocycles* **12**, 1463 (1979).
95. T. Fujii, M. Ohba, S. C. Pakrashi, and E. Ali, *Tetrahedron Lett.* 4955 (1979).
96. T. Fujii, K. Yamada, S. Yoshifuji, S. C. Pakrashi, and E. Ali, *Tetrahedron Lett.* 2553 (1976).
97. T. Fujii, S. Yoshifuji, S. Minami, S. C. Pakrashi, and E. Ali, *Heterocycles* **8**, 175 (1977).
98. T. Shioiri, Y. Yokoyama, Y. Kasai, and S. Yamada, *Tetrahedron* **32**, 2211 (1976).
99. T. Fujii and M. Ohba, *Heterocycles* **19**, 857 (1982).
100. T. Fujii, M. Ohba, and H. Suzuki, *Heterocycles* **19**, 705 (1982).
101. H. T. Openshaw and N. Whittaker, *Chem. Commun.* 131 (1966).
102. C. Szántay and G. Kalaus, *Acta Chim. Acad. Sci. Hung.* **49**, 427 (1966); *CA* **67**, 22072k (1967).
103. C. Szántay and G. Kalaus, *Chem. Ber.* **102**, 2270 (1969).

104. T. Kametani, Y. Suzuki, and M. Ihara, *Heterocycles* **11**, 415 (1978); *Can. J. Chem.* **57**, 1679 (1979).
105. A. Popelak, E. Haack, and H. Spingler, *Tetrahedron Lett.* 1081 (1966).
106. M. Ohba, M. Hayashi, and T. Fujii, *Heterocycles* **14**, 299 (1980).
107. T. Fujii, M. Ohba, A. Popelak, S. C. Pakrashi, and E. Ali, *Heterocycles* **14**, 971 (1980).
108. A. R. Battersby, J. R. Merchant, E. A. Ruveda, and S. S. Salgar, *Chem. Commun.* 315 (1965).
109. T. Fujii, S. Yoshifuji, and H. Kogen, *Tetrahedron Lett.* 3477 (1977).
110. T. Fujii, H. Kogen, and M. Ohba, *Tetrahedron Lett.* 3111 (1978).
111. J. R. Merchant and S. S. Salgar, *Indian J. Chem.* **13**, 100 (1975); *CA* **82**, 140361m (1975).
112. P. Bellet, *Ann. Pharm. Fr.* **10**, 81 (1952).
113. P. Bellet, *Ann. Pharm. Fr.* **12**, 466 (1954).
114. A. R. Battersby, B. Gregory, H. Spencer, J. C. Turner, M.-M. Janot, P. Potier, P. Francois, and J. Levisalles, *Chem. Commun.* 219 (1967).
115. O. Kennard, P. J. Roberts, N. W. Isaacs, F. H. Allen, W. D. S. Motherwell, K. H. Gibson, and A. R. Battersby, *Chem. Commun.* 899 (1971); P. J. Roberts, N. W. Isaacs, F. H. Allen, W. D. S. Motherwell, and O. Kennard, *Acta Crystallogr., Sect. B* **30**, 133 (1974).
116. A. R. Battersby, A. R. Burnett, and P. G. Parsons, *Chem. Commun.* 1280 (1968); *J. Chem. Soc. C* 1187 (1969).
117. R. S. Kapil and R. T. Brown, *in* "The Alkaloids" (R. H. F. Manske and R. G. A. Rodrigo, eds.), Vol. XVII, Chapter 5. Academic Press, New York, 1979.
118. T. Naito, O. Miyata, I. Ninomiya, and S. C. Pakrashi, *Heterocycles* **16**, 725 (1981).
119. A. R. Battersby, H. T. Openshaw, and H. C. S. Wood, *Experientia* **5**, 114 (1949).
120. H. T. Openshaw and H. C. S. Wood, *J. Chem. Soc.* 391 (1952).
121. H. T. Openshaw, *Spec. Publ.—Chem. Soc.* **3**, 28 (1955).
122. Y. Ban and M. Terashima, *Chem. Pharm. Bull.* **13**, 775 (1965).
123. A. Brossi, M. Gerecke, A. R. Battersby, R. S. Kapil, Y. Ban, and M. Terashima, *Experientia* **22**, 134 (1966).
124. R. F. Tietz and W. E. McEwen, *J. Am. Chem. Soc.* **75**, 4945 (1953).
125. K.-A. Kovar, H. Ehrhardt, and H. Auterhoff, *Arch. Pharm. Ber. Dtsch. Pharm. Ges.* **302**, 220 (1969).
126. K.-A. Kovar, *Arch. Pharm. Ber. Dtsch. Pharm. Ges.* **303**, 579 (1970).
127. A. R. Battersby, *Spec. Publ.—Chem. Soc.* **3**, 36 (1955).
128. F. L. Pyman, *J. Chem. Soc.* **111**, 419 (1917).
129. K.-A. Kovar, P. Andreas, and H. Auterhoff, *Arch. Pharm. Ber. Dtsch. Pharm. Ges.* **305**, 940 (1972).
130. M. Gerecke and A. Brossi, *Helv. Chim. Acta* **47**, 1117 (1964).
131. D. E. Clark, P. G. Holton, R. F. K. Meredith, A. C. Ritchie, T. Walker, and K. D. E. Whiting, *J. Chem. Soc.* 2479 (1962).
132. N. Whittaker, *J. Chem. Soc. C* 94 (1969).
133. A. Brossi, M. Baumann, F. Burkhardt, R. Richle, and J. R. Frey, *Helv. Chim. Acta* **45**, 2219 (1962).
134. E. Merck, A.-G., *Br. Pat.* **1,122,212** (1968); *CA* **70**, 47319w (1969).
135. C. Szántay and J. Rohaly, *Acta Chim. Acad. Sci. Hung.* **96**, 55 (1978); *CA* **89**, 129768y (1978).
136. C. Szántay, J. Rohaly, and I. Jelinek, *U. S. Pat.* **4,133,812** (1979); *CA* **90**, 152032h (1979).
137. A. C. Ritchie, D. R. Preston, T. Walker, and K. D. E. Whiting, *J. Chem. Soc.* 3385 (1962); A. C. Ritchie, J. D. Cocker, and G. B. Webb, *U. S. Pat.* **3,481,934** (1969); *CA* **72**, 55733h (1970).
138. A. Buzas, J. P. Finet, J. P. Jacquet, and G. Lavielle, *Tetrahedron Lett.* 2433 (1976).

139. J. P. Yardley, *Synthesis* 543 (1973).
140. A. Brossi and O. Schnider, *Helv. Chim. Acta* **45**, 1899 (1962).
141. T. Fujii and S. Yoshifuji, *Chem. Pharm. Bull.* **27**, 2497 (1979).
142. G. R. Pettit and S. K. Gupta, *Can. J. Chem.* **45**, 1561 (1967).
143. G. R. Pettit and S. K. Gupta, *Can. J. Chem.* **45**, 1600 (1967).
144. M. Nacken, P. Pachaly, and F. Zymalkowski, *Arch. Pharm. Ber. Dtsch. Pharm. Ges.* **303**, 122 (1970).
145. J. Rohaly and C. Szántay, *Acta Chim. Acad. Sci. Hung.* **96**, 45 (1978); *CA* **89**, 147115n (1978).
146. J. Gilbert, C. Gansser, C. Viel, R. Cavier, E. Chenu, and M. Hayat, *Farmaco, Ed. Sci.* **33**, 237 (1978); *CA* **89**, 43891w (1978).
147. H. T. Openshaw, N. C. Robson, and N. Whittaker, *J. Chem. Soc. C* 101 (1969).
148. M. Pailer, K. Schneglberger, and W. Reifschneider, *Monatsh. Chem.* **83**, 513 (1952).
149. M. Pailer and H. Strohmayer, *Monatsh. Chem.* **83**, 1198 (1952).
150. Y. Tomimatsu, *Yakugaku Zasshi* **73**, 75 (1953); *CA* **47**, 10543d (1953).
151. S. Sugasawa and K. Oka, *Pharm. Bull.* **2**, 85 (1954).
152. M. Kirisawa, *Chem. Pharm. Bull.* **7**, 38 (1959).
153. M. Barash and J. M. Osbond, *J. Chem. Soc.* 2157 (1959).
154. A. Buzas, R. Cavier, F. Cossais, J.-P. Finet, J.-P. Jacquet, G. Lavielle, and N. Platzer, *Helv. Chim. Acta* **60**, 2122 (1977).
155. S. Sugasawa and K. Sakurai, *Yakugaku Zasshi* **62**, 82 (1942); S. Sugasawa and H. Shigehara, *ibid.* **65B**, 369 (1945); M. Kirisawa, *Chem. Pharm. Bull.* **7**, 35 (1959).
156. S. Sugasawa, *Yakugaku Zasshi* **69**, 8 (1949); *CA* **44**, 1514a (1950).
157. S. Sugasawa and K. Kobayashi, *Proc. Jpn. Acad.* **24**(9), 23 (1948); K. Kobayashi, *Yakugaku Zasshi* **69**, 91 (1949); *CA* **44**, 1514f (1950).
158. R. F. Watts and F. D. Popp, *J. Heterocycl. Chem.* **15**, 1267 (1978).
159. E. Seguin and M. Koch, *C. R. Hebd. Seances Acad. Sci., Ser. C* **284**, 933 (1977); *CA* **87**, 136102p (1977); E. Seguin, M. Koch, E. Chenu, and M. Hayat, *Helv. Chim. Acta* **63**, 1335 (1980).
160. F. Sigaut-Titeux, L. Le Men-Olivier, and J. Le Men, *Heterocycles* **6**, 1895 (1977).
161. L. V. Feyns and L. T. Grady, *Anal. Profiles Drug Subst.* **10**, 289 (1981); *CA* **96**, 24834h (1982).
162. H. A. Dingjan, S. M. Dreyer -Van Der Glas, and G. T. Tjan, *Pharm. Weekbl.* **115**, 445 (1980); *CA* **93**, 245534a (1980).
163. S. Rolski, Z. Zakrzewski, and Z. Neumann, *Farm. Pol.* **22**, 571 (1966); *CA* **66**, 79617a (1967).
164. S. Rolski, Z. Zakrzewski, and J. Sochon, *Farm. Pol.* **22**, 641 (1966); *CA* **66**, 79618b (1967).
165. S. Rolski, Z. Zakrzewski, and D. Zawistowska, *Farm. Pol.* **22**, 725 (1966); *CA* **67**, 57269g (1967).
166. S. Rolski, Z. Zakrzewski, and B. Kowalewska, *Farm. Pol.* **24**, 901 (1968); *CA* **70**, 109183h (1969).
167. Z. Zakrzewski and J. Jarzebinski, *Farm. Pol.* **29**, 517 (1973); *CA* **79**, 118289k (1973).
168. G. Huesmann, *PTA Prakt. Pharm.* **7**, 332, 334, 336 (1978); *CA* **90**, 29071t (1979).
169. L. Larini, P. E. de T. Salgado, R. Teixcira de Camargo, and M. das G. Vilela, *An. Farm. Quim. Sao Paulo* **17**, 37 (1977); *CA* **89**, 208825t (1978).
170. M. Noguchi, M. Kubo, T. Hayashi, and M. Ono, *Chem. Pharm. Bull.* **26**, 3652 (1978); *CA* **90**, 127465u (1979).
171. J. Pasich, M. Kowalczuk, and W. Felinska, *Herba Pol.* **25**, 223 (1979); *CA* **92**, 28643k (1980).

172. J. Jarzebinski and K. Gryz, *Acta Pol. Pharm.* **34**, 637 (1977); *CA* **89**, 30839k (1978).
173. B. A. Persson and G. Schill, *Acta Pharm. Suec.* **3**, 281 (1966); *CA* **67**, 76687h (1967).
174. Cl. Moussion, H. Corneteau, H. L. Boiteau, and C. Boussicault, *J. Eur. Toxicol.* **2**, 98 (1969); *CA* **71**, 68941d (1969).
175. D. Westerlund, K. O. Borg, and P. O. Lagerstrom, *Acta Pharm. Suec.* **9**, 47 (1972); *CA* **77**, 13822j (1972).
176. V. Das Gupta and H. B. Herman, *J. Pharm. Sci.* **62**, 311 (1973); *CA* **78**, 88557f (1973).
177. E. Stahl and E. Willing, *Planta Med.* **34**, 192 (1978); *CA* **90**, 43705j (1979).
178. S. Tsurubou, N. Ohno, and T. Sakai, *Nippon Kagaku Kaishi* 828 (1980); *CA* **93**, 245568q (1980).
179. M. Przyborowska, *Acta Pol. Pharm.* **26**, 325 (1969); *CA* **72**, 107658j (1970).
180. L. Jusiak, *Acta Pol. Pharm.* **28**, 423 (1971); *CA* **76**, 27911d (1972).
181. L. Jusiak, *Acta Pol. Pharm.* **29**, 277 (1972); *CA* **77**, 156356w (1972).
182. L. Jusiak, *Acta Pol. Pharm.* **30**, 49 (1973); *CA* **79**, 18905q (1973).
183. L. Jusiak, *Acta Pol. Pharm.* **31**, 635 (1974); *CA* **83**, 15537t (1975).
184. L. Jusiak, B. Szabelska, and B. Mazurek, *Acta Pol. Pharm.* **35**, 201 (1978); *CA* **89**, 204917b (1978).
185. A. Caron dos Anjos, *Trib. Farm.* **35**, 53 (1967); *CA* **75**, 25300j (1971).
186. I. Andrei, *Farmacia (Bucharest)* **21**, 735 (1973); *CA* **81**, 96372u (1974).
187. G. K. Rao, *Labdev, Part B* **7**, 207 (1969); *CA* **72**, 6257x (1970).
188. I. Makauskas, R. Gonzalez, B. Soler, and L. Fernandez, *Rev. Cubana Farm.* **10**, 243 (1976); *CA* **88**, 11788f (1978).
189. A. Affonso and D. M. Shingbal, *Can. J. Pharm. Sci.* **8**, 57 (1973); *CA* **79**, 57721w (1973).
190. K. Macek, *Ann. Ist. Super. Sanita* **2**, 133 (1966); *CA* **66**, 40732j (1967).
191. Z. I. El-Darawy and Z. M. Mobarak, *Pharmazie* **29**, 391 (1974); *CA* **81**, 130750c (1974).
192. H. Nerlo and K. Koziejewska, *Acta Pol. Pharm.* **23**, 531 (1966); *CA* **66**, 108189v (1967).
193. S. Janicki, *Herba Pol.* **15**, 239 (1969); *CA* **73**, 28993g (1970).
194. C. Fernandez, A. Alessandri, and M. E. Fernandez, *Rev. Bras. Farm.* **52**, 181 (1971); *CA* **76**, 158397p (1972).
195. R. Tulus and G. Iskender, *Istanbul Univ. Eczacilik Fak. Mecm.* **5**, 130 (1969); *CA* **73**, 69887u (1970).
196. S. Zadeczky, D. Kuttel, and M. Szigetvary, *Acta Pharm. Hung.* **42**, 7 (1972); *CA* **76**, 103792h (1972).
197. F. Sita, V. Chmelova, and K. Chmel, *Cesk. Farm.* **22**, 234 (1973); *CA* **79**, 97020y (1973).
198. E. Novakova and J. Vecerkova, *Cesk. Farm.* **22**, 347 (1973); *CA* **80**, 74362t (1974).
199. J. Bertram, *Dtsch. Apoth.-Ztg.* **114**, 1395 (1974); *CA* **82**, 35097g (1975).
200. M. Sobiczewska, *in* "Chromatografia Cienkowarstwowa w Analizie Farmaceutycznej" (B. Borkowski, ed.), p. 186. Panst. Zakl. Wydawn. Lek., Warsaw, 1973; *CA* **82**, 103199u (1975).
201. K. F. Ahrend and D. Tiess, *Wiss. Z. Univ. Rostock, Math.-Naturwiss. Reihe* **22**, 951 (1973); *CA* **83**, 72815f (1975).
202. E. Stahl and W. Schmitt, *Arch. Pharm. Ber. Dtsch. Pharm. Ges.* **308**, 570 (1975); *CA* **83**, 197866e (1975).
203. L. Lepri, P. G. Desideri, and M. Lepori, *J. Chromatogr.* **123**, 175 (1976); *CA* **85**, 124206w (1976).
204. C. Schuijt, G. M. J. Beijersbergen van Henegouwen, and K. W. Gerritsma, *Analyst (London)* **102**, 298 (1977); *CA* **87**, 141319u (1977).
205. M. S. Dahiya and G. C. Jain, *Indian J. Criminol.* **4**, 131 (1976); *CA* **89**, 37467w (1978).
206. R. A. Egli and S. Tanner, *Fresenius' Z. Anal. Chem.* **295**, 398 (1979); *CA* **91**, 96665y (1979).

207. B. B. Wheals, *J. Chromatogr.* **187**, 65 (1980); *CA* **92**, 153234d (1980).
208. H. J. Huizing, F. De Boer, and T. M. Malingre, *J. Chromatogr.* **195**, 407 (1980); *CA* **93**, 143968c (1980).
209. M. C. Dutt and T. T. Poh, *J. Chromatogr.* **195**, 133 (1980); *CA* **93**, 245528b (1980).
210. S. El-Masry, M. G. El-Ghazooly, and A. A. Omar, *Sci. Pharm.* **48**, 330 (1980); *CA* **94**, 127430m (1981).
211. M. C. Dutt and T. T. Poh, *J. Chromatogr.* **206**, 267 (1981); *CA* **94**, 145409r (1981).
212. A. Brantner and J. Vamos, *Acta Pharm. Hung.* **50**, 185 (1980); *CA* **94**, 71604b (1981).
213. M. Petkovic and J. Tomin, *Arh. Farm.* **30**, 3 (1980); *CA* **94**, 127422k (1981).
214. M. S. Habib and K. J. Harkiss, *J. Pharm. Pharmacol.* **21**, Suppl. 57 (1969); *CA* **72**, 63612h (1970).
215. M. S. Habib and K. J. Harkiss, *Planta Med.* **18**, 270 (1970); *CA* **73**, 69889w (1970).
216. V. Massa, F. Gal, P. Susplugas, and G. Maestre, *Trav. Soc. Pharm. Montpellier* **30**, 301 (1970); *CA* **75**, 25448p (1971).
217. M. S. Habib, *Planta Med.* **27**, 294 (1975); *CA* **83**, 39628z (1975).
218. N. P. Sahu and S. B. Mahato, *J. Chromatogr.* **238**, 525 (1982); *CA* **97**, 11906d (1982).
219. W. Kamp and W. J. M. Onderberg, *Pharm. Weekbl.* **101**, 1077 (1966);*CA* **66**, 31977r (1967).
220. A. Kaess and C. Mathis, *Ann. Pharm. Fr.* **24**, 753 (1966); *CA* **67**, 14874h (1967).
221. D. Ghosh, D. D. Datta, and P. C. Bose, *J. Chromatogr.* **32**, 774 (1968); *CA* **68**, 72273t (1968).
222. V. Massa, F. Gal, and P. Susplugas, *Int. Symp. Chromatogr. Electrophor., Lect. Pap., 6th, 1970* 470 (1971); *CA* **78**, 25964r (1973).
223. C. Van Hulle, P. Braeckman, and R. Van Severen, *Farm. Tijdschr. Belg.* **55**, 150 (1978); *CA* **89**, 152769h (1978).
224. K. Takahashi and M. Ono, *Eisei Shikensho Hokoku* **97**, 79 (1979); *CA* **93**, 54054j (1980).
225. K. Takahashi and M. Ono, *Eisei Shikensho Hokoku* **97**, 21 (1979); *CA* **93**, 54055k (1980).
226. G. Bayraktar-Alpmen, *Eczacilik Bul.* **13**, 1 (1971); *CA* **75**, 67447t (1971).
227. F. Eiden and G. Khammash, *Pharm. Ztg.* **118**, 638 (1973); *CA* **79**, 97034f (1973).
228. B. Borkowski and M. Sobiczewska, *Herba Pol.* **14**, 35 (1968); *CA* **70**, 14436d (1969).
229. M. Medianu, I. Cruceanu, and E. Covalschi. *Rom. RO* **69,891** (1981); *CA* **96**, 11745e (1982).
230. M. Sobiczewska and B. Borkowski, *Acta Pol. Pharm.* **27**, 469 (1970); *CA* **74**, 79726a (1971).
231. C. Mathis and A. Kaess, *J. Pharm. Belg.* **21**, 561 (1966); *CA* **66**, 79613w (1967).
232. A. Kaess and C. Mathis, *Chromatogr., Electrophor., Symp. Int., 4th, 1966* 531 (1968); *CA* **72**, 82993z (1970).
233. S. Kori and M. Kono, *Yakugaku Zasshi* **82**, 1211 (1962); *CA* **59**, 1945a (1963).
234. M. F. Sharkey, E. Smith, and J. Levine, *J. Assoc. Off. Anal. Chem.* **54**, 614 (1971); *CA* **75**, 40484r (1971).
235. R. W. Frei, W. Santi, and M. Thomas, *J. Chromatogr.* **116**, 365 (1976); *CA* **84**, 95668v (1976).
236. E. Graf and W. Rosenberg, *Arch. Pharm. Ber. Dtsch. Pharm. Ges.* **303**, 209 (1970); *CA* **73**, 28950r (1970).
237. R. Verpoorte and A. Baerheim Svendsen, *J. Chromatogr.* **100**, 227 (1974); *CA* **82**, 73250j (1975).
238. J. C. Gfeller, G. Frey, J. M. Huen, and J. P. Thevenin, *J. Chromatogr.* **172**, 141 (1979); *CA* **91**, 82828s (1979).

42 TOZO FUJII AND MASASHI OHBA

239. C. E. Werkhoven-Goewie, U. A. T. Brinkman, and R. W. Frei, *Anal. Chim. Acta* **114**, 147 (1980); *CA* **92**, 185952c (1980).
240. G. M. Hatfield, L. Arteaga, J. D. Dwyer, T. D. Arias, and M. P. Gupta, *J. Nat. Prod.* **44**, 452 (1981); *CA* **95**, 138480z (1981).
241. R. Gimet and A. Filloux, *J. Chromatogr.* **177**, 333 (1979); *CA* **92**, 11282u (1980).
242. S. J. Bannister, J. Stevens, D. Musson, and L. A. Sternson, *J. Chromatogr.* **176**, 381 (1979); *CA* **91**, 204075s (1979).
243. A. S. C. Wan, *J. Chromatogr.* **60**, 371 (1971); *CA* **75**, 144032h (1971).
244. M. Yoshimura, M. Sugii, Y. Fujimura, and Y. Tomita, *Bunseki Kagaku* **23**, 280 (1974); *CA* **82**, 47776r (1975).
245. H. J. Uhlmann, *Dtsch. Apoth.-Ztg.* **115**, 1097 (1975); *CA* **83**, 120954z (1975).
246. T. Paal, *Acta Pharm. Hung.* **45**, 66 (1975); *CA* **83**, 168531m (1975).
247. C. Luca, C. Baloescu, G. Semenescu, T. Tolea, and E. Semenescu, *Rev. Chim. (Bucharest)* **30**, 72 (1979); *CA* **90**, 210186z (1979).
248. Zh. Tencheva, G. Velinov, and O. Budevski, *Farmatsiya (Sofia)* **31**, 12 (1981); *CA* **95**, 156655q (1981).
249. A. I. Popov and G. I. Oleshko, *Farmatsiya (Moscow)* **30**, 30 (1981); *CA* **94**, 180761w (1981).
250. A. I. Popov and G. I. Oleshko, *Farmatsiya (Moscow)* **30**, 73 (1981); *CA* **95**, 49323h (1981).
251. H. Reinicke and K. Schiemann, *Dtsch. Apoth.-Ztg.* **109**, 2007 (1969); *CA* **72**, 136461v (1970).
252. P. P. Suprun, *Farm. Zh. (Kiev)* **30**, 68 (1975); *CA* **83**, 168511e (1975).
253. A. I. Popov, *Deposited Doc.* **VINITI 410–478** (1978); *CA* **91**, 216855j (1979).
254. K. Acel and M. M. Tuckerman, *J. Pharm. Sci.* **59**, 1649 (1970); *CA* **74**, 15744x (1971).
255. C.-C. Ch'en and J.-L. Tai, *Yao Hsueh Hsueh Pao* **12**, 803 (1965); *CA* **68**, 98662u (1968).
256. R. M. Pinyazhko, *Aptechn. Delo* **15**, 42 (1966); *CA* **66**, 79627d (1967).
257. A. M. Taha, A. K. S. Ahmad, C. S. Gomaa, and H. M. El-Fatatry, *J. Pharm. Sci.* **63**, 1853 (1974); *CA* **82**, 64588y (1975).
258. A. D. Thomas, *J. Pharm. Pharmacol.* **28**, 838 (1976); *CA* **86**, 96062w (1977).
259. M. R. I. Saleh, S. El-Masry, and N. El-Shaer, *J. Assoc. Off. Anal. Chem.* **62**, 1113 (1979); *CA* **91**, 199004m (1979).
260. M. A. H. Elsayed, M. A. Abdel Salam, N. A. Abdel Salam, and Y. A. Mohammed. *Planta Med.* **34**, 430 (1978); *CA* **90**, 157104t (1979).
261. A. M. Taha and C. S. Gomaa, *J. Assoc. Off. Anal. Chem.* **59**, 683 (1976); *CA* **85**, 37305j (1976).
262. T. Sasaki, I. Hara, and M. Tsubouchi, *Chem. Pharm. Bull.* **24**, 1254 (1976); *CA* **85**, 83279b (1976).
263. R. K. Seth and G. K. Ray, *Indian J. Pharm.* **29**, 130 (1967); *CA* **67**, 67623g (1967).
264. C. N. Carducci, G. R. Barcic, and I. G. Stringa, *Rev. Asoc. Bioquim. Argent.* **232**, 23 (1978); *CA* **90**, 110053k (1979).
265. P. Dubois, J. Lacroix, R. Lacroix, P. Levillain, and C. Viel, *J. Pharm. Belg.* **36**, 203 (1981); *CA* **95**, 121218x (1981).
266. W. Wisniewski and A. Pietura, *Acta Pol. Pharm.* **24**, 393 (1967); *CA* **68**, 16174z (1968).
267. R. K. Seth and G. K. Ray, *Indian J. Pharm.* **29**, 203 (1967); *CA* **67**, 94029z (1967).
268. D. M. Shingbal, *Indian J. Pharm.* **36**, 83 (1974); *CA* **82**, 77136z (1975).
269. H. Wullen, E. Stainier, and M. Luyckx, *J. Pharm. Belg.* **21**, 409 (1966); *CA* **66**, 22265e (1967).
270. F. Nachtmann, H. Spitzy, and R. W. Frei, *Anal. Chim. Acta* **76**, 57 (1975); *CA* **83**, 53111g (1975).

271. M. Carlassare, *Boll. Chim. Farm.* **113**, 270 (1974); *CA* **81**, 140813q (1974).
272. J. M. Rouzioux and A. Badinand, *J. Eur. Toxicol.* **4**, 509 (1972); *CA* **77**, 147406k (1972).
272a. Y. Minami, T. Mitsui, and Y. Fujimura, *Bunseki Kagaku* **30**, 811 (1981); *CA* **96**, 62356m (1982).
273. J. Sagel, *Pharm. Weekbl.* **107**, 119 (1972); *CA* **76**, 117533m (1972).
274. I. Arenas de Castano and G. S. Veloza, *Rev. Colomb. Cienc. Quim.-Farm.* **2**, 105 (1973); *CA* **81**, 41384u (1974).
275. G. Dusinsky and L. Faith, *Pharmazie* **22**, 475 (1967); *CA* **68**, 53280b (1968).
276. P. Balatre, J. C. Guyot, and M. Traisnel, *Ann. Pharm. Fr.* **24**, 425 (1966); *CA* **66**, 5733n (1967).
277. A. De Marco and E. Mecarelli, *Boll. Chim. Farm.* **109**, 516 (1970); *CA* **74**, 130415c (1971).
278. H. Budzikiewicz, S. C. Pakrashi, and H. Vorbrüggen, *Tetrahedron* **20**, 399 (1964).
279. A. Popelak, E. Haack, and H. Spingler, *Tetrahedron Lett.* 5077 (1966).
280. M. C. Koch, M. M. Plat, N. Préaux, H. E. Gottlieb, E. W. Hagaman, F. M. Schell, and E. Wenkert, *J. Org. Chem.* **40**, 2836 (1975).
281. G. Höfle, N. Nagakura, and M. H. Zenk, *Chem. Ber.* **113**, 566 (1980).
282. G. Van Binst and D. Tourwe, *Heterocycles* **1**, 257 (1973); D. Tourwe and G. Van Binst, *ibid.* **9**, 507 (1978).
283. M. Sugiura, N. Takao, K. Iwasa, and Y. Sasaki, *Chem. Pharm. Bull.* **26**, 1168 (1978).
284. D. W. Hughes and D. B. MacLean, *in* "The Alkaloids" (R. H. F. Manske and R. G. A. Rodrigo, eds.), Vol. XVIII, Chapter 3. Academic Press, New York, 1981.
285. T. Fujii, H. Kogen, S. Yoshifuji, and K. Iga, *Chem. Pharm. Bull.* **27**, 1847 (1979).
286. T. Date, K. Aoe, M. Ohba, and T. Fujii, *Yakugaku Zasshi* **99**, 865 (1979); *CA* **91**, 192699h (1979).
287. A. I. Scott, S.-L. Lee, M. G. Culver, W. Wan, T. Hirata, F. Guéritte, R. L. Baxter, H. Nordlöv, C. A. Dorschel, H. Mizukami, and N. E. MacKenzie, *Heterocycles* **15**, 1257 (1981).
288. A. R. Battersby and B. Gregory, *Chem. Commun.* 134 (1968).
289. J. R. Gear and A. K. Garg, *Tetrahedron Lett.* 141 (1968); A. K. Garg and J. R. Gear, *ibid.* 4377 (1969); *Phytochemistry* **11**, 689 (1972).
290. A. R. Battersby, *Pure Appl. Chem.* **14**, 117 (1967).
291. A. R. Battersby, R. Binks, W. Lawrie, G. V. Parry, and B. R. Webster, *J. Chem. Soc.* 7459 (1965).
292. A. R. Battersby and R. J. Parry, *Chem. Commun.* 901 (1971).
293. N. Nagakura, G. Höfle, and M. H. Zenk, *J. Chem. Soc., Chem. Commun.* 896 (1978); N. Nagakura, G. Höfle, D. Coggiola, and M. H. Zenk, *Planta Med.* **34**, 381 (1978).
294. A. R. Battersby, N. G. Lewis, and J. M. Tippett, *Tetrahedron Lett.* 4849 (1978).
295. P. Synek and V. Synek, *Int. Z. Klin. Pharmakol., Ther. Toxikol.* **2**, 371 (1969); *CA* **72**, 64930x (1970).
296. A. P. Grollman, *Ohio State Med. J.* **66**, 257 (1970); *CA* **72**, 130666u (1970).
297. P. E. Thompson, *Arch. Invest. Med., Supl.* 245 (1971); *CA* **77**, 48m (1972).
298. H. H. Anderson, *in* "Drill's Pharmacology in Medicine" (J. R. DiPalma, ed.), 4th ed., p. 1793. McGraw-Hill, New York, 1971; *CA* **77**, 134969w (1972).
299. B. R. Manno and J. E. Manno, *Clin. Toxicol.* **10**, 221 (1977); *CA* **86**, 25758s (1977).
300. A. P. Grollman and Z. Jarkovsky, *Antibiotics (N.Y.)* **3**, 420 (1975).
301. W. C. T. Yang and M. Dubick, *Pharmacol. Ther.* **10**, 15 (1980); *CA* **93**, 160729t (1980).
302. I. M. Rollo, *in* "The Pharmacological Basis of Therapeutics" (A. G. Gilman, L. S. Goodman, and A. Gilman, eds.), 6th ed., Chapter 46. Macmillan, New York, 1980.
303. R. I. Yusupova and L. M. Gordeeva, *Med. Zh. Uzb.* 66 (1969); *CA* **72**, 129612s (1970).

304. I. De Carneri, *Riv. Parassitol.* **31,** 1 (1970); *CA* **74,** 29260y (1971).
305. S. A. Imam, *Indian J. Biochem.* **7,** 206 (1970); *CA* **75,** 45920q (1971).
306. R. J. Duma, *Antimicrob. Agents Chemother. 1970* 109 (1971); *CA* **75,** 137960f (1971).
307. C. J. Flickinger, *Exp. Cell Res.* **68,** 381 (1971); *CA* **76,** 21509g (1972).
308. L. S. Diamond and I. L. Bartgis, *Arch. Invest. Med., Supl.* 339 (1971); *CA* **77,** 233t (1972).
309. N. Trevino-Garcia Manzo, I. Ruiz de Chavez, and M. De La Torre, *Arch. Invest. Med., Supl.* 187 (1971); *CA* **77,** 1209b (1972).
310. K.-M. Cho, J.-Y. Lee, and C.-T. Soh, *Kisengchunghak Chapji* **7,** 121 (1969); *CA* **77,** 83928x (1972).
311. P. Myjak, *Bull. Inst. Mar. Med. Gdansk* **23,** 75 (1972); *CA* **77,** 160511k (1972).
312. C. J. Flickinger, *Exp. Cell Res.* **74,** 541 (1972); *CA* **78,** 12050q (1973).
313. G. P. Dutta and J. N. S. Yadava, *Indian J. Med. Res.* **60,** 1156 (1972); *CA* **78,** 38619g (1973).
314. N. Entner and A. P. Grollman, *J. Protozool.* **20,** 160 (1973); *CA* **78,** 80561g (1973).
315. L. Cerva, *Adv. Antimicrob. Antineoplastic Chemother., Proc. Int. Congr. Chemother., 7th, 1971* **1,** 431 (1972); *CA* **79,** 87831z (1973).
316. A. K. Chatterjee, S. K. Dey, and A. D. Ray, *Acta Histochem.* **48,** 193 (1974); *CA* **80,** 91284m (1974).
317. C. Pareyre and G. Deysson, *C. R. Hebd. Seances Acad. Sci., Ser. D* **277,** 2689 (1973); *CA* **80,** 116663z (1974).
318. M. M. Husain and V. K. M. Rao, *Indian J. Microbiol.* **13,** 37 (1973); *CA* **80,** 117551y (1974).
319. R. Campos, *Rev. Brasil. Clin. Ter.* 587 (1973); *CA* **81,** 10098z (1974).
320. S. Azhar and V. K. M. Rao, *Zentralbl. Bakteriol., Parasitenkd., Infektionskr. Hyg., Abt. 1: Orig., Reihe A* **230,** 270 (1975); *CA* **82,** 108054c (1975).
321. B. N. K. Prasad, *Indian J. Exp. Biol.* **12,** 578 (1974); *CA* **82,** 149902w (1975).
322. W. Kasprzak, *Wiad. Parazytol.* **21,** 195 (1975); *CA* **83,** 157998t (1975).
323. G. P. Dutta and J. N. S. Yadava, *Indian J. Med. Res.* **64,** 224 (1976); *CA* **84,** 116335b (1976).
324. V. C. Pandey, G. P. Dutta, and V. K. M. Rao, *Indian J. Exp. Biol.* **14,** 142 (1976); *CA* **84,** 144930k (1976).
325. N. Entner, *U. S. NTIS, AD Rep.* **AD-A046924** (1977); *CA* **88,** 164831q (1978).
326. G. P. Dutta and L. Narain, *Indian J. Exp. Biol.* **16,** 838 (1978); *CA* **89,** 141171f (1978).
327. U. S. Singh, M. K. Raizada, and V. K. M. Rao, *Zentralbl. Bakteriol., Parasitenkd., Infektionskr. Hyg., Abt. 1: Orig., Reihe A* **241,** 358 (1978); *CA* **89,** 210941w (1978).
328. R. A. Neal, *Arch. Invest. Med.* **9,** Suppl. 1, 387 (1978); *CA* **90,** 49007e (1979).
329. N. Entner, *J. Protozool.* **26,** 324 (1979); *CA* **91,** 84058q (1979).
330. A. L. Yudin, *Acta Protozool.* **18,** 571 (1979); *CA* **93,** 38197c (1980).
331. W. B. Eubank and R. E. Reeves, *Am. J. Trop. Med. Hyg.* **30,** 900 (1981); *CA* **95,** 144553s (1981).
332. B. N. K. Prasad and R. Srivastava, *Curr. Sci.* **51,** 199 (1982); *CA* **97,** 3437m (1982).
332a. F. D. Gillin and L. S. Diamond, *J. Antimicrob. Chemother.* **8,** 305 (1981); *CA* **96,** 3549d (1982).
333. R. A. Neal, *Ann. Trop. Med. Parasitol.* **64,** 159 (1970); *CA* **74,** 2385f (1971).
334. D. Seth and C. D. Lovekar, *Indian J. Med. Res.* **60,** 1251 (1972); *CA* **78,** 24183y (1973).
335. B. M. Stammers, *Z. Parasitenkd.* **47,** 145 (1975); *CA* **84,** 25796m (1976).
336. E. Schwartz, *Zentralbl. Bakteriol., Parasitenk., Infektionskr. Hyg., Abt. 1: Orig.* **212,** 115 (1969); *CA* **72,** 97671s (1970).
337. E. Schwartz and L. Badalik, *Zentralbl. Bakteriol., Parasitenk., Infektionskr. Hyg., Abt. 1: Orig.* **215,** 90 (1970); *CA* **74,** 63076c (1971).

338. E. Schwartz, *Bratisl. Lek. Listy* **54**, 32 (1970); *CA* **77**, 135573z (1972).
339. I. Rosztoczy, *Acta Microbiol.* **16**, 227 (1969); *CA* **72**, 64993v (1970).
340. N. Z. Khazipov and R. P. Tyurikova, *Uch. Zap. Kazan. Gos. Vet. Inst. im. N.E. Baumana* **109**, 54 (1971); *CA* **77**, 14765e (1972).
341. N. G. Khalitova, *Uch. Zap. Kazan. Gos. Vet. Inst. im. N.E. Baumana* **112**, 73 (1972); *CA* **81**, 72578q (1974).
342. T. V. Ramabhadran and R. E. Thach, *J. Virol.* **34**, 293 (1980); *CA* **93**, 38227n (1980).
343. F. Dubini, R. Mattina, and M. Falchi, *Chemioter. Antimicrob.* **3**, 5 (1980); *CA* **95**, 73539d (1981).
344. G. Vajda and J. Hanisch, *Acta Physiol. Acad. Sci. Hung.* **36**, 171 (1969); *CA* **73**, 2530d (1970).
345. I. Kantemir, *Ankara Univ. Tip Fak. Mecm.* **23**, 329 (1970); *CA* **74**, 97734f (1971).
346. W. R. Jondorf, B. J. Abbott, N. H. Greenberg, and J. A. R. Mead, *Chemotherapy* **16**, 109 (1971); *CA* **74**, 110058k (1971).
347. R. K. Morrison, D. E. Brown, E. K. Timmens, M. A. Nieglos, and C. D. Molins, *U. S. C. F. S. T. I., PB Rep.* **195705** (1970); *CA* **75**, 47357k (1971).
348. R. K. Johnson and W. R. Jondorf, *Biochem. J.* **126**, 22 (1972); *CA* **76**, 135942y (1972).
349. R. K. Johnson and W. R. Jondorf, *Biochem. J.* **140**, 87 (1974); *CA* **81**, 130847q (1974).
350. M. Gosalvez, M. Blanco, J. Hunter, M. Miko, and B. Chance, *Eur. J. Cancer* **10**, 567 (1974); *CA* **82**, 118796k (1975).
351. E. Schwartz, M. Vincurova, and L. Badalik, *Bratisl. Lek. Listy* **67**, 583 (1977); *CA* **87**, 193969a (1977).
352. G. D. Stoner, M. B. Shimkin, A. J. Kniazeff, J. H. Weisburger, E. K. Weisburger, and G. B. Gori, *Cancer Res.* **33**, 3069 (1973); *CA* **81**, 34254y (1974).
353. National Cancer Institute, *U. S. NTIS, PB Rep.* **PB-278891** (1977); *CA* **89**, 123248g (1978).
354. St. C. Bartsokas, D. G. Papadimitriou, C. Blassopoulos, J. Hatziminas, D. Kannas, and Z. Katapoti, *Arzneim.-Forsch.* **21**, 1543 (1971); *CA* **76**, 68135x (1972).
355. A. De Micheli, G. A. Medrano, A. Villarreal, and D. Sodi-Pallares, *Arch. Inst. Cardiol. Mex.* **45**, 469 (1975); *CA* **84**, 69335q (1976).
356. G. Zbinden, R. Kleinert, and B. Rageth, *J. Cardiovasc. Pharmacol.* **2**, 155 (1980); *CA* **93**, 37172d (1980).
357. R. S. Gupta and L. Siminovitch, *Cell* **10**, 61 (1977); *CA* **86**, 134423z (1977).
358. R. S. Gupta and L. Siminovitch, *Cell* **9**, 213 (1976); *CA* **87**, 16486v (1977).
359. R. S. Gupta and L. Siminovitch, *Biochemistry* **16**, 3209 (1977); *CA* **87**, 81991v (1977).
360. R. S. Gupta and L. Siminovitch, *Somatic Cell Genet.* **4**, 77 (1978); *CA* **88**, 186968s (1978).
361. R. S. Gupta and L. Siminovitch, *J. Biol. Chem.* **253**, 3978 (1978); *CA* **89**, 57158z (1978).
362. D. Boersma, S. M. McGill, J. W. Mollenkamp, and D. J. Roufa, *J. Biol. Chem.* **254**, 559 (1979); *CA* **90**, 101034c (1979).
363. D. Boersma, S. M. McGill, J. W. Mollenkamp, and D. J. Roufa, *Proc. Natl. Acad. Sci. U. S. A.* **76**, 415 (1979); *CA* **90**, 116563y (1979).
364. V. E. Reichenbecher, Jr. and C. T. Caskey, *J. Biol. Chem.* **254**, 6207 (1979); *CA* **91**, 121298r (1979).
365. J. J. Wasmuth, J. M. Hill, and L. S. Vock, *Somatic Cell Genet.* **6**, 495 (1980); *CA* **93**, 129605a (1980).
366. J. J. Wasmuth, J. M. Hill, and L. S. Vock, *Mol. Cell. Biol.* **1**, 58 (1981); *CA* **94**, 100579e (1981).
367. J. J. Madjar, K. Nielsen-Smith, M. Frahm, and D. J. Roufa, *Proc. Natl. Acad. Sci. U. S. A.* **79**, 1003 (1982); *CA* **96**, 175258z (1982).
368. L. A. Salako, *J. Pharm. Pharmacol.* **22**, 69 (1970); *CA* **72**, 53535w (1970).

369. L. A. Salako, *Eur. J. Pharmacol.* **11**, 342 (1970); *CA* **73**, 118902f (1970).
370. G. Achari, P. K. Banerji, and M. K. Kapoor, *Indian J. Med. Res.* **60**, 273 (1972); *CA* **77**, 135044c (1972).
371. L. A. Salako, *West Afr. J. Pharmacol. Drug Res.* **1**, 13 (1974); *CA* **82**, 38626q (1975).
372. D. Mitolo-Chieppa and A. Marino, *Boll. Soc. Ital. Biol. Sper.* **50**, 1349 (1974); *CA* **83**, 71554h (1975).
373. W. G. Bradley, J. D. Fewings, J. B. Harris, and M. A. Johnson, *Br. J. Pharmacol.* **57**, 29 (1976); *CA* **85**, 72154f (1976).
374. L. Bindoff and M. J. Cullen, *J. Neurol. Sci.* **39**, 1 (1978); *CA* **90**, 132748t (1979).
375. S. Perlman and S. Penman, *Biochem. Biophys. Res. Commun.* **40**, 941 (1970); *CA* **73**, 105388t (1970).
376. A. K. Chatterjee, A. D. Roy, S. C. Datta, and B. B. Ghosh, *Experientia* **26**, 1077 (1970); *CA* **74**, 11750y (1971).
377. P. S. Lietman, *Mol. Pharmacol.* **7**, 122 (1971); *CA* **74**, 109962u (1971).
378. H. R. Mahler, L. R. Jones, and W. J. Moore, *Biochem. Biophys. Res. Commun.* **42**, 384 (1971); *CA* **74**, 137620z (1971).
379. B. Hogan and P. R. Gross, *J. Cell Biol.* **49**, 692 (1971); *CA* **75**, 32071p (1971).
380. H. F. Lodish, D. Housman, and M. Jacobsen, *Biochemistry* **10**, 2348 (1971); *CA* **75**, 58729j (1971).
381. W. F. Bridgers, R. D. Cunningham, and G. Gressett, *Biochem. Biophys. Res. Commun.* **45**, 351 (1971); *CA* **76**, 880f (1972).
382. R. K. Johnson, J. D. Donahue, and W. R. Jondorf, *Xenobiotica* **1**, 131 (1971); *CA* **76**, 10324s (1972).
383. Z. Gilead and Y. Becker, *Eur. J. Biochem.* **23**, 143 (1971); *CA* **76**, 68071y (1972).
384. E. Battaner and D. Vazquez, *Biochim. Biophys. Acta* **254**, 316 (1971); *CA* **76**, 108559x (1972).
385. R. I. Glazer and A. C. Sartorelli, *Biochem. Biophys. Res. Commun.* **46**, 1418 (1972); *CA* **76**, 123073c (1972).
386. T. M. Cashman, K. A. Conklin, and S. C. Chou, *Experientia* **28**, 520 (1972); *CA* **77**, 57053p (1972).
387. D. K. Dube, S. Chakrabarti, and S. C. Roy, *Cancer (Philadelphia)* **29**, 1575 (1972); *CA* **77**, 99450w (1972).
388. D. K. Dube, S. Chakrabarti, and S. C. Roy, *Sci. Cult.* **38**, 139 (1972); *CA* **77**, 99456c (1972).
389. S. Chakrabarti, D. K. Dube, and S. C. Roy, *Biochem. J.* **128**, 461 (1972); *CA* **77**, 122499e (1972).
390. S. Chakrabarti, D. K. Dube, and S. C. Roy, *Biochem. Pharmacol.* **21**, 2539 (1972); *CA* **77**, 147502p (1972).
391. R. Parenti-Rosina and F. Parenti, *Boll. Soc. Ital. Biol. Sper.* **48**, 248 (1972); *CA* **77**, 147980t (1972).
392. C. Elson, E. Shrago, E. Sondheimer, and M. Yatvin, *Biochim. Biophys. Acta* **297**, 125 (1973); *CA* **78**, 81936v (1973).
393. K. M. Hwang, L. C. Yang, C. K. Carrico, R. A. Schulz, J. B. Schenkman, and A. C. Sartorelli, *J. Cell Biol.* **62**, 20 (1974); *CA* **81**, 86497z (1974).
394. C. T. Roberts, Jr. and E. Orias, *J. Cell Biol.* **62**, 707 (1974); *CA* **81**, 163942v (1974).
395. C. E. Brinckerhoff and M. Lubin, *JNCI, J. Natl. Cancer Inst.* **53**, 567 (1974); *CA* **81**, 167428e (1974).
396. G. Farkas, F. Antoni, M. Staub, and P. Piffko, *Acta Biochim. Biophys.* **9**, 63 (1974); *CA* **82**, 102c (1975).
397. N. G. Ibrahim, J. P. Burke, and D. S. Beattie, *J. Biol. Chem.* **249**, 6806 (1974); *CA* **82**, 39730f (1975).

398. A. Jimenez, B. Littlewood, and J. Davies, *Mol. Mech. Antibiot. Action Protein Biosynth. Membr., Proc. Symp., 1971* 292 (1972); *CA* **83**, 53856d (1975).
399. E. Gravela, G. Bertone, and G. Poli, *Boll. Soc. Ital. Biol. Sper.* **50**, 1674 (1974); *CA* **83**, 109369s (1975).
400. J. L. McCullough and G. D. Weinstein, *J. Invest. Dermatol.* **65**, 394 (1975); *CA* **83**, 172344v (1975).
401. L. Carrasco, E. Battaner, and D. Vazquez, *Methods Enzymol.* **30**, Pt. F, 282 (1974); *CA* **83**, 203473t (1975).
402. M. N. Gadaleta, M. Greco, G. Del Prete, and C. Saccone, *Arch. Biochem. Biophys.* **172**, 238 (1976); *CA* **84**, 39817f (1976).
403. R. A. Lansman and D. A. Clayton, *J. Mol. Biol.* **99**, 777 (1975); *CA* **84**, 72005n (1976).
404. I. W. Sherman, *Comp. Biochem. Physiol. B* **53**(4B), 447 (1976); *CA* **85**, 43519y (1976).
405. C. Ceccarini and H. Eagle, *In Vitro* **12**, 346 (1976); *CA* **85**, 154574g (1976).
406. U.-B. Westerberg, G. Bolcsfoldi, and E. Eliasson, *Biochim. Biophys. Acta* **447**, 203 (1976); *CA* **85**, 157110p (1976).
407. B. Emmerich, H. Hoffman, V. Erben, and J. Rastetter, *Biochim. Biophys. Acta* **447**, 460 (1976); *CA* **86**, 25896k (1977).
408. J. S. Tscherne and S. Pestka, *Antimicrob. Agents Chemother.* **8**, 479 (1975); *CA* **86**, 134284e (1977).
409. C. E. Brinckerhoff and M. Lubin, *JNCI, J. Natl. Cancer Inst.* **58**, 605 (1977); *CA* **86**, 165170u (1977).
410. A. D. Ray, A. K. Chatterjee, and S. C. Datta, *Jpn. J. Pharmacol.* **27**, 165 (1977); *CA* **87**, 284s (1977).
411. A. J. Dunn, H. E. Gray, and P. M. Iuvone, *Pharmacol., Biochem. Behav.* **6**, 1 (1977); *CA* **87**, 15897t (1977).
412. N. L. Oleinick, *Arch. Biochem. Biophys.* **182**, 171 (1977); *CA* **87**, 111326b (1977).
413. S. Ohi, *Experientia* **33**, 1184 (1977); *CA* **87**, 177576q (1977).
414. V. Sundararaman and D. J. Cummings, *Mech. Ageing Dev.* **6**, 393 (1977); *CA* **88**, 34371d (1978).
415. F. M. Baccino, G. Cecchini, F. Palmucci, V. Sverko, L. Tessitore, and M. F. Zuretti, *Biochim. Biophys. Acta* **479**, 91 (1977); *CA* **88**, 44952h (1978).
416. M. P. Chitnis and R. K. Johnson, *JNCI, J. Natl. Cancer Inst.* **60**, 1049 (1978); *CA* **89**, 84569r (1978).
417. A. Contreras, D. Vazquez, and L. Carrasco, *J. Antibiot.* **31**, 598 (1978); *CA* **89**, 122980c (1978).
418. L. N. Drozdovskaya, *in* "Khimicheskii Mutagenez i Gibridizatsiya" (I. A. Rapoport, ed.), pp. 194–200. Izd. Nauka, Moscow, 1978; *CA* **91**, 158m (1979).
419. E. Walker and D. N. Wheatley, *J. Cell. Physiol.* **99**, 1 (1979); *CA* **91**, 2414r (1979).
420. J. A. Bilello, G. Warnecke, and G. Koch, *Haematol. Bluttransfus.* **23**, 303 (1978); *CA* **92**, 1062h (1980).
421. L. M. Pike, J. X. Khym, M. H. Jones, W. H. Lee, and E. Volkin, *J. Biol. Chem.* **255**, 3340 (1980); *CA* **93**, 1487w (1980).
422. R. S. Gupta, J. J. Krepinsky, and L. Siminovitch, *Mol. Pharmacol.* **18**, 136 (1980); *CA* **93**, 160962p (1980).
423. E. Walker and L. H. Chappell, *Comp. Biochem. Physiol. C* **67C**, 129 (1980); *CA* **94**, 77596x (1981).
424. A. K. Chatterjee and A. D. Ray, *Jpn. J. Exp. Med.* **52**, 27 (1982); *CA* **96**, 173972d (1982).
425. C. Pareyre, *C. R. Hebd. Seances Acad. Sci., Ser. D* **273**, 143 (1971); *CA* **75**, 106227t (1971).
426. M. L. Murphy, R. T. Bulloch, and M. B. Pearce, *Am. Heart J.* **87**, 105 (1974); *CA* **80**, 128330b (1974).

427. M. A. Goldsmith, M. Slavik, and S. K. Carter, *Cancer Res.* **35**, 1354 (1975); *CA* **83**, 53490e (1975).
428. J. C. Murphy, P. Skierkowski, E. S. Watson, R. M. Folk, and C. L. Litterst, *U. S. NTIS, PB Rep.* **PB-274082** (1977); *CA* **88**, 164350g (1978).
429. B. R. Jones and G. A. Ofosu, *Cytobios* **19**, 109 (1977); *CA* **89**, 173715c (1978).
430. M. A. Dubick, G. Hanasono, and W. C. T. Yang, *Proc. West. Pharmacol. Soc.* **22**, 401 (1979); *CA* **92**, 34365n (1980).
431. G. Renna, G. Siro-Brigiani, R. Cagiano, and V. Cuomo, *Atti Relaz.—Accad. Pugliese Sci., Parte 2* **35**, 35 (1977); *CA* **92**, 174292m (1980).
432. M. A. Dubick and W. C. T. Yang, *Toxicol. Appl. Pharmacol.* **54**, 311 (1980); *CA* **93**, 107164d (1980).
433. M. A. Dubick and W. C. T. Yang, *J. Pharm. Sci.* **70**, 343 (1981); *CA* **94**, 150376g (1981).
434. K. Fujii, H. Jaffe, Y. Bishop, E. Arnold, D. Mackintosh, and S. S. Epstein, *Toxicol. Appl. Pharmacol.* **16**, 482 (1970); *CA* **73**, 2132a (1970).
435. L. Vlckova, V. Vondrejs, and J. Necasek, *Folia Microbiol. (Prague)* **15**, 76 (1970); *CA* **73**, 11643v (1970).
436. W. G. Levine, *J. Pharmacol. Exp. Ther.* **175**, 301 (1970); *CA* **73**, 128907a (1970).
437. A. Daroczy and T. Valyi-Nagy, *Prog. Antimicrob. Anticancer Chemother., Proc. Int. Congr. Chemother., 6th, 1969* **1**, 190 (1970); *CA* **74**, 74945m (1971).
438. P. Lensky, *Muench. Med. Wochenschr.* **113**, 231 (1971); *CA* **74**, 98037t (1971).
439. P. Ponka and J. Neuwirt, *Biochim. Biophys. Acta* **230**, 381 (1971); *CA* **74**, 109193u (1971).
440. Y. H. Tan, J. A. Armstrong, and M. Ho, *Virology* **44**, 503 (1971); *CA* **75**, 61355q (1971).
441. R. K. Johnson, P. Mazel, J. D. Donahue, and W. R. Jondorf, *Biochem. Pharmacol.* **20**, 955 (1971); *CA* **75**, 61871e (1971).
442. R. A. Raff, *Exp. Cell Res.* **71**, 455 (1972); *CA* **77**, 14955s (1972).
443. F. P. Kovalenko, *Tr. Vses. Inst. Gel'mintol. im. K. I. Skryabina* **17**, 71 (1971); *CA* **77**, 56334u (1972).
444. A. K. Chatterjee, S. C. Datta, and B. B. Ghosh, *Br. J. Pharmacol.* **44**, 810 (1972); *CA* **77**, 70174y (1972).
445. S. A. Imam, V. K. M. Rao, and K. Kar, *Biochem. Pharmacol.* **21**, 3089 (1972); *CA* **78**, 24147q (1973).
446. S. Lal, T. L. Sourkes, K. Missala, and G. Belendiuk, *Eur. J. Pharmacol.* **20**, 71 (1972); *CA* **78**, 92509j (1973).
447. S. Azhar and V. K. Mohan, *Zentralbl. Bakteriol., Parasitenk., Infektionskr. Hyg., Abt. 1: Orig., Reihe A* **225**, 553 (1973); *CA* **80**, 105227j (1974).
448. A. K. Chatterjee and A. D. Ray, *J. Pharm. Pharmacol.* **25**, 827 (1973); *CA* **80**, 128160w (1974).
449. D. J. Miletich, A. D. Ivankovic, R. F. Albrecht, and E. T. Toyooka, *J. Pharm. Pharmacol.* **26**, 101 (1974); *CA* **81**, 20967w (1974).
450. G. Rez and J. Kovacs, *Acta Biol. Acad. Sci. Hung.* **24**, 237 (1973); *CA* **81**, 99481q (1974).
451. E. Gravela and G. Poli, *IRCS Libr. Compend.* **2**, 1534 (1974); *CA* **82**, 11871d (1975).
452. S. C. Datta, A. K. Chatterjee, and B. B. Ghosh, *J. Pharm. Pharmacol.* **26**, 547 (1974); *CA* **82**, 38526g (1975).
453. J. Kovacs and G. Rez, *Virchows Arch. B* **15**, 209 (1974); *CA* **82**, 51482j (1975).
454. C. S. Henney, J. Gaffney, and B. R. Bloom, *J. Exp. Med.* **140**, 837 (1974); *CA* **82**, 55881m (1975).
455. P. Synek, *Cas. Lek. Cesk.* **113**, 856 (1974); *CA* **82**, 92862b (1975).
456. H. Schellekens, J. H. P. M. Huffmeyer, and L. J. L. D. Van Griensven, *J. Gen. Virol.* **26**(Pt. 2), 197 (1975); *CA* **83**, 646d (1975).

457. J. T. Rico, J. M. G. T. Rico, M. T. Zambelli de Almeida, A. C. Cravo, and J. M. C. Ferreira, *C. R. Seances Soc. Biol. Ses Fil.* **169**, 749 (1975); *CA* **83**, 201905m (1975).
458. J. Kovacs, G. Rez, and A. Kiss, *Cytobiologie* **11**, 309 (1975); *CA* **84**, 242q (1976).
459. D. L. Moyer, R. S. Thompson, and I. Berger, *Contraception* **16**, 39 (1977); *CA* **87**, 178189c (1977).
460. A. Jimenez, L. Carrasco, and D. Vazquez, *Biochemistry* **16**, 4727 (1977); *CA* **87**, 179198s (1977).
461. Y. Kase, T. Yakushiji, H. Seo, M. Sakata, G. Kito, K. Takahama, and T. Miyata, *Nippon Yakurigaku Zasshi* **73**, 605 (1977); *CA* **88**, 15657k (1978).
462. P. D. Brown-Woodman and I. G. White, *Theriogenology* **8**, 199 (1977); *CA* **88**, 58644w (1978).
463. D. K. Dwivedi, *Aust. J. Pharm. Sci.* **7**, 29 (1978); *CA* **89**, 85004q (1978).
464. S. G. Deans and J. E. Smith, *Trans. Br. Mycol. Soc.* **72**, 201 (1979); *CA* **91**, 69179m (1979).
465. M. A. Dubick and W. C. T. Yang, *Proc. West. Pharmacol. Soc.* **22**, 411 (1979); *CA* **92**, 34366p (1980).
466. U. N. Das and B. V. Chainlu, *IRCS Med. Sci.: Libr. Compend.* **8**, 499 (1980); *CA* **93**, 106973e (1980).
467. C. Agostini, D. Castelluccio, and F. Corbetta, *IRCS Med. Sci.: Libr. Compend.* **8**, 571 (1980); *CA* **93**, 180449e (1980).
468. C. Agostini, M. Secchi, and D. Venturelli, *Experientia* **36**, 1067 (1980); *CA* **93**, 198661c (1980).
469. A. D. D'Angeac and A. H. Hale, *Cell. Immunol.* **55**, 342 (1980); *CA* **93**, 236731k (1980).
470. P. Y. W. Chow, M. K. Holland, D. A. I. Suter, and I. G. White, *Int. J. Fertil.* **25**, 281 (1980); *CA* **94**, 114918b (1981).
471. R. R. Dalvi and A. Peeples, *J. Pharm. Pharmacol.* **33**, 51 (1981); *CA* **94**, 186576m (1981).
472. C. Agostini, *IRCS Med. Sci.: Libr. Compend.* **9**, 405 (1981); *CA* **95**, 36590t (1981).
473. R. L. Hallberg, P. G. Wilson, and C. Sutton, *Cell (Cambridge, Mass.)* **26**(Pt. 1), 47 (1981); *CA* **95**, 215700a (1981).
474. J. B. Weissberg and J. J. Fischer, *Radiat. Res.* **88**, 597 (1981); *CA* **95**, 215210r (1981).
475. I. D. Paul, *Med. Lab. Sci.* **39**, 15 (1982); *CA* **96**, 159091j (1982).
475a. N. P. Trifunac and G. S. Bernstein, *Contraception* **25**, 69 (1982); *CA* **96**, 155734y (1982).
476. N. Farjaudon, C. Pareyre, and G. Deysson, *C. R. Seances Soc. Biol. Ses Fil.* **171**, 34 (1977); *CA* **87**, 96749w (1977).
477. M. Mitrovic, *U. S. Pat.* **3,488,422** (1970); *CA* **72**, 97883n (1970).
478. H.-J. Rim, D.-S. Chang, Il Hyun, and S.-D. Song, *Kisaengch'ung Hak Chapchi* **13**, 123 (1975); *CA* **85**, 40895g (1976).
479. G. H. Al-Khateeb, T. I. Al-Jeboori, and K. A. Al-Janabi, *Chemotherapy (Basel)* **23**, 230 (1977); *CA* **86**, 150938b (1977).
480. W. Peters, E. R. Trotter, and B. L. Robinson, *Ann. Trop. Med. Parasitol.* **74**, 289 (1980); *CA* **93**, 179601s (1980).
481. G. H. Al-Khateeb and A. L. Molan, *Chemotherapy (Basel)* **27**, 117 (1981); *CA* **94**, 150178u (1981).
482. H.-J. Rim, S.-W. Jo, K.-H. Joo, and S.-S. Kim, *Kisaengch'ung Hak Chapchi* **18**, 185 (1980); *CA* **95**, 18912v (1981).
483. E. Grunberg and H. N. Prince, *Ann. N. Y. Acad. Sci.* **173** (Art. 1), 122 (1970); *CA* **73**, 75294q (1970).
484. L. A. Salako and A. O. Durotoye, *Eur. J. Pharmacol.* **14**, 200 (1971); *CA* **75**, 18365f (1971).
485. A. O. Durotoye and L. A. Salako, *Life Sci.* **10**(Pt. 1), 623 (1971); *CA* **75**, 74474c (1971).

486. A. O. Durotoye and L. A. Salako, *Br. J. Pharmacol.* **44**, 723 (1972); *CA* **77**, 70009y (1972).
487. L. A. Salako, *Br. J. Pharmacol.* **46**, 725 (1972); *CA* **78**, 66871b (1973).
488. L. A. Salako and A. O. Durotoye, *Eur. J. Pharmacol.* **23**, 6 (1973); *CA* **79**, 132980k (1973).
489. L. A. Salako, *Eur. J. Pharmacol.* **12**, 124 (1970); *CA* **73**, 118788y (1970).
490. K. K. F. Ng and Y. T. Ng, *J. Pharm. Pharmacol.* **22**, 787 (1970); *CA* **73**, 118802y (1970).
491. L. A. Salako, *J. Pharm. Pharmacol.* **22**, 938 (1970); *CA* **74**, 85994c (1971).
492. L. A. Salako, *Ghana J. Sci.* **11**, 12 (1971); *CA* **76**, 135729j (1972).
493. O. D. Gulati, R. Makol, and D. S. Shah, *Br. J. Pharmacol.* **48**, 314 (1973); *CA* **79**, 100436w (1973).
494. J. D. Donahue, R. K. Johnson, and W. R. Jondorf, *Br. J. Pharmacol.* **43**, 456 (1971); *CA* **76**, 81223u (1972).
495. R. K. Johnson, W. T. Wynn, and W. R. Jondorf, *Biochem. J.* **125**, 26 (1971); *CA* **76**, 80966b (1972).
496. R. K. Johnson and W. R. Jondorf, *Xenobiotica* **3**, 85 (1973); *CA* **79**, 49071u (1973).
497. A. O. Durotoye, *West Afr. J. Pharmacol. Drug Res.* **3**, 119 (1976); *CA* **87**, 145804w (1977).
498. D. S. Shah, S. P. Rathod, and M. P. Patel, *Indian J. Pharmacol.* **11**, 189 (1979); *CA* **92**, 104129q (1980).
499. A. Pedrazzoli, B. Gradnik, and L. Dall'Asta, *J. Med. Chem.* **14**, 255 (1971); *CA* **74**, 112277e (1971).
500. B. Gradnik, L. Dall'Asta, and A. Pedrazzoli, *Ger. Offen.* **1,952,873** (1970); *CA* **73**, 15064e (1970).
501. T. Shoji and K. Kisara, *Oyo Yakuri* **10**, 407 (1975); *CA* **88**, 115366h (1978).
502. K. Akiba, K. Onodera, K. Kisara, and H. Fujikura, *Nippon Yakurigaku Zasshi* **75**, 201 (1979); *CA* **91**, 32924z (1979).
503. M. C. Benbadis, F. Levy, C. Pareyre, and G. Deysson, *C. R. Hebd. Seances Acad. Sci.,* Ser. D **272**, 707 (1971); *CA* **74**, 95592j (1971).
504. C. Pareyre, *Mem. Soc. Bot. Fr.* 167 (1970); *CA* **75**, 4426q (1971).
505. L. Carrasco, A. Jimenez, and D. Vazquez, *Eur. J. Biochem.* **64**, 1 (1976); *CA* **84**, 174566p (1976).
506. C. Pareyre, *Bull. Soc. Bot. Fr.* **121**, 3 (1974); *CA* **82**, 68774d (1975).
507. K. R. Kirtikar and B. D. Basu, "Indian Medicinal Plants," 2nd ed., Vol. II, pp. 1236–1239. L. M. Basu, Allahabad, India, 1933.
508. R. N. Chopra, I. C. Chopra, K. L. Handa, and L. D. Kapur, "Indigenous Drugs of India," 2nd ed., pp. 270–271. U. N. Dhar and Sons, Calcutta, 1958.
509. B. Dasgupta, *J. Pharm. Sci.* **54**, 481 (1965).
510. Raymond-Hamet, *C. R. Seances Soc. Biol. Ses Fil.* **135**, 1011 (1941); *CA* **39**, 1929_8 (1945).
511. M. Hayat, G. Mathe, E. Chenu, H. P. Husson, T. Sevenet, C. Kan, and P. Potier, *C.R. Hebd. Seances Acad. Sci., Ser. D* **285**, 1191 (1977); *CA* **88**, 32117b (1978); G. Mathé, M. Hayat, E. Chenu, H. P. Husson, T. Thevenet, C. Kan, and P. Potier, *Cancer Res.* **38**, 1465 (1978).
512. E. Ali, R. R. Sinha, B. Achari, and S. C. Pakrashi, *Heterocycles,* **19**, 2301 (1982).
513. T. Fujii, M. Ohba, H. Suzuki, S. C. Pakrashi, and E. Ali, *Heterocycles* **19**, 2305 (1982).
514. S. C. Pakrashi, R. Mukhopadhyay, P. P. Ghosh Dastidar, A. Bhattacharjya, and E. Ali, *Tetrahedron Lett.* **24**, 291 (1983).

—— CHAPTER 2 ——

ELUCIDATION OF STRUCTURAL FORMULA, CONFIGURATION, AND CONFORMATION OF ALKALOIDS BY X-RAY DIFFRACTION

ISABELLA L. KARLE

Laboratory for the Structure of Matter, Naval Research Laboratory, Washington, D.C.

I. Introduction

A classic procedure in the chemical study of alkaloids includes extraction from a plant or animal source, chemical isolation and purification of the active principal, elucidation of the molecular formula by means of chemical degradation, mass spectrometry, infrared, ultraviolet, and nuclear magnetic resonance analyses, and laboratory synthesis of the product. Serious diffi-

THE ALKALOIDS, VOL. XXII

ISBN 0-12-469522-1

culties arise in establishing the molecular formula when the amount of purified product is extremely minute, precluding extensive chemical experimentation, when the chemical linkages are new or unusual so that comparison of NMR spectra of the unknown with established structures cannot be made, or when the number of asymmetric centers is large. Under these circumstances, X-ray diffraction analysis of a single crystal is not only particularly useful but often indispensable.

One of the more notable examples is provided by batrachotoxin, a potent neurotoxin contained in the skin of frogs from tropical American jungles (1). Native Indians impale the tiny frogs on twigs and hold them over a fire until they exude a milky liquid that contains the deadly toxin. Blow darts used in hunting are then tipped with the toxin. The amount of purified congeners obtained from ethanolic extracts of skins from 8000 frogs was miniscule and quite insufficient to determine the structural formulae by standard procedures. A very few, extremely minute crystals were grown of the O-p-bromobenzoate derivative of batrachotoxinin A, one of the congeners. The largest of the needle-shaped crystals (0.05 × 0.03 × 1.0 mm and a mass of only 2 μg) was selected for the X-ray analysis that established the structural formula for this steroidal alkaloid with several novel structural features (2,3). Other immediate advantages of the crystal structure analysis were the additional types of information obtained simultaneously such as the stereoconfiguration at the nine asymmetric centers, the absolute stereoconfiguration, the conformations of the six rings, particularly the new seven-membered ring containing the alkaloid function, values of bond lengths, bond angles and torsional angles, the geometry of intermolecular hydrogen bonds, and the mode of packing in the unit cell.

Structure analysis by X-ray diffraction has played a vital role in the alkaloid discipline for many years. One of the reasons has been that many alkaloids have a propensity for forming salts with Br^- and I^- and thus provided crystals ideal for the application of the earlier method of structure analysis that was strongly dependent upon the presence of a heavy atom. A more recent method of deriving structures from X-ray data relies on the determination of phases associated with structure factor magnitudes directly from the measured intensities of the scattered X rays (4,5). The direct method, as it is commonly called, does not require the presence of a heavy atom, and thus most structures of crystals composed of only light atoms such as C, H, N, and O can be solved readily. One of the first applications of the direct method of structure analysis was to the structure of reserpine, $C_{33}H_{40}N_2O_9$ (6). The stereochemistry had already been established by chemical means but there remained the problem of finding the spatial arrangement of the atoms in the molecule (Fig. 1), where it was shown that the trimethoxybenzoxy group is nearly perpendicular to the indole moiety and

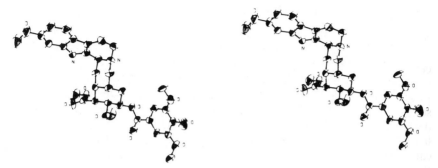

FIG. 1. A stereodiagram showing the structure and conformation of reserpine (6), one of the first crystal-structure determinations without the presence of a heavy atom. The ellipsoids represent the thermal motions of the atoms, drawn at the 50% probability level.

that only the methyl group of the middle methoxy moiety is out of the plane containing the remainder of the atoms in the trimethoxybenzoxy group. Since that time the molecular structures of hundreds of complex alkaloids have been elucidated or confirmed by X-ray analysis and their stereoconfigurations and conformations established. This chapter is not intended to be a review or compendium of alkaloid structures. Bibliographic references to work published since 1935 on crystal structures of organic compounds, including a division on alkaloids, have been compiled by the Cambridge Data Center. The center publishes a series of reference books (7) containing the names of the compounds, formulas, names of authors, and literature references. They also maintain a series of computer-based files containing bibliographic data, structural data, and chemical connectivity information.

The objectives of this chapter are to indicate in a brief form the procedures used in crystal-structure analysis, to provide a guide to the evaluation of the published results, to show examples of the different types of information available from diffraction experiments, and to describe a few alkaloids with unusual structural features.

II. X-Ray Diffraction of Single Crystals

A. EXPERIMENTAL PROCEDURE

The first requirement for X-ray diffraction analysis is a single crystal. A crystal is a solid composed of a group of atoms or molecules that is repeated regularly in three dimensions. The basic repeating unit is called a unit cell characterized by three axial directions, usually all different in length, and the three angles formed by the axes, not necessarily equal to 90°. Within a unit cell of a particular cyrstal, there may be symmetry elements such as rotation axes, screw axes, mirror planes, glide planes, and/or inversion centers that produce one or more additional exact images of a subgroup of atoms. Such a subgroup is called an asymmetric unit. In publications of crystal structure determinations, the coordinates of only the atoms in an asymmetric unit are listed, because the locations of the remainder of the atoms in the unit cell can be readily deduced by applying the symmetry operations associated with that particular cell. In crystals of alkaloids, an asymmetric unit usually consists of the alkaloid molecule or ion, a counterion, if present, and associated solvent molecules, such as water or ethanol; although in some crystals there are two or more alkaloid molecules in an asymmetric unit, not necessarily with identical conformations. There are 232 different arrangements of symmetry elements and axes, called space groups [listed in *International Tables for X-Ray Crystallography,* Vol. 1 (*8*)], however, alkaloids generally crystallize in one of two space groups, monoclinic $P2_1$ and orthorhombic $P2_12_12_1$. The former has a cell with two asymmetric units exactly related by an operation of a twofold screw axis, whereas the latter has a cell with four asymmetric units exactly related by three nonintersecting twofold screw axes. The molecules in the unit cells described by both these space groups possess a chirality, consistent with the fact that almost all alkaloid molecules are chiral and occur naturally with a particular handedness.

Crystals of alkaloids are usually grown by slow evaporation of a solution of the alkaloid in common solvents such as ethanol, ethyl acetate, acetonitrile, or mixture of solvents or by slowly cooling a supersaturated solution obtained by heating. Too rapid crystal growth produces microcrystals that are too small for diffraction purposes. Suitable crystals have dimensions of 0.1–0.5 mm in cross section. Good single crystals usually have regular shapes, such as prismatic or acicular, good faces, and are clear, not opaque. Intergrown crystals, clusters, bundles, crystals shaped like stars or vees are not single crystals and cannot be used unless a single crystal fragment can be cleaved or cut out of the composite. Faces of single crystals when viewed through crossed Nicols of a polarizing microscope will have optical extinctions every 90° of rotation (except for those with cubic symmetry). If optical extinctions occur more often than at 90° intervals, or only part of a face is darkened, the "crystal" is not single but is either "twinned" or a conglomerate of crystals.

The repetitions of the atoms or molecules in all directions in the crystal form many sets of parallel planes that act as diffraction gratings. When a beam of X rays impinges on a crystal that has a set of planes in a proper orientation with respect to the direction of the beam, diffraction will take place, i.e., some of the X rays will be bent away from the main beam, forming a new beam. The angle at which scattering takes place depends on the wave length of the X rays, 1.5418 Å if X rays are generated from a Cu anode, and inversely on the spacing between the parallel planes involved. The intensity of a diffracted beam is a function of the number of atoms and the kinds of atoms present in the particular scattering planes, among other factors. Each row of spots in a diffraction pattern corresponds to a different family of planes, and thus to a different view of the atomic distribution in the cyrstal. In order to bring each different set of planes into diffracting position, the orientation of the crystal with respect to the impinging X-ray beam must be changed. With present-day computer-controlled, automatic diffracto-meters, the task of measuring and recording intensity data is simplified immensely. From a crystal of a medium-sized alkaloid molecule, the inten-sities of 2000–4000 diffraction spots are measured. Thus, there is an abundance of experimental data from which to derive the precise location of each atom in the unit cell.

B. INTERPRETATION OF DIFFRACTION DATA

Each diffracted spot has four numbers associated with it, $h,k,l,$ and F_{hkl}. The quantity F_{hkl} is related to the experimentally measured intensity, whereas the h,k,l indices (small integers) designate the diffracting plane in the crystal that produced the diffracted beam. The h,k,l are inversely proportional to the intercepts of the corresponding planes on the chosen axes. It is quite simple to index the diffraction pattern, that is, assign the proper h,k,l to each diffracted spot, using geometrical considerations and having a knowledge of the symmetry and dimensions of the unit cell (9).

The structure of the unit cell of a crystal may be described in terms of the electron density distribution in the cell. The x,y,z coordinates of the maxima of the electron density function correspond to the positions of the atoms. The electron density distribution function $\rho(x,y,z)$ may be represented by a three-dimensional Fourier series

$$\rho(x,y,z) = \frac{1}{V} \sum_{h,k,l=-\infty}^{\infty} F_{hkl} \, e^{-2\pi i(hx+ky+lz)} \tag{1}$$

where the coefficients $F_{hkl} = |F_{hkl}|e^{-i\phi_{hkl}}$. \hfill (2)

From experiment, the values for the $|F_{hkl}|$ and h,k,l are readily derived, but Eq. (1) cannot be calculated without the knowledge of individual ϕ_{hkl} values, that is, the phases.

For a long time it was thought that the phase problem was unsolvable and that special devices had to be used to determine crystal structures. One of these which proved very useful in the past, and is still used fairly extensively, depends on the presence of a heavy atom in the asymmetric unit. The

Patterson function (*10,11*),

$$P(x,y,z) = \sum_{h,k,l=-\infty}^{\infty} |F_{hkl}|^2 e^{-2\pi i(hx+ky+lz)} \qquad (3)$$

is very similar to Eq. (1), except that the coefficients are $|F_{hkl}|^2$ instead of F_{hkl} and therefore the need to know the individual phase values is obviated. The maxima of the Patterson function represent interatomic vectors in the crystal of which there are $N(N-1)$ where N is the number of atoms in the unit cell of the crystal. A vector map calculated with Eq. (3) is usually extremely difficult to interpret unless there is one atom, or a very few atoms, considerably heavier than all the others in the asymmetric unit. In that case, the vectors associated with the heavy atom can be readily identified and the x,y,z coordinates established. Phases calculated on the basis of the position of the heavy atom or atoms are generally a good approximation to the true phases for the crystal and, when they are used in Eq. (1), the positions of the light atoms can usually be determined. The HI salt of sewarine (*12*) can serve as an example of this process, where the positions of all light atoms were readily derived on the basis of the known coordinates for the iodide ion (Fig. 2).

Although the "heavy atom" method has been successful in establishing many structures, particularly those of alkaloids since alkaloids have a propensity for crystallizing as halide salts, there had been an urgent need to develop a procedure for phase determination that was not dependent on the presence of a heavy atom in a crystal. Such a procedure, now commonly called the "direct method" of phase determination, has been devised. Karle and Hauptman (*13*) recognized that the number of unique reflections measured in an X-ray pattern is 25–50 times greater than the number of unknowns in a crystal, the unknown quantities being the three coordinates

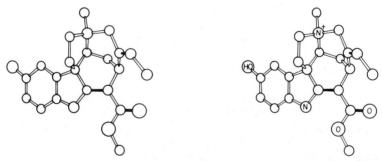

FIG. 2. A stereodiagram of the sewarine cation (*12*). The structure analysis was based on the position of the iodide ion. The absolute configuration was determined from the anomalous scattering of the iodide ion.

for each atom in an asymmetric unit, and consequently, that there existed a large overdeterminancy in a mathematical sense. For practical purposes, it has been more expedient to determine the phases associated with the measured intensities as an intermediate step and then to use them in Eq. (1) to derive the atomic positions. On the basis that the electron density function [Eq. (1)] must be a nonnegative function, Karle and Hauptman (*13*) derived the complete set of inequalities relating the structure factors F_{hkl} in any crystal. Relatively simple relationships among the phases ϕ_{hkl}, dependent on the experimentally measured magnitudes of $|F_{hkl}|$, were readily extracted from the inequalities . The following are examples of the most useful phase relationships*:

$$\phi_{h_{1+2}k_{1+2}l_{1+2}} \approx \phi_{h_1k_1l_1} + \phi_{h_2k_2l_2} \tag{4}$$

(valid for reflections with large $|E_{hkl}|$, where $|E_{hkl}|$ are the normalized values for $|F_{hkl}|$), and

$$\tan \phi_{h_{1+2}k_{1+2}l_{1+2}} = \frac{\sum \partial E_{h_1k_1l_1}E_{h_2k_2l_2} \mid \sin(\phi_{h_1k_1l_1} + \phi_{h_2k_2l_2})}{\sum \partial E_{h_1k_1l_1}E_{h_2k_2l_2} \mid \cos(\phi_{h_1k_1l_1} + \phi_{h_2k_2l_2})} \tag{5}$$

A modus operandi developed for applying the above relationships to experimental data in order to derive phases, and, in turn, the structures has been described by Karle and Karle (*4*). The phase derivation is initiated with very few phases, up to three known phases from assigning an origin for the unit cell and several unknown phases represented by algebraic symbols. The process cascades quickly and the phase values are refined and extended with the tangent formula Eq. (5). Subsequently, the direct method for phase determination has been formulated into a number of commonly used computer programs, such as MULTAN (*14*), SHELX (*15*), and XRAY 76 (*16*) by means of which it is often possible to obtain the solution of a crystal structure quite automatically. Thousands of structures of crystals containing only light atoms, as well as crystals containing heavier atoms, have been solved by the direct method of phase determination.

C. EVALUATION OF RESULTS

After the positions of the atoms in an asymmetric unit have been located in electron density maps, the best values for the x,y,z coordinates are obtained by means of a least-squares refinement in which the squared values of the differences between the values of the experimentally observed F_{hkl} and the calculated F_{hkl}, based on the determined coordinates of each atom, are minimized. A measure of the minimization and agreement is an R factor,

*Note that $h_{1+2} = h_1 + h_2$, $k_{1+2} = k_1 + k_2$ and $l_{1+2} = l_1 + l_2$.

where

$$R = \sum (|F_{\text{obs}}| - |F_{\text{calc}}|) \Big/ \sum |F_{\text{obs}}| \qquad (6)$$

Other mathematical expressions for R values are also used. A well-refined, correct structure will yield an R value in the range 3–10%. An incorrect structure or one in which atoms are missing will have much higher R values, perhaps 25% or greater.

Thermal vibration of atoms in organic crystals is appreciable at room temperature and its effect on the calculated F_{hkl} must be incorporated. During the least-squares refinement, not only are the best values of the coordinates sought to minimize the sum of $(|F_{\text{obs}}| - |F_{\text{calc}}|)^2$, but the six parameters describing the thermal ellipsoids of each atom are also varied to obtain their best values. Figure 3 shows a computer drawing of a derivative of the diterpene alkaloid lycoctonine with the thermal parameters of the C, N, and O atoms indicated by the ellipsoids (17). Cooling an organic crystal to liquid nitrogen temperatures during X-ray data collection will decrease considerably the thermal motion, and consequently the size of the thermal ellipsoids.

There are a number of factors that can affect the final value of the agreement factor R, such as the amount of thermal motion. Some other factors are the quality and size of the crystal used for data collection. Too small a specimen will produce weak reflections that are difficult to measure accurately. A soft crystal is readily susceptible to physical distortions that will result in attenuations in the magnitudes of the reflections. Some molecules, usually large ones, have some small variation in their conformation, such as rotations about single bonds, from cell to cell in the crystal that produce a positional disorder and have the effect of increasing the R factor.

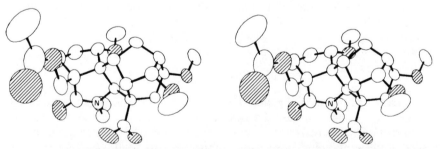

FIG. 3. A stereodiagram of an acetoxylactam acid from 4-amino-4-deoxymethylene anhydrolycoctonam (17). The thermal ellipsoids, at the 50% probability level, show the large increase in thermal motion of the atoms on the periphery of the molecule as compared with the atoms in the more rigid, fused-ring system. The O atoms are crosshatched.

Even with all of these possible problems, it is generally easy to achieve an R factor of less than 10%, which guarantees that the structure is quite correct and that the geometric parameters, such as bond lengths and torsional angles, are quite precise. In older structure determinations, perhaps before 1970–1975, the experimental X-ray intensities were recorded on film and estimated by visual comparison with a standard film density strip, rather than being recorded by a scintillation counter on an automatic diffractometer. In those publications, R factors are sometimes higher, perhaps even 12–15%, but, even so, the results are entirely reliable, except possibly for the precise evaluation of the thermal parameters, information that generally is of lesser concern to alkaloid chemists.

An individual discussion and examples of the various types of information available from a crystal structure determination follows.

D. STRUCTURAL FORMULA AND RELATIVE CONFIGURATION

A crucial problem in organic chemistry is the elucidation of the structural formula. One of the first unknown structures containing only light atoms that was established by X-ray diffraction was that of the alkaloid jamine from *Ormosia jamaicensis,* where only the chemical composition was known, $C_{21}H_{35}N_3$ (*18*). The three-dimensional electron density map, shown in Fig. 4, shows that the molecule is composed of six 6-membered rings, five of which have the chair conformation and one the boat conformation. The three N atoms were identified by the heavier densities associated with them, as compared with the C atoms, and by shorter bond lengths where the 9 C—N bonds were found to be near 1.47 Å whereas the 20 C—C bonds averaged 1.55 Å. Furthermore, the relative stereoconfiguration at all the asymmetric centers was established simultaneously (Fig. 5).

Jamine is most unusual in that it occurs as a racemate in *Ormosia jamaicensis.* Recently, jamine isolated from *Ormosia costulata* was found crystallized in a noncentrosymmetric space group, $P2_12_12_1$, where it necessarily must be optically active. The structure determination (*19*) yielded a molecule identical to that found in the earlier publication (*18*).

E. ABSOLUTE CONFIGURATION

A unique contribution of X-ray analysis to stereochemistry is the ability to determine directly the absolute configuration of a molecule providing that in the noncentrosymmetric crystal there is an atom present that scatters X rays anomalously (*20,21*). Actually, all atoms scatter X rays anomalously, but the effect is much more noticeable and more easily measured for the

ISABELLA L. KARLE

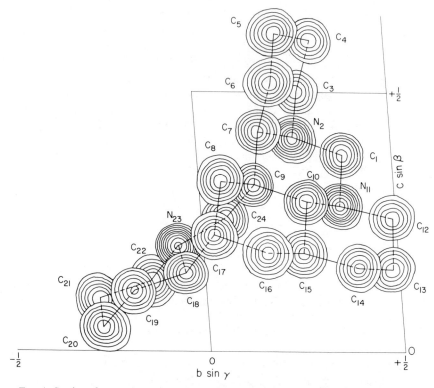

FIG. 4. Sections from a three-dimensional electron density map for jamine (*18*). Contours are spaced by $1e/\text{Å}^3$.

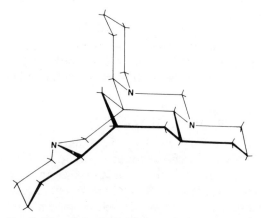

FIG. 5. The relative stereoconfiguration of jamine (*18*).

heavier atoms. When the wavelength of the incident X radiation is close to an absorption edge of one particular kind of atom in a crystal (e.g., a Br. atom) the intrinsic phase change associated with scattering by this atom is slightly different from that of the other atoms (e.g., C and N). The slight phase lag leads to a small difference in the intensities of reflections hkl and those in the opposite direction, that is, $\bar{h}\bar{k}\bar{l}$. Originally it had been assumed that the intensities of hkl and $\bar{h}\bar{k}\bar{l}$ reflections were equal (Friedel's Law). The differences in the intensities of reflections hkl and $\bar{h}\bar{k}\bar{l}$ lead to the determination of the absolute configuration of a molecule if the structure is already known except for its chirality. Some early direct determinations of the absolute configurations of alkaloids and related molecules were those of D-methadone·HBr by Hanson and Ahmed in 1958 (22) and codeine·HBr·$2H_2O$ by Kartha et al. (23). With improvements in accurate intensity measurements, absolute configurations have been determined for several molecules based only on the anomalous dispersion effects of oxygen atoms, where the differences in the intensities for hkl and $\bar{h}\bar{k}\bar{l}$ reflections are very small. Obviously, extremely careful measurements have to be made on crystals of very good quality. An example is the determination of the absolute configuration of chaetoglobosin A (24).

F. CONFORMATION

Alkaloid molecules are usually composed of fused ring systems that allow relatively little flexibility. Their fixed conformation, for example, is determined to be chair or boat as shown in Fig. 5. Side groups, however, are often connected to the ring nucleus by single bonds about which there is the possibility of rotation. The stable or "preferred" conformation is automatically obtained from a crystal structure analysis. In addition to such descriptive terms as trans, gauche, chair, or envelope, the amount of rotation about each bond can be calculated from the experimentally determined coordinates to give a quantitative measure. The torsional angle about bond B—C in Fig. 6 is calculated from the positions of atoms A, B, C, and D and represents the angle through which atom A must be turned to superimpose atom D. A value of 0° corresponds to the cis conformation. Alternating values near $+60°$ and $-60°$ in a saturated six-membered ring correspond to the chair conformation. The values of torsional angles are usually determined with estimated standard deviations (e.s.d. values) of less than 1°.

Knowledge of the conformations of molecules is important for making structure–activity comparison and in developing a drug-receptor site theory. The preferred conformation may indicate a lock and key arrangement or a probable distribution of charge on the surface of the molecule.

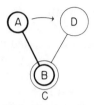

FIG. 6. The torsion angle along bond B—C is measured by the number of degrees A must be turned to superimpose D. Clockwise (right-handed) rotation is positive.

G. BOND LENGTHS AND BOND ANGLES

Precise bond lengths and bond angles are an additional dividend from crystal structure analyses. Typical e.s.d. values for organic molecules are 0.005 Å for bond distances and 0.5° for bond angles. Structures with precise bond values for a sufficient number of related compounds in certain classes of substances have been reported so that statistical analyses can be made. For example, it has been possible to establish reaction paths from a study of the systematic small changes in bond lengths (*25*) or to order electron attractive forces of moieties labeled X that affect the bond/no-bond distances between S and O in systems containing $C{=}O \cdots S{-}X$ groups (*26*).

H. CRYSTAL PACKING AND HYDROGEN BONDING

Usually the final type of information obtained from a crystal structure analysis is the arrangement of the molecules in the cell with respect to neighboring molecules. The placement of the molecules must be consistent, of course, with the symmetry operations of the crystallographic space groups. The formation of the crystal during crystal growth is influenced not only by the shape of the molecule but also by the attractive forces between molecules, such as possible hydrogen bond formation between polar groups.

In the crystal of 6-hydroxycrinamine, for example, there are two independent molecules of the alkaloid in each asymmetric unit, that is, the two molecules are not related by symmetry operations although the conformations of the two molecules are very similar (*26*). These pairs of molecules, shown in Fig. 7, occur as dimers, and each pair is held together by two hydrogen bonds between atoms N-5 · · · O*-6 and N*-5 · · · O-6, where the respective N · · · O distances are 2.88 and 2.83 Å. The hydrogen bonds to the nitrogen atoms N-5 and N*-5 provide the fourth bond linkage in a tetrahedral configuration.

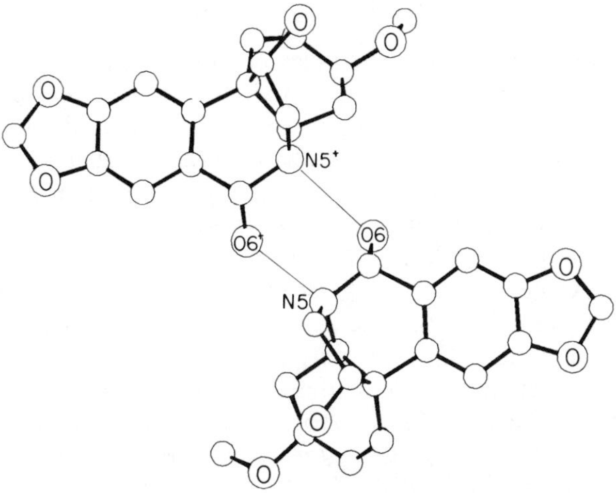

FIG. 7. Dimers of 6-hydroxycrinamine formed by a pair of hydrogen bonds between two independent molecules (26). The hydrogen bonds are indicated by thin lines.

In addition to identifying the donors and acceptors in the hydrogen bonds present in a crystal, all the geometric parameters of the hydrogen bonds are readily derived from the coordinates of the atoms involved.

III. Alkaloids from Neotropical Poison Frogs

A unique source of poisons for arrows and blow darts is the skin secretions of certain brightly colored frogs native to the rain forests of tropical America. The nature and the action of the poisons has attracted the attention of toxicologists, pharmacologists, biologists, chemists, and crystallographers, and through their cooperative efforts the structures, syntheses, and biological activities of these unusual alkaloids have been well characterized and documented. Two articles by Witkop and Gössinger (27) and by Daly (28) present comprehensive reviews and bibliographies.

The indispensable contribution of crystallography has been to establish the molecular formulas, the stereoconfigurations, and the absolute configurations of five quite different toxins representative of the five major classes of dendrobatid alkaloids. Each of the five substances had unexpected and novel chemical bonding. Knowledge of the structures from crystal structure analyses led to reevaluations of the spectral properties of analogs and congeners and to the definitions of the structures of more than 200 dendrobatid alkaloids.

A. Batrachotoxinin A

Extracts from the skins of the species *Phyllobates aurotaenia* contained the major alkaloids batrachotoxin, homobatrachotoxin, and batrachotoxinin A. The *O-p*-bromobenzoate derivative of batrachotoxinin A, recrystallized from acetone, formed extremely minute, needle-shaped crystals. The very small size of the crystal used to collect X-ray diffraction data produced only 830 measurable intensities. Even so, it proved possible to eke out the molecular formula and stereoconfiguration of the unknown toxin (Fig. 8). The C, N, and O atoms were distinguished initially by weights of peaks in the electron density map (Fig. 9), and confirmed by values of the thermal factors and bond lengths derived from the least-squares refinement (2,3). Final proof of structure came with the synthesis of batrachotoxinin A. (29) and the comparison with the natural product. The absolute configuration, as shown in Fig. 8, was established directly by measuring the effects of the anomalous scattering of the Br atom (30).

Unique structural features of this steroidal alkaloid are the cis junctions between rings A/B and C/D, the ether linkage between C-3 and C-9, the OH group on C-11, and the alkaloid function in a seven-membered ring formed by a methylethylamine moiety bridging the methyl on C-13 and OH on C-14 of a steroid nucleus. A reanalysis of the spectral data showed that batrachotoxin is the 20-α-dimethylpyrrole carboxylate and homobatrachotoxin the 20-α-ethylmethylpyrrole carboxylate of batrachotoxinin A (31).

The above three alkaloids are found in the skins of dendrobatid frogs of the genus *Phyllobates terribilis, P. aurotaenia,* and *P. bicolor,* all of Colombia, and to a very minor extent in *P. vittatus* of Costa Rica and *P. lugubris* of Panama (28).

B. Pumiliotoxin C

An active principle from the skin of a small, brightly colored, dendrobatid frog occurring in Panama, *Dendrobates pumilio,* was found to be relatively

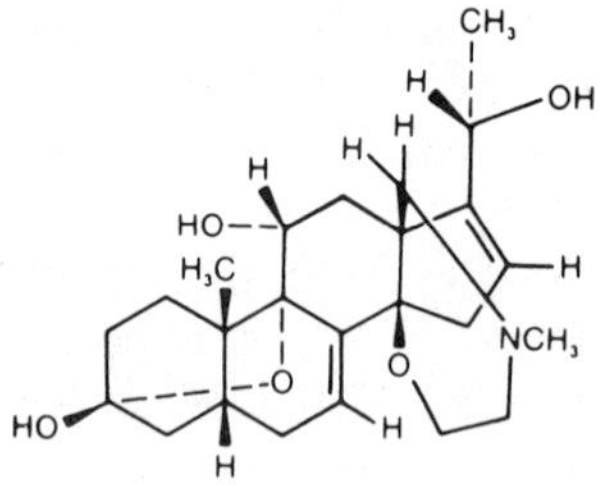

FIG. 8. Structure of batrachotoxinin A (2,3).

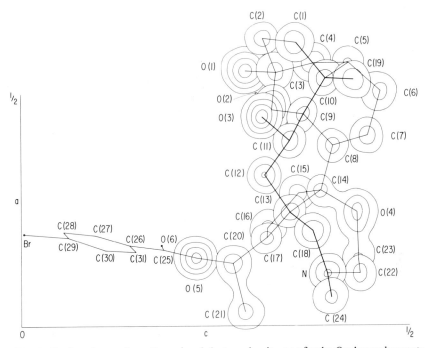

FIG. 9. Sections from a three-dimensional electron-density map for the O-p-bromobenzoate derivative of batrachotoxinin A (2). The contours are spaced by $2e/Å^3$.

nontoxic as compared with batrachotoxin, but at high concentration it blocks neuromuscular transmission. The chemical formula, established by crystal structure analysis of the HCl salt, is quite simple (32). A later analysis, in which the X-ray intensity data were measured on an automatic diffractometer, rather than by visual estimation, established the absolute configuration (33). The disubstituted cis-perhydroquinoline (Fig. 10) has been proposed as a parent member of the pumiliotoxin C class of dendrobatid alkaloids.

C. HISTRIONICOTOXINS

Another small, brightly colored, dendrobatid frog, *Dendrobates histrionicus,* occurring in Colombia and Ecuador, yields two major toxic alkaloids from defensive skin secretions, histrionicotoxin ($C_{19}H_{25}NO$) and dihydroisohistrionicotoxin, ($C_{19}H_{27}NO$). These alkaloids have a quite different molecular structure than batrachotoxin and pumiliotoxin C. They are quite unique in having a spiropiperidine ring system with acetylenic and/or allenic moieties in the side chains.

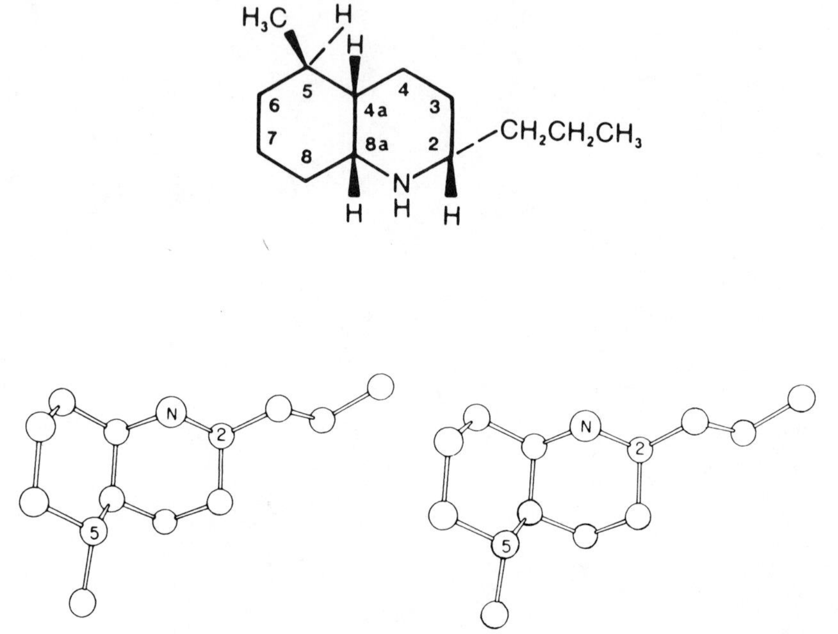

FIG. 10. Pumiliotoxin C formula, structure, and absolute configuration determined by X-ray analysis (32,33).

The molecular structures, stereoconfigurations, and absolute configurations were established by X-ray diffraction analyses of single crystals (34,35) and are displayed in Fig. 11. The HCl salt of histrionicotoxin was used for the crystal structure analysis, but the anomalous scattering from the Cl⁻ ion in this crystal did not exhibit sufficiently large differences to determine the absolute configuration confidently. Accordingly, the analysis was repeated with an isomorphous crystal of the HBr salt and the absolute configuration, shown in Fig. 11, was definitively determined and in agreement with the weaker indication from the HCl salt. The absolute configuration of dihydroisohistrionicotoxin was assumed to be the same as that of histrionicotoxin.

The histrionicotoxins probably represent the first examples of the occurrence of the acetylene and allene moieties in the animal kingdom. They presented the opportunity to establish the bond lengths and angles for these unsaturated moieties in the crystalline state. Subsequently, other alkaloids occurring in smaller quantities in the skins of *Dendrobates histrionicus* were shown, by means of mass spectra and NMR spectra, to be dihydro-, tetrahydro-, octahydro-, neodihydro-, allodihydro-, and allotetrahydrohistrionicotoxins, among others (36). The congeners differ only in the

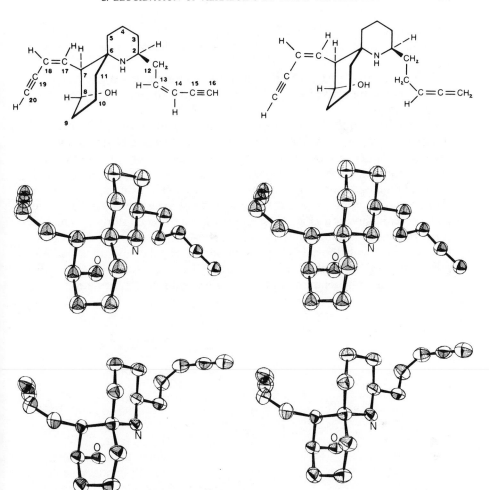

FIG. 11. Stereodiagrams of histrionicotoxin (bottom) and dihydroisohistrionicotoxin (middle). Histrionicotoxin on left and dihydroisohistrionicotoxin on right (top) (*34,35*).

saturation of the two side chains and all of them can be reduced to perhydrohistrionicotoxin (dodecahydro). The spiro ring system, with the internal NH · · · O hydrogen bond, remains the same.

D. GEPHYROTOXIN

An additional compound isolated from the skin of *Dendrobates histrionicus,* referred to originally as HTX-D, was obviously not a histrionicotoxin, although it contained a five-carbon chain (*cis* $CH_2CH = CHC \equiv CH$) iden-

tical to one of the side chains in histrionicotoxin. It forms the parent
compound of yet another class of dendrobatid alkaloids called gephyro-
toxins.

Gephyrotoxin ($C_{19}H_{29}NO$), in contrast to histrionicotoxin and pumiliotoxin C which are active in the acetylcholine receptor channel, is a muscar-
inic antagonist.

A crystallographic analysis of the HBr salt of gephyrotoxin (33) revealed
the structure to be a novel tricyclic alkaloid with two side chains, shown in
Fig. 12. A comparison of Fig. 12 with Fig. 10 shows that both gephyrotoxin
and pumiliotoxin C contain a cis-decahydroquinoline ring system. It should
be noted that for a cis-decahydroquinoline ring system, with each ring in the
chair conformation, there are two distinct conformations for the same
absolute configuration (Fig. 13(a) and (b)). The two conformers are not
superimposable although they are interconvertible. Gephyrotoxin has conformation (a) with the side chain axial on the cyclohexane ring, whereas
pumiliotoxin C has conformation (b) with a CH_3 group in the same position
as the side chain in gephyrotoxin but in the equatorial conformation.

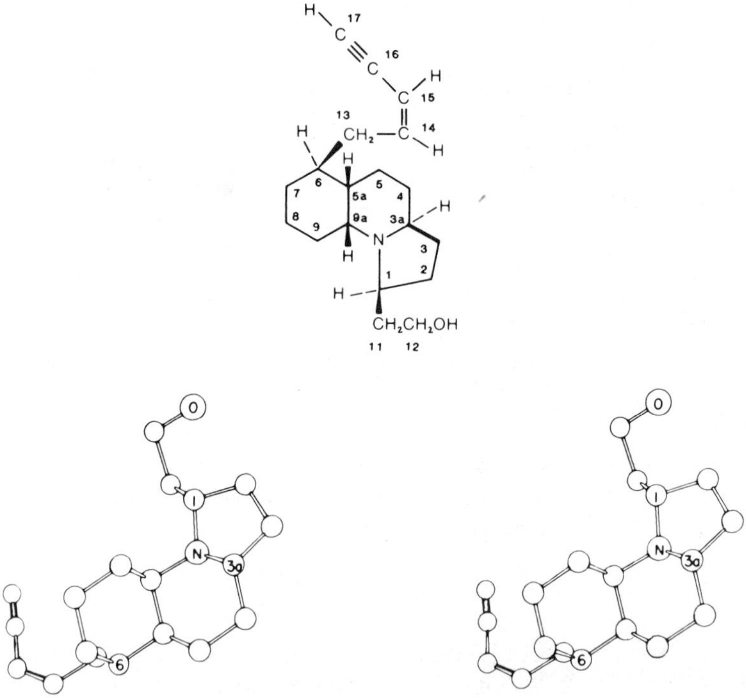

FIG. 12. Gephyrotoxin (33).

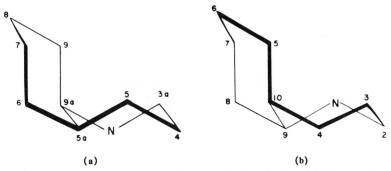

(a)   (b)

FIG. 13. Two different conformations for *cis*-decahydroquinoline with the same absolute configuration: (a) as in gephyrotoxin, Fig. 12, and (b) as in pumiliotoxin C, Fig. 10. The numbering coincides with the numbering in the respective molecules.

Gephyrotoxin does not contain the spiro ring system of histrionicotoxin but does retain one side chain with the vinylacetylene group. The third ring is a five-membered ring that incorporates the N atom and has a $-CH_2CH_2OH$ side chain. The pyramidal N atom is readily protonated and participates in hydrogen bonding with the counterion.

E. INDOLIZIDINE ALKALOID, MEMBER OF PUMILIOTOXIN A CLASS

The Panamanian frog *Dendrobates pumilio* yielded not only pumiliotoxin C (*vide ante*) but also pumiliotoxins A and B. The instability of pumiliotoxins A and B under acid conditions has interfered with the preparation of a crystalline salt and many years after the isolation of these alkaloids the structural formulas remained unknown. Pumiliotoxins A and B and many related alkaloids are widely distributed in frogs of the genus *Dendrobates*. Serendipitously, in searching for a different alkaloid, a major alkaloid with a molecular weight of 251 (pumiliotoxin 251D) was isolated from the skin of the Ecuadorian poison frog *Dendrobates tricolor*. It proved to be a simpler analog of pumiliotoxins A and B, but most importantly, it was possible to crystallize the HCl salt and consequently to derive its structural formula by X-ray diffraction analysis (*37*).

The structure of pumiliotoxin 251D is shown in Fig. 14. Positional disorder in the crystal, as evidenced by the increasing size of the experimentally determined thermal parameters for each succeeding atom in the side chain, was sufficiently large that reliable values for the bond lengths and bond angle for the three terminal atoms could not be determined. However, the mass spectrum of the compound, including data recorded from the specific crystal used for the X-ray analysis, shows that the side chain does

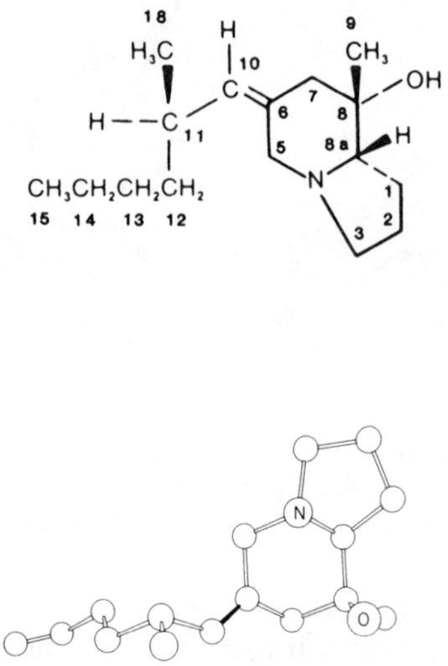

FIG. 14. Pumiliotoxin 251D (*37*).

have a saturated $CH_2CH_2CH_2CH_3$ terminal group. In addition to the high mobility of the terminus of the long side chain, there are very weak attractive forces in the part of the crystal occupied by the hydrocarbon side chains. The two nearest intermolecular approaches between chain carbon atoms are 4.0 Å and 4.2 Å while others are larger, whereas normal van der Waals approaches for $C \cdot \cdot \cdot C$ interactions between neighboring molecules have values of 3.6–3.8 Å. The high mobility of the hydrocarbon side chain is also correlated with the remarkable volatility of the free base.

Pumiliotoxin 251D contains an indolizidine moiety in common with gephyrotoxin, but in other respects it is quite different from the other classes

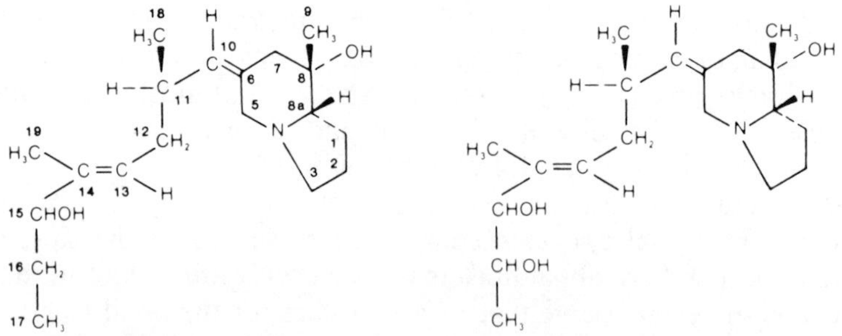

FIG. 15. Structures of pumiliotoxin A (left) and pumiliotoxin B (right) based on the structure of pumiliotoxin 251D, as derived from X-ray analysis and spectral data (*37*).

of dendrobatid alkaloids. Unlike the preceding three classes, this dendrobatid alkaloid could not be derived simply by biosynthetic ring closure of a
postulated precursor 2,6-disubstituted piperidine. It seems that a 2,5-disub-
stituted piperidine precursor is needed (28).

Once the structure of pumiliotoxin 251D was known from crystal structure analysis, the structures of a number of other alkaloids of the pumiliotoxin A class were derived from mass spectra and NMR spectra (see e.g., Fig.
15). A problem still remains in assigning the stereoconfigurations of the
hydroxyl groups in the long side chains. This kind of problem could be easily
resolved by crystallography, if it were possible to grow a crystal.

IV. Weberine and Polymethoxylated Analogs

The aromatic, tetramethoxy-substituted, tetrahydroisoquinoline alkaloid
weberine has been isolated recently from extracts of the Mexican cactus
Pachycereus weberi (38). The rare occurrence of four adjacent methoxy
groups on an aryl moiety prompted the synthesis and examination by
crystal-structure analyses of analogs and derivatives with four and five
adjacent methoxy groups on the phenyl ring. In contrast to the problem with

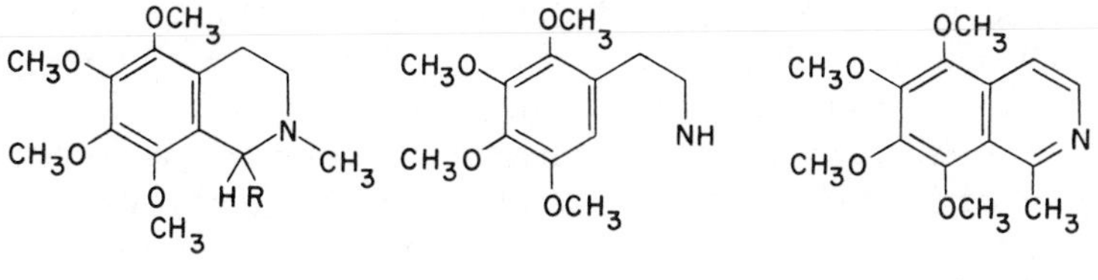

(a) R = H
(b) R = CH$_3$

(c)

(d)

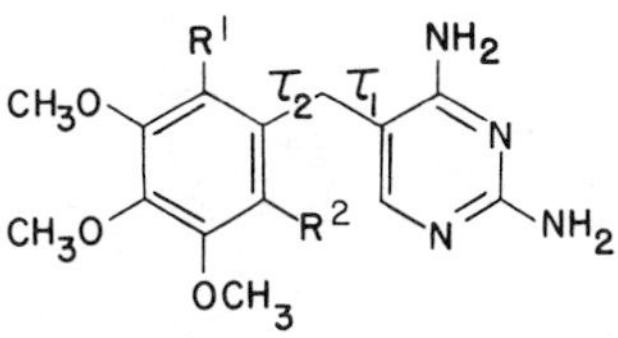

(e) R^1 = R^2 = H
(f) R^2 = H, R^1 = OCH$_3$
(g) R^1 = R^2 = OCH$_3$

FIG. 16. Weberine (a) and polymethoxylated analogs.

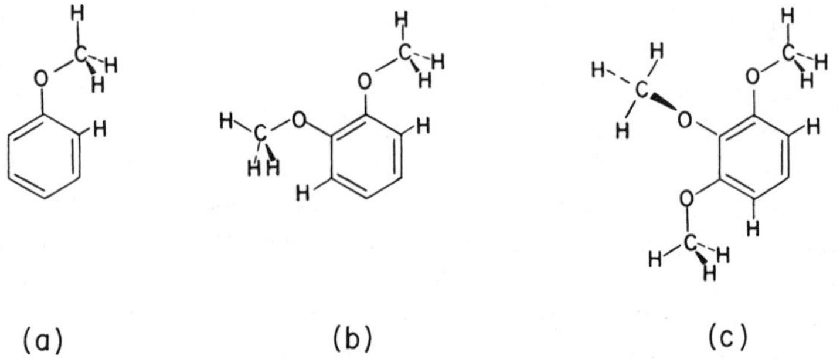

72   ISABELLA L. KARLE

(a)   (b)   (c)

FIG. 17. Conformation of unhindered OCH$_3$ groups on phenyl rings. (a) Monomethoxy; (b) adjacent dimethoxy; (c) adjacent trimethoxy.

poison frog alkaloids (Section III) where the structural formulas of the toxins were completely unknown, in this group of compounds all the formulas were known. The task for the crystal-structure analyses was to establish the conformations of the methyl groups in the polymethoxylated phenyl compounds shown in Fig. 16.

A survey of crystal structures of numerous methoxy-substituted aromatics showed that, for 30 unhindered monomethoxy derivatives, the methoxy groups are nearly coplanar with the phenyl ring (see Anderson *et al.* (*39*), and reference therein). One H atom of the terminal methyl group is coplanar and trans with respect to the C$_{ring}$—O bond, whereas the remaining two H atoms straddle the H atom on the adjacent ring carbon (Fig. 17(a)). In the same survey, a planar orientation for methoxy groups was found in 30 of 32 unhindered *o*-dimethoxy derivatives; while for three adjacent methoxy groups, the outer methoxy groups have been found to be coplanar whereas the middle methoxy is nearly perpendicular (see Fig. 17(c) and reserpine in

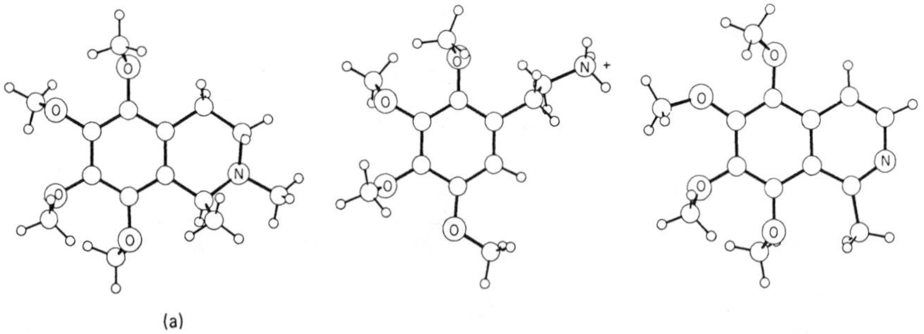

(a)

FIG. 18. A comparison of the conformations of 1-methylweberine (a) and two tetramethoxy derivatives. Hydrogen atoms are depicted by small spheres.

Fig. 1). In the case of three adjacent methoxy groups, for spatial reasons they cannot all be coplanar with the phenyl ring.

The conformations, as determined by crystal structure analyses, of 1-methylweberine, two tetramethoxy weberine analogs, and three polymethoprims, shown in Fig. 16, are displayed in Figs. 18 and 19. These compounds were synthesized by Takahashi and Brossi (*39a*) and the crystal structures were determined by J. L. Flippen-Anderson, J. F. Chiang, and I. L. Karle (*39b*), with the exception of trimethoprim (*40*) where the three adjacent methoxy groups have the same conformation as shown in Fig. 17(c).

In two tetramethoxy compounds (Figs. 18(b) and 19(b)), there is an H atom on the ring ortho to a methoxy group, and consequently, that methoxy group is coplanar with the ring, consistent with the observations for mono-, *o*-di-, and *o*-trimethoxy compounds shown in Fig. 17. Methoxy groups located between two other methoxy groups, or substituents other than hydrogen, are not coplanar with the phenyl ring; rather, their torsion angles about the C_{ring}—O bonds generally fall in the range $90 \pm 25°$. In compounds shown in Figs. 18(a), 18(b), and 19(c), there is a regular alternation of methyl groups above, below, above, below, etc., the plane of the phenyl ring, a pleasing and orderly arrangement. However, in compounds shown in Figs. 18(c) and 19(b), the noncoplanar methoxy groups are irregularly directed above, above, above, below and below, above, above the phenyl ring, respectively. Thus, there does not appear to be an orderly trend, even in these very similar compounds. The only steric consideration that may govern the conformations of the methoxy groups in these polymethoxy molecules is that the minimum distance between an O atom in one methoxy and a methyl C in an adjacent methoxy is at least 3.0 Å, whereas the minimum distance between two neighboring methyl C atoms is at least 3.5 Å.

The polymethoprims (Fig. 19) differ chemically only in the number of

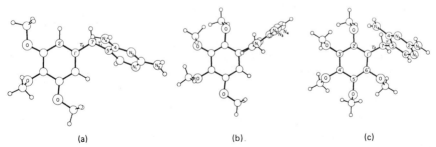

(a)   (b)   (c)

FIG. 19. A comparison of the conformations of (a) trimethoprim (*40*), (b) tetramethoprim, and (c) pentamethoprim.

methoxy groups, however, there are significant differences in the orientation of the two rings with respect to each other. The conformational freedom between the two rings can be described by the torsional angles about C-5–C-7 and C-7–C-1′, the two bonds joining the rings. If τ_1 is the torsional angle defined by C-4, C-5, C-7, C-1′ and τ_2 the torsional angle defined by C-5, C-7, C-1′, C-2′, then the values for τ_1 and τ_2 are $(-89°, +153°)$, $(-171°, +85°)$, $(-71°, +107°)$ for the tri-, tetra, and pentamethoprims, respectively. The orientation of the rings is fairly similar in the triand pentamethoprims but distinctly different in tetramethoprim.

V. Morphine, Agonists, and Antagonists

The stereodrawing of morphine plotted by a computer program (*41*) and based on the coordinates and thermal parameters experimentally determined by X-ray analysis of a crystal of the HBr salt (*42*) has been oriented to show the characteristic T shape of the molecule (or ion with the protonated N) in Fig. 20(a). Naloxone, a potent narcotic antagonist, differs from morphine chemically in the substitution of an allyl chain for the methyl group on the N atom, the substitution of OH for H at C-14, the saturation of the C-7–C-8 bond, and a carbonyl oxygen at C-6 rather than a hydroxyl. The stereodrawing of naloxone in Fig. 20(b) shows that morphine and naloxone (*43*) have identical molecular conformations, except for atoms C-6 and C-7 in ring D.

6-Ketomorphinans have been shown to have narcotic agonist properties, with the greatest effect on analgesic potency in the bay area encompassing atoms C-3 to C-6 (*44*). The rupture of the 4,5-O- bridge in morphine produces little change in conformation as can be seen by comparing 1-bromo-4-hydroxy-*N*-methylmorphinan-6-one (*45*) in Fig. 20(c) with naloxone, Fig. 20(b). Furthermore, a comparison of a 3,4-methylenedioxy-substituted morphinanone [solid lines in Fig. 20(d)] that is 20 times less potent than the analogous 4-methoxy derivative [dotted lines, Fig. 20(d)] shows the extremely close similarity in conformation despite a very different biological activity (*44*).

Cyclazocine, a nonaddictive analgesic antagonist of morphine, is used as a final example in this class of compounds. The presence of the *N*-cyclopropylmethyl side chain in cyclazocine and the *N*-allyl side chain in naloxone distinguishes the antagonists from the agonists that have simply an *N*-methyl moiety. In both antagonists, the side chains are disposed to the right of the T shape as drawn in Fig. 20(b) and (e), with torsional angles about C-9—N—C-17—C-18 having values of $-60°$ for cyclazocine (*46*) and comparably $-51°$ for naloxone (*43*).

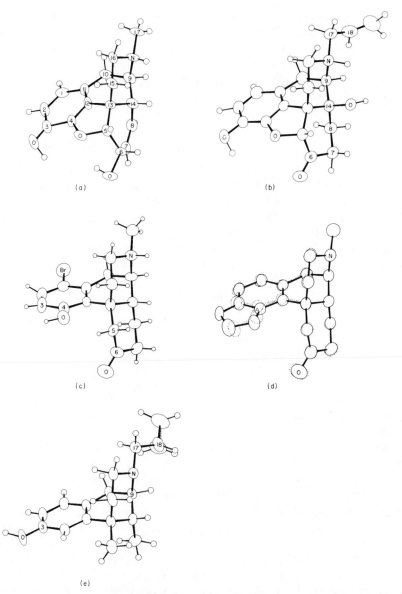

FIG. 20. A comparison of the conformations of (a) morphine (*42*), (b) naloxone (*43*), (c) 1-bromo-4-hydroxy-*N*-methylmorphinan-6-one (*45*), (d) a 3,4-methylenedioxy-substituted morphinanone (solid line) and a 4-methoxy-substituted morphinanone (dotted lines) (*44*), and (e) cyclazocine (*46*). In all molecules, the ellipsoids represent the experimentally determined thermal parameters at the 50% probability level. Hydrogen atoms have been omitted in (d).

The ring system in cyclazocine is abbreviated as compared with morphine or the morphinanones; nonetheless, the remaining atoms can be superimposed very closely over those of morphine, naloxone, and the agonists shown in Fig. 20(a–d). In other words, crystal structure analyses have shown that all six substances have an almost common geometrical arrangement. In this class of substances, the differences in their activities have to be attributed largely to factors other than conformation.

VI. Selected Bisindole and Bisditerpene Alkaloids

The major uses of crystallography in the province of alkaloids, namely, elucidating the structural formula, establishing the stereoconfiguration, and defining the preferred conformation, are illustrated further in the following examples. The alkaloid molecules were chosen mainly for their relatively large molecular weight and the unusual junctions between the two principal moieties.

A. STAPHISINE, A BISDITERPENE ALKALOID

The molecular formula of staphisine, isolated from *Delphinium staphisagria,* was corrected to $C_{43}H_{60}N_2O_2$ by X-ray diffraction studies (47). The first analysis was puzzling due to the diffuse appearance of a methoxy group in an electron-density map and the unusually low microanalyses for methoxy. Subsequently, it had been discovered that the crystal used for analysis was a mixed crystal of staphisine and the nonmethoxy alkaloid staphidine. A separation of the two alkaloids led to a pure staphisine crystal that was reexamined by X-ray diffraction.

Staphisine, shown in Fig. 21, is a bisditerpene alkaloid that consists of two atisine-type units (labeled A and B) that have joined to form a central six-membered ring. The two rather inflexible halves of the molecule are well separated by the central ring that contains oxygen OB and two spiro carbon atoms, C-16A and C-16B. Carbon-16B is also part of a three-membered ring, a rather unusual feature. The only other oxygen atom in the molecule, O_A, is part of a methoxy group. The absolute configuration has not been determined for this alkaloid, since the free base was used for X-ray analysis and there were no atoms present with suitable anomalous dispersion characteristics for determining the absolute configuration.

B. BISINDOLE ALKALOIDS: GEISSOSPERMINE, SUNGUCINE, AND GARDMULTINE

Geissospermine ($C_{40}H_{48}N_4O_3$) isolated from the bark of a Brazilian apocynacea (*Geissospermum laeve* Baillon) is composed of two fragments

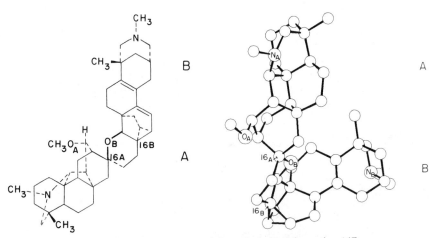

FIG. 21. Staphisine: structural formula and conformation (47).

containing indole moieties and hydrolyzes to one molecule each of geissoschizine and geissoschizoline. The main purpose of the crystal structure
analysis (48) was to establish the unknown stereochemistry around the bond
linking the two moieties, C-16′—C-17′. It was shown that at this junction
the stereoconfiguration is 16′R and 17′S. A projection along the
C-16′—C-17′ bond in Fig. 22 illustrates the arrangement of the substituents bonded to these atoms. The formula and a computer drawing of the
conformation of geissospermine are shown in Fig. 23. Note the parallel
stacking of the COOCH$_3$ moiety over the indole group of the geissoschizine
portion and the cis junction between rings C′ and D′.

Sungucine (C$_{42}$H$_{42}$N$_4$O$_2$), a related bisindole alkaloid extracted from the
roots of the *African Strychnos icaja* Baillon, has an unusual connection
[C-23—C-5] between the two similar parts of the molecule that have the
same stereochemistry. Crystal-structure analyses established the structural
formula (49) which is shown in Fig. 24 along with the conformation of the

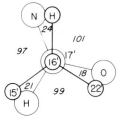

FIG. 22. The arrangement of substituents about the C-16′—C-17′ bond in geissospermine
(48).

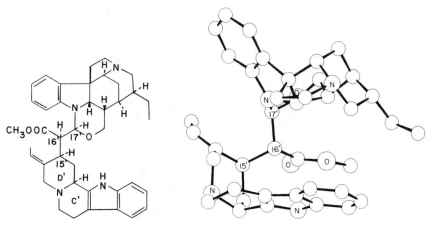

FIG. 23. Geissospermine: structural formula and conformation (48).

molecule. The ring junctions BC, CE, CF, and DE are all cis whereas the ring junction BF is trans.

Gardmultine ($C_{45}H_{54}N_4O_{10}$) a bisindole alkaloid isolated from *Gardneria multiflora* Makino, comprises gardneramine and chitosenine linked by a spiro five-membered ring. The structure proposed from chemical and spectroscopic properties (50) was confirmed by the crystal-structure analysis (51). The structural formula and the stereodrawing, shown in Fig. 25, illustrate the spiro junction at atom C-16 between the two component alkaloids of gardmultine.

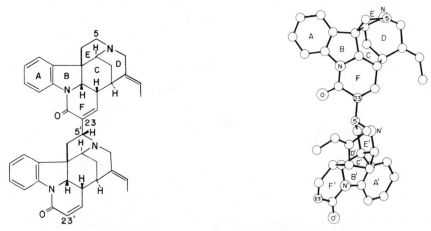

FIG. 24. Sungucine: structural formula and conformation (49).

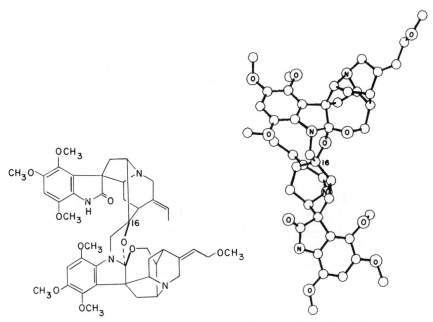

FIG. 25. Gardmultine: structural formula and conformation (51).

VII. Veratridine, a Sodium Channel Neurotoxin

Batrachotoxin (see Section III, A) binds to a single receptor site associated with sodium channels that mediate the electrical excitability of nerve, heart, and skeletal muscle. Although batrachotoxin is a full agonist of the receptor, there are three other types of neurotoxins with disparate chemical structures that act as partial agonists, among them veratridine ($C_{36}H_{52}NO_{11}$) isolated from the *Veratrum* genus. Crystal-structure analyses of drugs in this series can provide the topochemical information needed for the identification of the common structural features that are required for toxin action.

Figure 26 shows the formula, conformation, and absolute configuration of veratridine (52). The structure and conformation of the C_{27}-steroidal base is very similar to that determined for zygacine ($C_{32}H_{49}NO_8$) another *Veratrum* alkaloid (53). Variations in the pharmacological effects of *Veratrum* alkaloids appear to be dependent on the substituents attached to the essentially rigid molecular framework. The numerous intramolecular hydrogen bonds that enhance the molecular rigidity are indicated in Fig. 26. The positions of all the hydrogen atoms have been located for the veratridine molecule and the intramolecular donors and acceptors are identified as

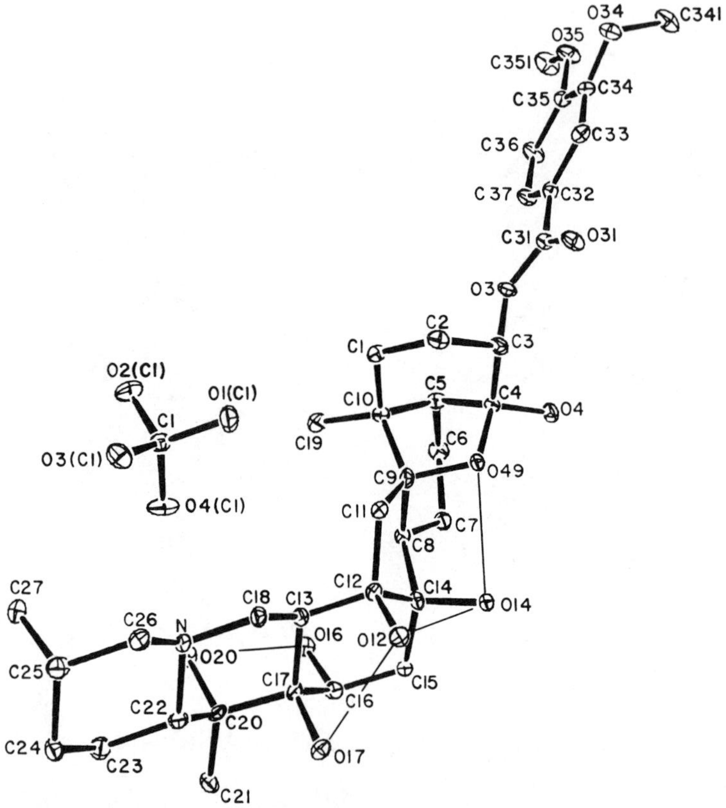

FIG. 26. Veratridine. Thin lines indicate the four internal OH · · · O hydrogen bonds (52) that contribute to the rigidity of the molecule.

O-17—H · · ·O-12—H · · ·O-14—H · · ·O-49 in a connected sequence and the isolated O-20—H · · ·O-16 bond.

Batrachotoxin and veratridine have quite different molecular structures, nevertheless they seem to bind to the same receptor. With precise structural data for the conformations of both molecules, it was possible to search for points of similarity for the attractive sites on the surfaces of these molecules. The points of similar activity may be approximated by a model requiring a triangle of reactive oxygen atoms placed at a distance in a specific range from the tertiary nitrogen atom. A least-squares fit of the three rigid oxygen atoms in batrachotoxin, O-3—H, O-11—H, and the 3α,9α-hemiketal, to three oxygen atoms with similar separations and similar distances to the nitrogen atom in veratridine showed that the hemiketal O-49, O-4—H, and O-14—H satisfied the above requirements. Figure 27 shows the distances in

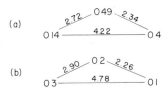

FIG. 27. Distances (in Å) between reactive oxygen atoms in veratridine (a) and batracho-toxin (b) (see Figs. 26, 8, and 9). The average distance between the oxygen triangle and the nitrogen is 5.7 Å in veratridine and 6.6 Å in batrachotoxin (52).

the oxygen triangles of the two molecules (52). Thus, the similar spatial arrangements of polar groups on the surfaces of these two quite different alkaloids are correlated with their similar binding propensities toward a particular receptor.

VIII. Summary

The objectives in this chapter are to illustrate the various kinds of information that are immediately derivable from a crystal-structure analysis by X-ray diffraction and their applications to alkaloid molecules. From unequivocally providing the structural formula, stereochemistry, and absolute configuration of an alkaloid of unknown formula to establishing the spatial arrangement of the atoms, particularly the disposition of the atoms that are the reactive centers or that provide a steric hindrance to reaction, crystal-structure analyses are extremely useful and often indispensable.

REFERENCES

1. F. Marki and B. Witkop, *Experientia* **19,** 329(1963).
2. I. L. Karle and J. Karle, *Acta Crystallogr., Sect. B* **B25,** 428(1969).
3. T. Tokuyama, J. Daly, B. Witkop, I. L. Karle, and J. Karle, *J. Am. Chem. Soc.* **90,** 1917(1968).
4. J. Karle and I. L. Karle, *Acta Crystallogr.* **21,** 849(1966).
5. J. Karle, *Adv. Chem. Phys.* **16,** 131(1969).
6. I. L. Karle and J. Karle, *Acta Crystallogr., Sect. B* **B24,** 81(1968).
7. O. Kennard, D. C. Watson, F. H. Allen, and S. M. Weeds, eds., "Molecular Structures and Dimensions," Vols. 1–12. Cambridge Univ. Press, London and New York, 1970–1981.
8. "International Tables for X-Ray Crystallography," Vol. I. Kynoch Press, Birmingham, England, 1962.
9. G. H. Stout and L. H. Jensen, "X-Ray Structure Determination." Macmillan, New York, 1968.
10. A. L. Patterson, *Phys. Rev.* **46,** 372(1935).
11. A. L. Patterson, *Z. Kristallogr.* **90,** 517(1935).
12. Jean M. Karle and P. W. LeQuesne, *Chem. Commun.* 416(1972).

13. J. Karle and H. Hauptman, *Acta Crystallogr.* **3**, 181(1950).
14. G. Germain, P. Main, and M. M. Woolfson, *Acta Crystallogr., Sect. A* **A27**, 368(1971).
15. G. M. Sheldrick, "SHELX 76. Program for Crystal Structure Determination." University of Cambridge, England, 1976.
16. J. M. Stewart, "The XRAY 76 System," Tech. Rep. TR-446. Computer Science Center, University of Maryland, College Park, 1976.
17. M. Cygler, M. Przybylska, and O. E. Edwards, *Acta Crystallogr., Sect. B* **B38**, 479(1982).
18. I. L. Karle and J. Karle, *Acta Crystallogr.* **17**, 1356(1964).
19. J. K. Frank, E. N. Duesler, N. N. Thayer, R. Heckendorn, K. L. Rinehart, Jr., and I. C. Paul, *Acta Crystallogr., Sect. B* **B34**, 2316(1978).
20. J. M. Bijvoet, *Proc. Acad. Sci. Amsterdam* **52**, 313(1949).
21. J. M. Bijvoet, A. F. Peerdeman, and A. J. van Bommel, *Nature (London)* **168**, London, 271(1951).
22. A. W. Hanson and F. R. Ahmed, *Acta Crystallogr.* **11**, 724(1958).
23. G. Kartha, F. R. Ahmed, and W. H. Barnes, *Acta Crystallogr.* **15**, 326(1962).
24. J. V. Silverton and C. Kabuto, *Acta Crystallogr., Sect. B* **B34**, 588(1978).
25. J. D. Dunitz, "X-Ray Analysis and the Structure of Organic Molecules," Chapter 7, p.337. Cornell Univ. Press, Ithaca, 1979.
26. J. Karle, J. A. Estlin, and I. L. Karle, *J. Am. Chem. Soc.* **89**, 6510(1967).
27. B. Witkop and E. Gössinger, *in* "The Alkaloids" (A.R. Brossi, ed.), Vol. XXI, Chapter 5. Academic Press, Inc., New York, 1983.
28. J. W. Daly, *Fortschr. Chem. Org. Naturst.* **41**, 205(1982).
29. R. Imhof, E. Gössinger, W. Graf, L. Berner-Fenz, H. Berner, R. S. Chanfelberger, and H. Wehrli, *Helv. Chim. Acta* **56**, 139(1973).
30. R. D. Gilardi, *Acta. Crystallogr., Sect. B* **B26**, 440(1970).
31. T. Tokuyama, J. Daly, and B. Witkop, *J. Am. Chem. Soc.* **91**, 3931(1969).
32. J. W. Daly, T. Tokuyama, G. Habermehl, I. L. Karle, and B. Witkop, *Justus Liebigs Ann. Chem.* **729**, 198(1969).
33. J. W. Daly, B. Witkop, T. Tokuyama, T. Nishikawa, and I. L. Karle, *Helv. Chim. Acta* **60**, 1128(1977).
34. J. W. Daly, I. Karle, C. W. Myers, T. Tokuyama, J. A. Waters, and B. Witkop. *Proc. Natl. Acad. Sci. U.S.A.* **68**, 1870(1971).
35. I. L. Karle, *J. Am. Chem. Soc.* **95**, 4036(1973).
36. T. Tokuyama, K. Uenoyama, G. Brown, J. W. Daly, and B. Witkop, *Helv. Chim. Acta* **57**, 2597(1974).
37. J. W. Daly, T. Tokuyama, T. Fujiwara, R. J. Highet, and I. L. Karle, *J. Am. Chem. Soc.* **102**, 830(1980).
38. R. Mata and J. L. McLaughlin, *Phytochemistry* **19**, 673(1980).
39. G. M. Anderson, III, P. A. Kollman, L. N. Domelsmith, and K. N. Houk, *J. Am. Chem. Soc.* **101**, 2344(1979).
39a. K. Takahashi and A. Brossi, *Heterocycles* **19**, 691 (1982).
39b. A. Brossi, P. N. Sharma, K. Takahashi, J. F. Chiang, I. L. Karle, and G. Siebert, *Helv. Chim. Acta* **66**, 795.
40. T. F. Koetzle and G. J. B. Williams, *J. Am. Chem. Soc.* **98**, 2074(1976).
41. C. K. Johnson, "ORTEP," Rep. No. ORNL-5138. Oak Ridge Nat. Lab., Oak Ridge, Tennessee, 1976.
42. L. Gylbert, *Acta Crystallogr., Sect. B* **B29**, 1630(1973).
43. I. L. Karle, *Acta Crystallogr., Sect. B* **B30**, 1682(1974).
44. A. Brossi, L. Atwell, A. E. Jacobson, M. D. Rozwadowska, H. Schmidhammer, J. L. Flippen-Anderson, and R. Gilardi, *Helv. Chim. Acta* **65**, 2394 (1982).

45. A. Brossi, F. L. Hsu, K. C. Rice, M. D. Rozwadowska, H. Schmidhammer, C. D. Hufford, C. C. Chiang, and I. L. Karle, *Helv. Chim. Acta* **64**, 1672(1981).
46. I. L. Karle, R. D. Gilardi, A. V. Fratini, and J. Karle, *Acta Crystallogr., Sect. B* **B25**, 1469(1969).
47. S. W. Pelletier, W. H. DeCamp, J. Finer-Moore and I. V. Micovic, *Acta Crystallogr., Sect. B* **B36**, 3040(1980).
48. A. Chiaroni and C. Riche, *Acta Crystallogr., Sect. B* **B35**, 1820(1979).
49. L. Dupont, O. Dideberg, J. Lamotte, K. Kambu, and L. Augenot, *Acta Crystallogr., Sect. B* **B36**, 1669(1980).
50. S. Sakai, N. Aimi, E. Yamaguchi, E. Yamanaka, and J. Haginiwa, *Tetrahedron Lett.* 719(1975).
51. J. V. Silverton and T. Akiyama, *J. Chem. Soc., Perkin Trans. 1* 1263(1982).
52. P. W. Codding, *J. Am. Chem. Soc.* (in publication).
53. R. F. Bryan, R. J. Restivo, and S. M. Kupchan, *J. Chem. Soc., Perkin Trans. 2* 386(1973).

CHAPTER 3

PUTRESCINE, SPERMIDINE, SPERMINE, AND RELATED POLYAMINE ALKALOIDS

ARMIN GUGGISBERG AND MANFRED HESSE

*Organisch-chemisches Institut der Universität Zürich,
Zürich, Switzerland*

I. Introduction

Diamines and polyamines occur in the plant and animal kingdoms as free bases (biogenic amines) as well as derivatives. The derivatives can be divided into several groups: One contains the di- or polyamine as part of a peptide [e.g., glutathionylspermidine, γ-glutamylcysteinylglycylspermidine *(1–3)*] or as part of an amino acid [e.g., putreanine, $NH_2(CH_2)_4NH(CH_2)_2COOH$ *(4)*]. Further, some antibiotics are known which contain a di- or polyamine. Examples are bleomycin *(5)* and tallysomycin *(6)*. In agreement with other chapters of "The Alkaloids," we will not discuss the aforementioned groups of natural products. In Section II the bases isolated from natural sources are listed and briefly summarized. The main part of this chapter will deal with the alkaloidal derivatives of di- and polyamines. These include fatty acids (including acetic acid) and cinnamic acid conjugates as well as the simple, methylated compounds.

II. Occurrence of Natural Diamines and Polyamines

In addition to the well-known compounds putrescine, spermidine, and spermine, many other simple di-, tri-, and tetraamine compounds are

THE ALKALOIDS, VOL. XXII

known in nature. Those that have been isolated as such from plants, animals, or microorganisms are presented in Table I. Review references for the individual bases are given where possible; otherwise the primary literature is cited. Some of these biogenic amines occur ubiquitously in nature. Their biochemistry, biogenesis, and function as well as analyses have been reviewed in several papers and books (7–17). The alkaloidal derivatives of these bases are summarized in Sections III and V.

III. Structure Elucidation and Synthesis of Polyamine Alkaloids

In contrast to the polycyclic indole, isoquinoline, or terpene alkaloids, the di- and polyamine alkaloids seem to be of much simpler construction. This first impression is misleading. Special structural features render this group of alkaloids even more difficult to handle than the above-mentioned ones. It should be noted that the structures of several polyamine alkaloids have had to be revised. Because of this, two main factors should be mentioned. (1) The alkaloids sometimes occur as mixtures that are very difficult to separate, and (2) the results from spectral or chemical analyses are equivocal (cf. references on structural elucidation in Section V). This group of alkaloids was the subject of several review papers (26–28) and especially covering the subject of synthesis (29–32). Some general aspects of the difficulties associated with the isolation and structure elucidation of the polyamine alkaloids are discussed.

Some of the alkaloids contain phenolic hydroxyl groups in addition to the basic amino nitrogen, so they are very polar, strong bases. In order to isolate these materials from plants and to purify the compounds to some extent, procedures must be applied that are different from those used in "normal" alkaloid chemistry. Extraction of plant sources under strongly basic or even strongly acidic conditions is dangerous because of the possibility of trans-amidation reactions. Products from this kind of rearrangement reaction are artifacts, and no one is interested in isolating artifacts.

Another difficulty is the determination of the homogenity of the isolated alkaloids. To illustrate this two extreme examples are given. By different chromatographic methods and because of its spectral properties (NMR, IR, MS), the natural, approximate 1:1 mixture of the spermidine alkaloids inandenin-12-one and inandenin-13-one appeared to be a pure compound. This was supported by the sharp melting point of its hydrochloride (150–151°C). Only after chemical degradation was the nature of the mixture determined by mass spectrometry. On the other hand, the behavior of the pure crystalline spermine alkaloid aphelandrine was that of a mixture. At least two criteria were in favor of a mixture. In the ^{1}H-NMR spectrum several singlets were registered for only one methyl group. Furthermore, several molecular ion peaks were found in the MS. The NMR data are in agreement with the existence of different conformations of medium and large ring compounds and the cis–trans isomerism of amide (lactam) bonds. The observation of several molecular ions in the mass spectrum of one compound must be interpreted in light of the characteristic behavior of the 1,3-diaminopropane unit. This structural element reacts with aldehydes (especially formaldehyde) that are present as impurities in solvents, even in those of high analytical purity. By this reaction, 1,3-diazacyclohexane derivatives are formed. In the mass spectrometer under electron impact conditions, the condensation products with formaldehyde show a strong $[M - 1]^+$ signal. The m/z value of this ion differs from that of the parent alkaloid by $+ 11$ mass units. Although the concentration of the formaldehyde condensation product is very small, its $[M - 1]^+$ signal is much more abundant than that of the molecular ion of the alkaloid itself. Other aldehydes give similar products (33).

Another major problem is the unequivocal determination of the connection points between the amino nitrogen atoms of the polyamine and the substituents. In the case of spermidine, for example, two primary amino groups exist. Both are very similar with respect to their neighboring groups but not identical (three and four methylene groups between two amino groups, respectively). When there are different substituents on the two nitrogen atoms, chemical degradation combined with spectral analyses often was necessary to determine the correct structure. From the natural occurrence of isomeric and homologous unsubstituted di- and polyamines (see Section II), it can be assumed that polyamine alkaloids will also be found that contain basic backbones other than those isolated to date. Spectroscopic and chemical degradation studies should keep this in mind.

In the analysis of ^{1}H-NMR spectra of compounds elaborated from aliphatic fatty acids and polyamines it is rather difficult to determine the linear or branched nature of the fatty acid. In order to differentiate, well-known oxidative degradation reactions (e.g., HNO_3) are recommended in addition to spectroscopic methods.

Some structure elucidation methods often have been used in the past and should also be recommended for future use. The Hofmann degradation is one of these reactions; another is the use of a KOH melt to hydrolyze (amide) bonds. This method is well-known for determining the nature of the polyamine. One must keep in mind, however, that in addition to the cleavage of polyamine substituent bonds, the polyamines themselves are degraded, and smaller amino fragments can be detected. Therefore, a careful analysis of the product is necessary.

An additional procedure for determining alkaloid structure is the mass

TABLE I

NATURALLY OCCURRING DI-, TRI-, AND TETRAAMINES

Structure	Trivial or systematic name	Reference
$H_2N—(CH_2)_3—NH_2$	1,3-Diaminopropane	7, 8, 13, 18
$H_2N—(CH_2)_4—NH_2$	Putrescine	7, 8, 13, 18
$H_2N—(CH_2)_5—NH_2$	Cadaverine	7, 8, 13, 18
$H_2N—CH_2—\underset{\underset{OH}{\mid}}{CH}—(CH_2)_2—NH_2$	2-Hydroxyputrescine	11–13
$\underset{HN}{\overset{H_2N}{>}}C—NH—(CH_2)_4—NH_2$	Agmatine	7, 8, 13, 18
$\underset{HN}{\overset{H_2N}{>}}C—NH—(CH_2)_4—NHC\overset{NH_2}{\underset{NH}{<}}$	Arcaine	7, 11, 13, 18
$\underset{HN}{\overset{H_2N}{>}}C—NH—(CH_2)_5—NH_2$	Homoagmatine	7, 8, 18, 19

Structure	Name	References
$H_2N\text{—}C(=NH)\text{—}NH\text{—}(CH_2)_5\text{—}NHC(=NH)(NH_2)$	Audouine	*11, 13*
$H_2N\text{—}(CH_2)_3\text{—}NH\text{—}(CH_2)_3\text{—}NH_2$	*sym*-Norspermidine	*8, 11, 13*
$H_2N\text{—}(CH_2)_3\text{—}NH\text{—}(CH_2)_4\text{—}NH_2$	Spermidine	*7, 8, 13, 18*
$H_2N\text{—}(CH_2)_3\text{—}NH\text{—}(CH_2)_5\text{—}NH_2$	*N*-(3-Aminopropyl)-1,5-diaminopentane	*11*
$H_2N\text{—}(CH_2)_4\text{—}NH\text{—}(CH_2)_4\text{—}NH_2$	*sym*-Homospermidine	*7, 8, 11, 18*
$H_2N\text{—}C(=NH)\text{—}NH\text{—}(CH_2)_3\text{—}NH\text{—}(CH_2)_4\text{—}NH_2$	*N*-(3-Guanidinopropyl)-1,4-diaminobutane	*13, 20*
$H_2N\text{—}C(=NH)\text{—}NH\text{—}(CH_2)_3\text{—}NH\text{—}(CH_2)_4\text{—}NHC(NH_2)(=NH)$	Hirudonine	*13, 21*
$H_2N\text{—}(CH_2)_3\text{—}NH\text{—}(CH_2)_3\text{—}NH\text{—}(CH_2)_3\text{—}NH_2$	*sym*-Norspermine (thermine)	*8, 12, 22*
$H_2N\text{—}(CH_2)_3\text{—}NH\text{—}(CH_2)_3\text{—}NH\text{—}(CH_2)_4\text{—}NH_2$	Thermospermine	*23–25*
$H_2N\text{—}(CH_2)_3\text{—}NH\text{—}(CH_2)_4\text{—}NH\text{—}(CH_2)_3\text{—}NH_2$	Spermine	*7, 8, 13, 18*

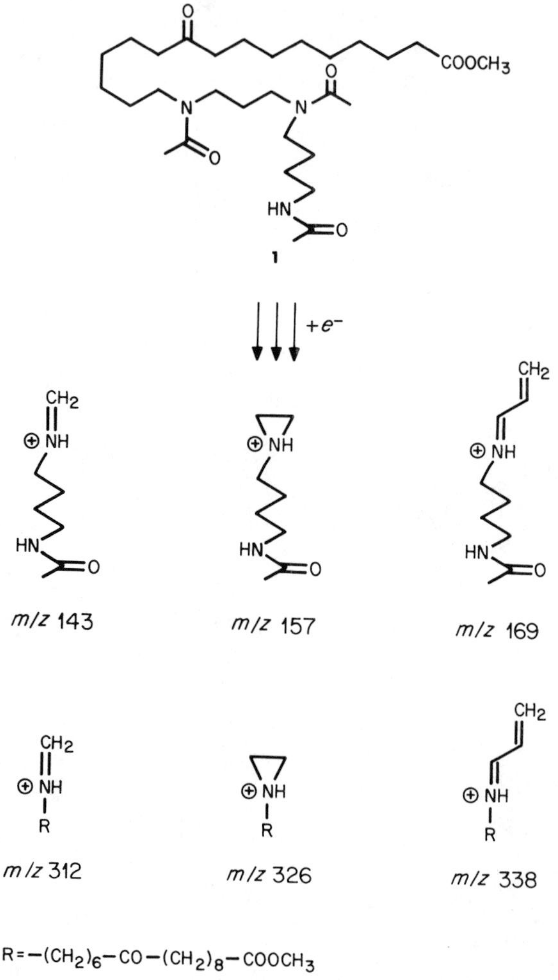

SCHEME 1. Mass spectral fragmentation of inandenin-12-one derivative **1**.

spectral analysis of the acetylated hydrolysates prepared from the alkaloids under mild conditions. This characteristic behavior is discussed using the degradation product **1** of inandenin-12-one as an example (Scheme 1). In the mass spectrum two peak triads are observed: m/z 143, 157, and 169 as well as m/z 312, 326, and 338. The first triad corresponds to ions containing the unbroken 1,4-diaminobutane unit while the 1,3-diaminopropane unit is degraded. The second group of signals corresponds to ions in which the 1,3-diaminopropane unit again is fragmented, but the positive charge is located at the other nitrogen atom. In both cases the fragmentation reactions take place in the smaller unit. By combining these two triads it is possible to determine the substitution pattern of the three (or four) nitrogen atoms and, by analyzing some other signals, the structure itself. A different substitution pattern at the spermidine residue will cause shifts of the peak triads. Similar observations can be made in the case of spermine derivatives (*34*).

A. PUTRESCINE TYPE

Until now only putrescine and 2-hydroxyputrescine have been found as basic skeletons of diaminoalkane alkaloids of this type.*

1. Simple Derivatives of Putrescine

A number of putrescine derivatives have been detected in nature for which one or two cinnamic acid analogs with amide linkages are known: 4-coumaroylputrescine (**2**), di-4-coumaroylputrescine (**3**), feruloylputrescine (**4**) (subaphylline), diferuloylputrescine (**5**), caffeoylputrescine (**6**) (paucine), dicaffeoylputrescine (**7**), sinapoylputrescine (**8**), and disinapoylputrescine (**9**). Paucine, one of the first diaminoalkane alkaloids known since 1894 as a component of the seed of *Pentaclethra macrophylla,* was hydrolyzed (40% KOH–H_2O) to give putrescine and caffeic acid. Structure **6** was deduced (*37*) from this data together with spectroscopic data, especially mass spectra. Another derivative of putrescine, subaphylline (**4**), first isolated from *Salsola subaphylla,* was shown to be the monoferuloyl derivative of putrescine by hydrolysis (30% KOH–H_2O) (*38*). The structure elucidation of the other derivatives of putrescine (**2, 3, 5, 7–9**) mentioned before was undertaken in a manner similar to that of **4** and **6**. Several compounds were synthesized and compared to the natural products and are summarized in Table II of Section V.

Two natural derivatives of 2-hydroxyputrescine have been found in wheat: *N*-(4-coumaroyl)- and *N*-feruloyl-2-hydroxyputrescine (**10** and **11**, respectively). By acid-catalyzed hydrolysis (*S*)-(+)-2-hydroxyputrescine and ferulic acid and 4-coumaric acid were formed, respectively. By comparison to synthetic *N*-carbamoylputrescine (**12**), the structure of a compound isolated from barley seedlings was confirmed (*39*).

Three methylated derivatives of putrescine are known. Tetramethylpu-

*In the hydrolysate of toxic components in the venom of bird-catching spiders (suborder Orthognatha, family Aviculariidae, subfamilies Grammostolinae and Theraphosinae), diaminopropane, spermine, 4-hydroxyphenylpyruvic acid, 4-hydroxybenzoic acid, 4-hydroxyphenylacetic acid, and 4-hydroxyphenyllactic acid have been identified. Separation and structure elucidation of the individual components of the original extract were not performed (*35,36*).

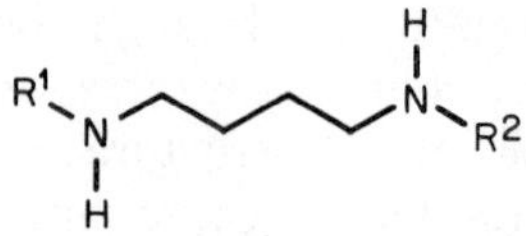

2 $R^1 = CO-CH\overset{E}{=}CH-$⟨benzene⟩$-OH$, $R^2 = H$

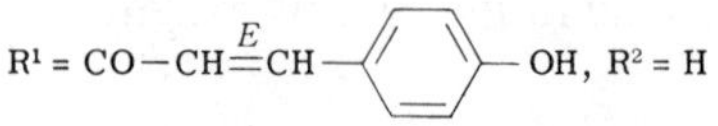

3 $R^1 = R^2 = CO-CH\overset{E}{=}CH-$⟨benzene⟩$-OH$

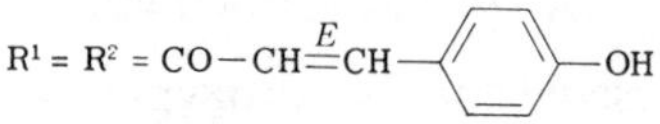

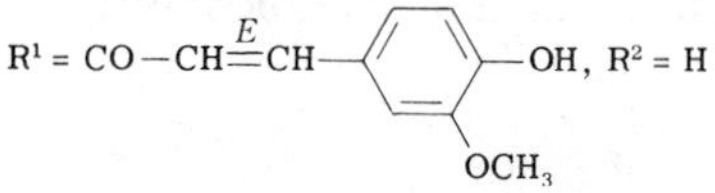

4 $R^1 = CO-CH\overset{E}{=}CH-$⟨benzene⟩$-OH$, $R^2 = H$
OCH_3

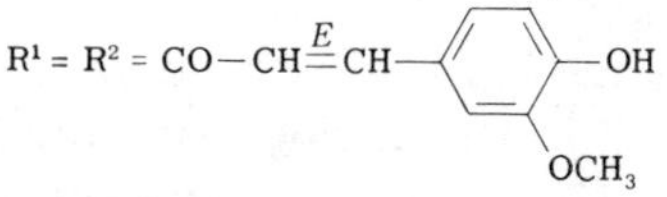

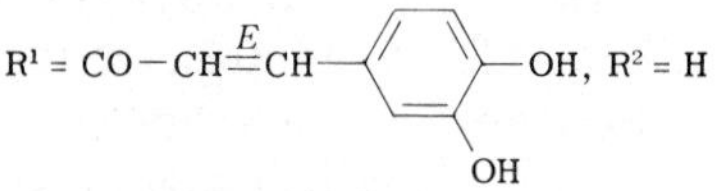

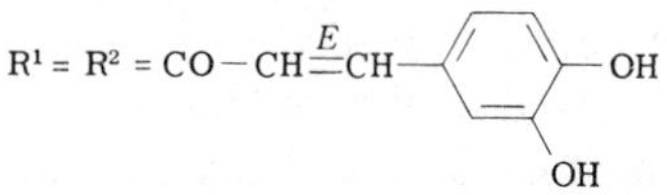

5 $R^1 = R^2 = CO-CH\overset{E}{=}CH-$⟨benzene⟩$-OH$
OCH_3

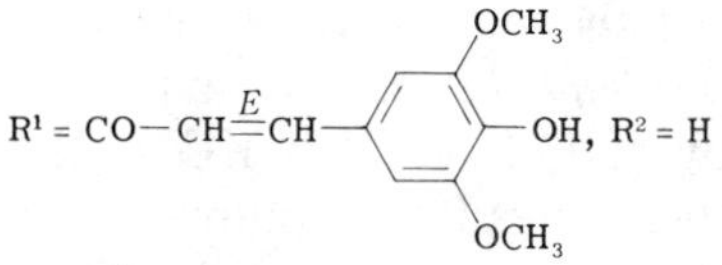

6 $R^1 = CO-CH\overset{E}{=}CH-$⟨benzene⟩$-OH$, $R^2 = H$
OH

7 $R^1 = R^2 = CO-CH\overset{E}{=}CH-$⟨benzene⟩$-OH$
OH

8 $R^1 = CO-CH\overset{E}{=}CH-$⟨benzene⟩$-OH$, $R^2 = H$
OCH_3 / OCH_3

9 $R^1 = R^2 = CO-CH\overset{E}{=}CH-$⟨benzene⟩$-OH$
OCH_3 / OCH_3

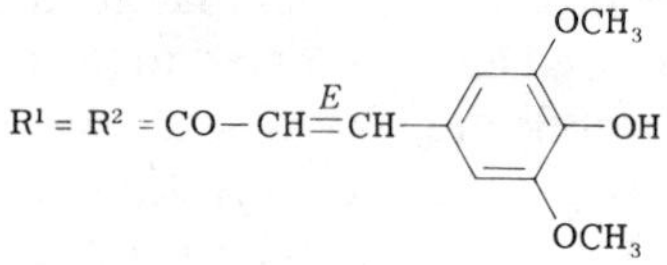

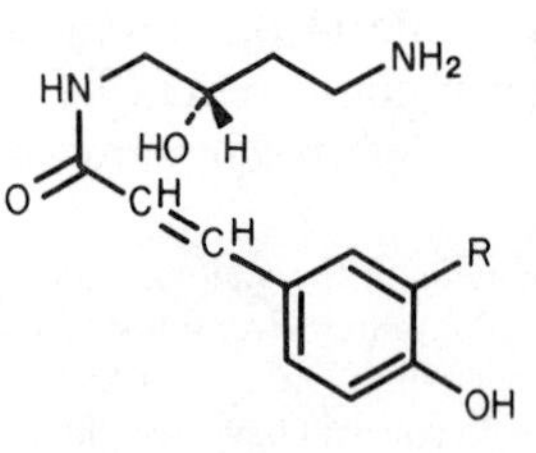

10 R = H   **11** R = OCH_3

12

trescine (**13**) was the first methylated derivative to be isolated from *Hyoscyamus muticus* in 1907 (*40*). The structure elucidation was performed in a classical way by Hofmann degradation to yield butadiene. Comparison of a synthetic sample to the quaternized natural product proved the identity of both tetramethylene-1,4-ditrimethylammonium diiodides (*40*). *N,N,N'*-Trimethyl-*N'*-(4-hydroxy-*Z*-cinnamoyl)putrescine (**14**) and *N,N,N'*-trimethyl-*N'*-(4-methoxy-*Z*-cinnamoyl)putrescine (**15**) were isolated from three *Kniphofia* species (*41*). Methylation of **14** afforded the methylester **15**. Structure elucidation was done by means of IR, ¹H-NMR, and mass spectral analyses, as well as by a direct comparison of the dihydro derivative of **15** to a synthetic sample (*41*).

13 R = CH₃
14 R = CO—CH=CH—⟨benzene⟩—OH
15 R = CO—CH=CH—⟨benzene⟩—OCH₃

2. Agmatine Derivatives

Several agmatine (**16**) (Scheme 2) derivatives have been isolated from barley seedlings. All are conjugates of coumaric acid and possess antifungal activity.

4-Courmaroylagmatine (**17**) picrate was hydrolyzed (2 *N* HCl, 100°C, 22 hr) to form agmatine (**16**); the coumaric acid moiety itself was apparently destroyed under the hydrolytic conditions employed. Synthesis of **17** established its structure (*9*).

The other compounds are hordatine A (**18**) and hordatine B (**19**). A mixture of hordatine A and B glucosides, called hordatine M (**20**), has not

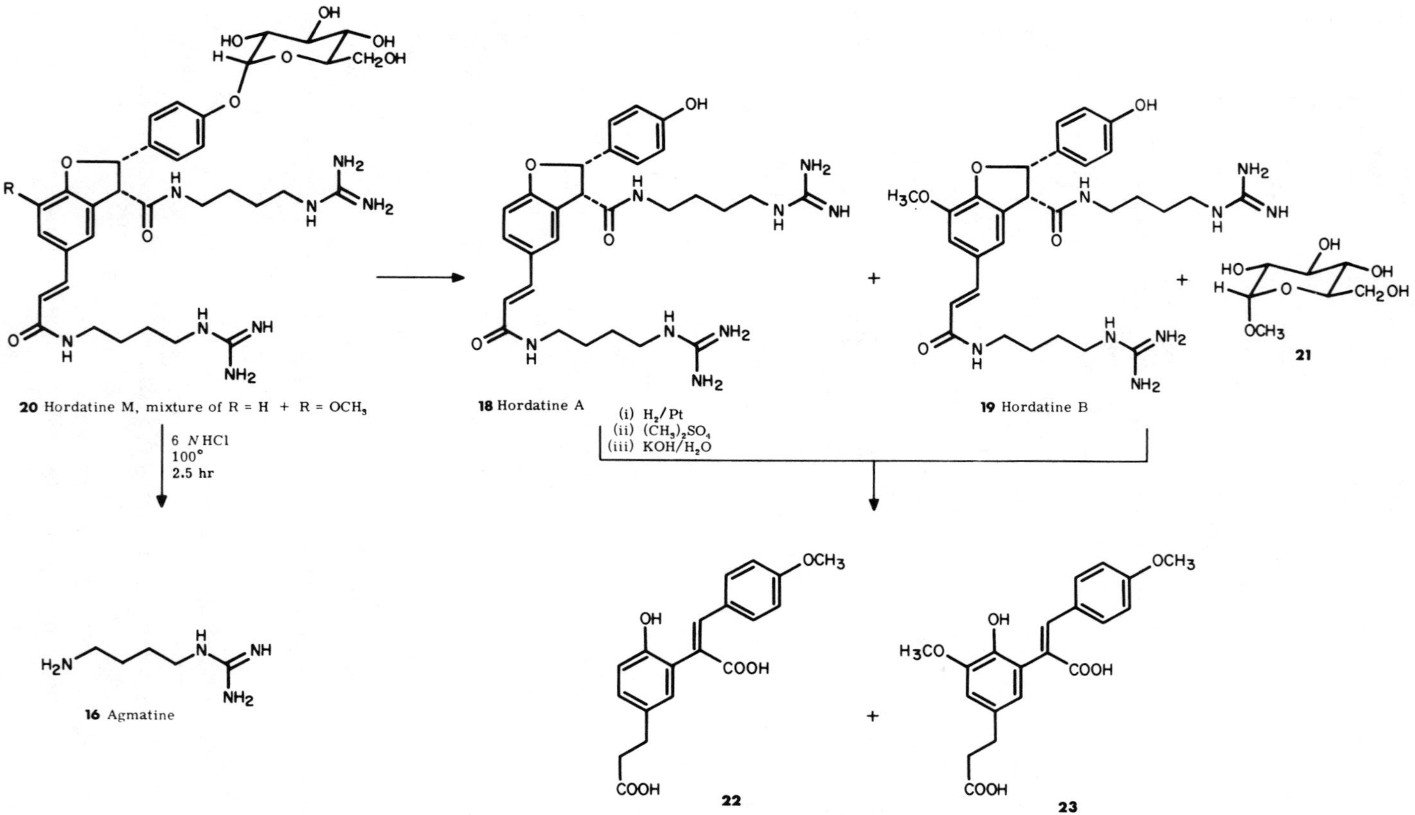

SCHEME 2. Structural work on the barley seedling alkaloids hordatine A, B, and M.

17

yet been separated. Most of the structure elucidation work has been done on this mixture (**20**) (*42–44*).

The most important degradation steps are summarized in Scheme 2. Under normal acetylation reaction conditions, hordatine M forms an O,O',O'',O'''-tetraacetyl derivative in the form of its dihydroacetate (because of the strongly basic nature of the two guanidine moieties). The basic backbone, agmatine (**16**) was found among the hydrolysates of **20**. Identification was made by direct comparison to an authentic sample and by hydrolysis (KOH) to putrescine. In addition to a variety of oxidative degradation reactions, the transformation of hordatine M (**20**) to hordatine A (**18**) and hordatine B (**19**) as well as α-D-methylglucopyranoside (**21**), by methanolic HCl was one of the most important reactions. Both compounds (**18** and **19**) found in the hydrolysis reaction product were identified with the naturally occurring materials. Further degradation of the mixture (**18/19**) was performed by catalytic hydrogenation of the "cinnamic acid" double bond, followed by methylation of the phenolic hydroxyl group and amide cleavage with KOH–H_2O. Two acids resulted (**22** and **23**) representing the chromophores of hordatine A and B, respectively. The final structure of the acids was confirmed by synthesis (*42*). In combination with spectroscopic data the results of the degradation reactions led to the proposed structures of the hordatines. ($\pm$)-Hordatine A was synthesized by oxidative coupling of 4-coumaroylagmatine (**17**) as its hydroacetate with horseradish peroxidase–H_2O_2 or with $K_3[Fe(CN)_6]$. These results confirmed the structure of **18** (*42*).

Recently, it has been shown that an enzyme present in extracts of the shoots of barley seedlings synthesizes both 4-coumaroylagmatine and the hordatines from 4-coumaroyl coenzyme A and [U-^{14}C]agmatine (*45*).*

---

*Other than this investigation, no further work has been published giving experimental results on the biosynthesis of the alkaloids that are the subject of this chapter.

3. Aerothionin

Aerothionin, a tetrabromo derivative, has been isolated from sponges *Aplysina aerophoba* and *Verongia thiona;* the proposed structure is **24** (*46*). The oxime **25** was formed by treatment of aerothionin with dilute alkali. The relative positions of the aromatic substituents in **25** were established by cleavage in hot 6 *N* HCl to give **26** and a lactone subsequently converted by methylation to **27**. The tetramethyl derivative of **25** was hydrolyzed to form

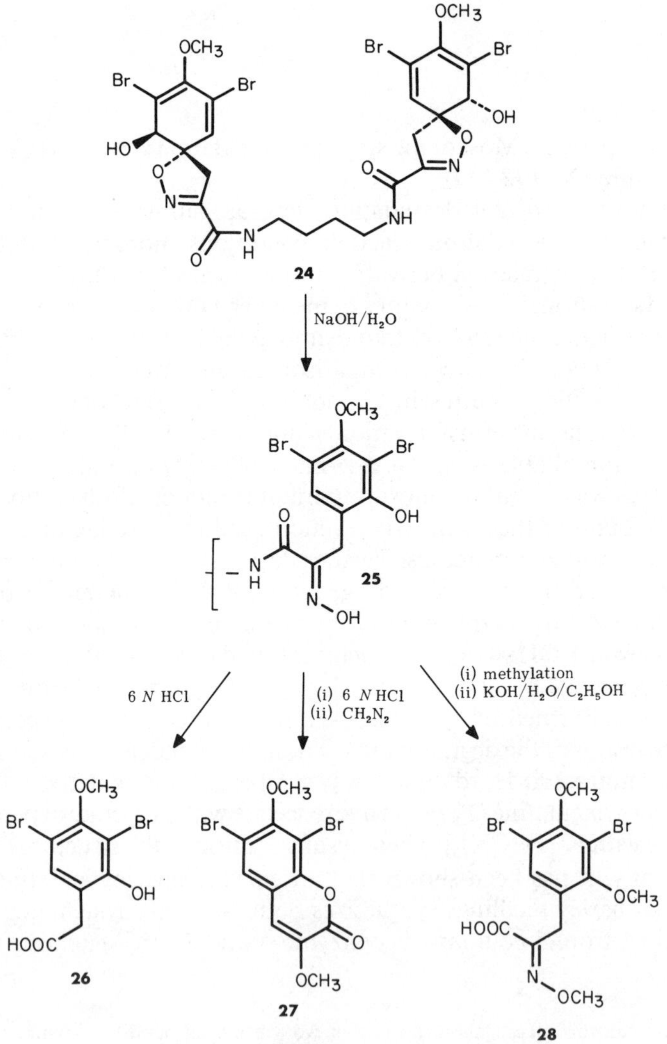

SCHEME 3. Degradation of aerothionin (**24**).

28. Condensation of the acid chloride of **28** with putrescine gave a product identical to the tetramethyl derivative of **25** (*46*) (see Scheme 3).

Ten years later the same aerothionin was isolated from a different sponge. X-Ray structure determination in addition to CD data were employed to assign its absolute configuration (*47*).

B. SPERMIDINE TYPE AND RELATED ALKALOIDS

1. The Simple Open-Chain Spermidine Derivative of Natural Origin

Spermidines substituted with cinnamic acid derivatives seem to be widely distributed in the plant kingdom. Cinnamic acid (alkaloid maytenine), caffeic acid (caffeoylspermidine, dicaffeoylspermidine), 4-coumaric acid (coumaroylspermidine, dicoumaroylspermidine, tricoumaroylspermidine), ferulic acid (feruloylspermidine, diferuloylspermidine), and sinapic acid (sinapoylspermidine, disinapoylspermidine) are known as aromatic amide substituents of spermidine. Occurrence, structure elucidation, and syntheses are summarized in Section V.

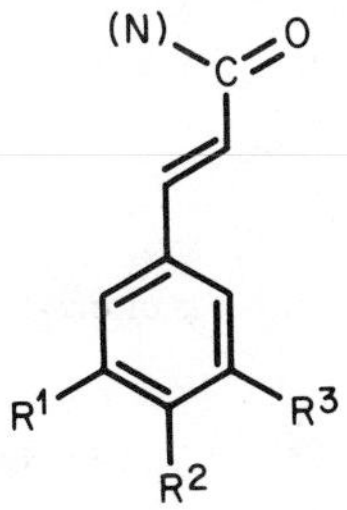

$R^1 = R^2 = OH$, $R^3 = H$	Caffeoyl
$R^1 = R^3 = H$, $R^2 = OH$	4-Coumaroyl
$R^1 = OCH_3$, $R^2 = OH$, $R^3 = H$	Feruloyl
$R^1 = R^3 = OCH_3$, $R^2 = OH$	Sinapoyl
$R^1 = R^2 = R^3 = H$	Cinnamoyl

Maytenine (**31**) was the first of these so-called simple alkaloids to be isolated and identified structurally by use of mass spectrometry, NMR, and UV spectroscopy (*48*). Several syntheses of **31** have been published in the past and are summarized in Scheme 4.

2. Siderophores

Three siderophores (microbial iron-transport agents) belong to the class of spermidine alkaloids: agrobactine (**32**) from *Agrobacterium tumefaciens*

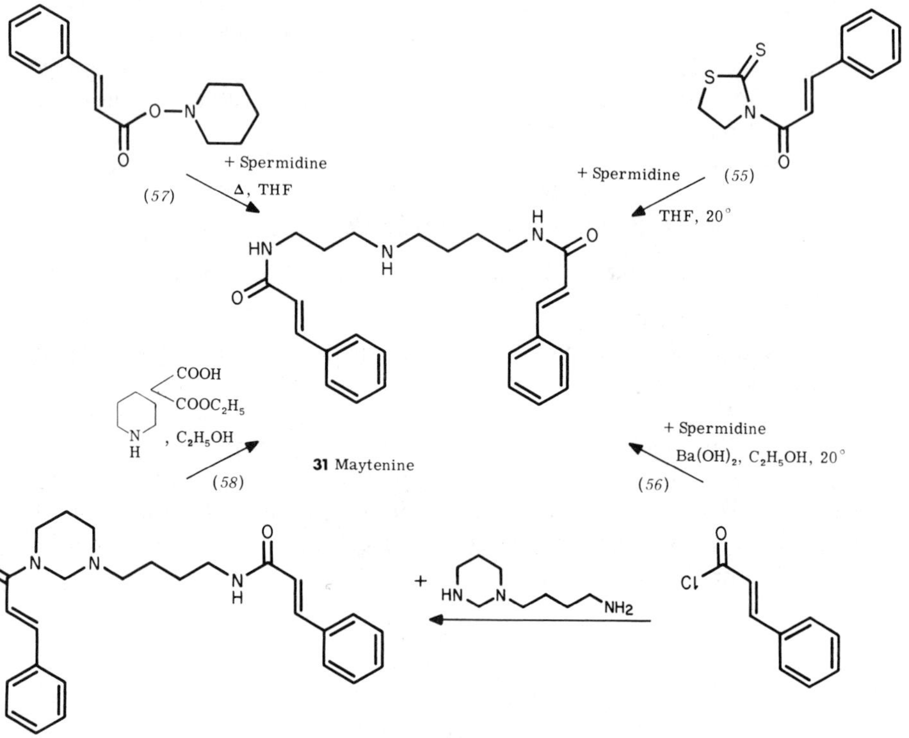

SCHEME 4. Syntheses of maytenine (**31**).

and parabactine (**33**) and 1,8-bis(2,3-dihydroxybenzamido)-4-azaoctane (compound II, **34**), both isolated from *Paracoccus denitrificans* (*49,50*).

The structure elucidation of **34** was based, in principle, on two facts: color reactions and hydrolysis. Compound **34** does not react with ninhydrin (therefore, no primary amino group is present) but it does react with 1-fluoro-2,4-dinitrobenzene. By exact determination of the absorption ratio E_{350}/E_{390} after the reaction with the latter reagent, Tait concluded that **34** contains a secondary amino group (*50*). Hydrolysis (6 N HCl, 110°C, 24 hr) of the so-called compound II afforded 2,3-dihydroxybenzoic acid and spermidine (*50*). The presence of two 2,3-dihydroxybenzoyl residues in **34** was demonstrated by its enzymatic (*50*) and chemical synthesis (*51–54*).

Hydrolysis experiments with the second siderophore, parabactin, gave four products: 2,3-dihydroxybenzoic acid, 2-hydroxybenzoic acid, L-threonine, and spermidine, in a ratio of 2:1:1:1. It was shown that parabactin is enzymatically built up from **34**, L-threonine, and 2-hydroxy-benzoic acid. Furthermore, it was shown that the threonine amino group is

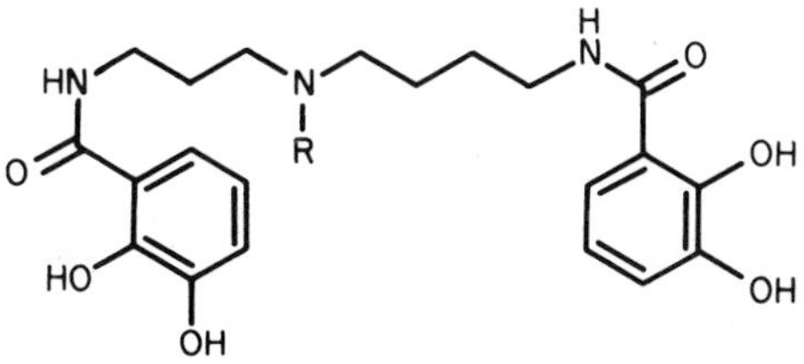

32   R = 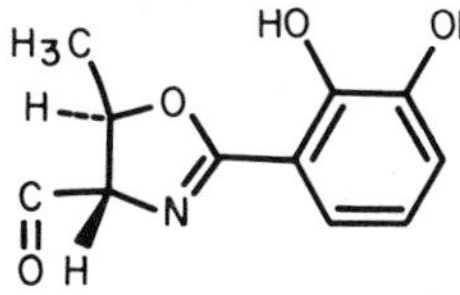   Agrobactine

33   R = 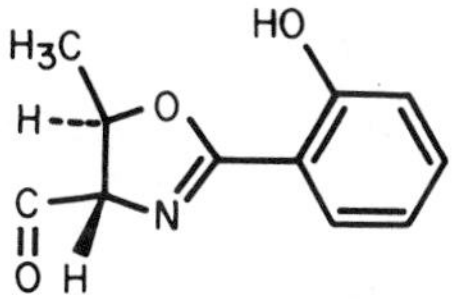   Parabactine

34   R = H   Compound II

part of a peptide bond. On the basis of these results, the structure **33a** for parabactin was proposed (*50*). Extensive UV spectral analysis of parabactin confirmed that the siderophore contains, instead of the peptide linkage between the threonine and the 2-hydroxybenzoic acid moieties, an oxazoline ring that is derived from the proposed structure by loss of water. Thus, the structure of parabactine was revised to **33** (*59*). Agrobactine (**32**), on the other hand, was found by hydrolysis to consist of 2,3-dihydroxybenzoic acid, L-threonine, and spermidine, in the ratio of 3:1:1. From special data and comparison to analogous compounds, structure **32** was proposed (*49,60*).

R =

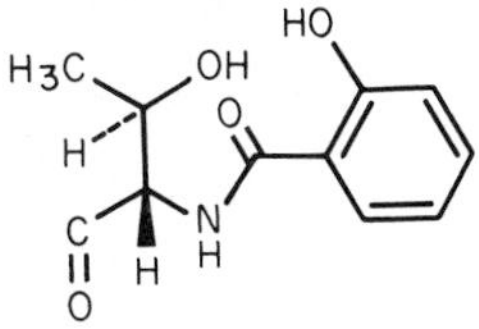

33a

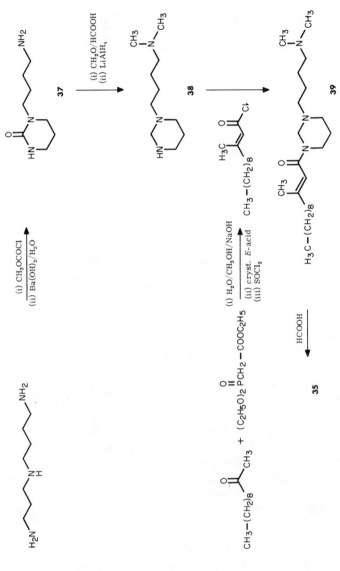

SCHEME 5. Synthesis of 5,12-dimethyl-1-dimethylamino-5,9-diazaheneicos-11-en-10-one (35).

3. Spermidines of Soft Corals

Two cytotoxic spermidine derivatives were isolated as a mixture from the Pacific soft coral *Sinularia brongersmai:* 5,12-dimethyl-1-dimethylamino-5,9-diazaheneicos-11-en-10-one (35) and its 11,12-dihydroderivative 36 (ratio 9:1). By hydrogenation of the mixture, a single compound was obtained that corresponded by GC/MS analysis to the minor component of the original mixture. Hydrolysis followed by esterification of the hydrogenated product 36 afforded methyl 3-methyldodecanoate and *N,N,N*-trimethylspermidine. "The results of a Hofmann degradation of 36 confirmed the conclusion on mass spectral evidence that the primary amide nitrogen is bonded to the trimethylene chain of the unsymmetrical spermidine molecule" (*61*).

35

36 11,12-Dihydro

MS and ^{1}H-NMR spectral analyses of the 35/36 mixture in addition to the chemical evidence led to their structures (*61*).

Synthesis of 35 (and 36) was accomplished according to the scheme depicted in Scheme 5 (*62*). Interesting to note is the ambivalence of the C_1 unit first introduced into spermidine: In compounds 37 and 38 it is used as a protecting group for one primary amino group and for the secondary amino group of spermidine. In the transformation of 39 to 35, the "protecting group" becomes a substituent of the final product.

4. The Oncinotines

The isomeric macrocyclic spermidine alkaloids oncinotine (40), neooncinotine (41), and isooncinotine (42) occur together in the bark of *Oncinotis nitida*. Whereas purification of isooncinotine was readily achieved, the separation of oncinotine and neooncinotine has not yet been possible. In contrast, their *N*-acetyl derivatives have been separated by TLC (*63*). The essential degradation reactions for structure elucidation therefore had to be made with the natural mixture of 40 and 41 (ratio ~ 7:3). Small amounts of pure oncinotine have been obtained by treating the mixture with potassium *tert*-butoxide in boiling toluene. Under these conditions, neooncinotine can

40 Oncinotine   **41** Neooncinotine

42 Isooncinotine

be transformed to isooncinotine (**42**), while oncinotine (**40**) remains stable. The physical properties of oncinotine and the natural mixture of oncinotine and neooncinotine show only slight differences.

A reaction pathway important for the structure elucidation of oncinotine and neooncinotine is the acid-catalyzed opening of the lactam ring to the amino acid, which, following esterification with methanolic HCl, is acetylated and the basic tertiary nitrogen atom methylated with CH_3I. The thus-formed quaternary ester acetate mixture of **43** and **44** (as the fluorides) is then pyrolyzed, whereby a Hofmann degradation reaction occurs to form the three compounds **45**, **46**, and **47**, shown in Scheme 6.

All three compounds have been synthesized and compared to the degradation products. The determination of absolute configuration has been carried out by means of the CD measurement of the LAH-reduction products of **46** in comparison with that of (*R*)-(−)-*N*-methylconiine (*61*). Mass spectral analysis of the nonquaternary ester acetate led to the same structure of the main component (**43**) of the mixture (*63,64*).

The structures of oncinotine, neooncinotine, and isooncinotine, as well as that of pseudo-oncinotine (so far not found in nature), have been confirmed by synthesis (*65,66*). First, the piperidine ring with the long aliphatic side chain was synthesized and then the rest of the spermidine chain was added prior to ring closure of the macrocyclic lactam. Synthesis of the piperidine moiety is shown in Scheme 7. Compound **48** was coupled with the protected bromide **49**, prepared by combined tosylation and acetylation of putrescine, followed by reaction with 1,3-dibromopropane (Scheme 8). The resulting alkylation product **50** was saponified with KOH and detosylated using electrolytic conditions. Cyclization to the macrocyclic *N*-acetylonci-

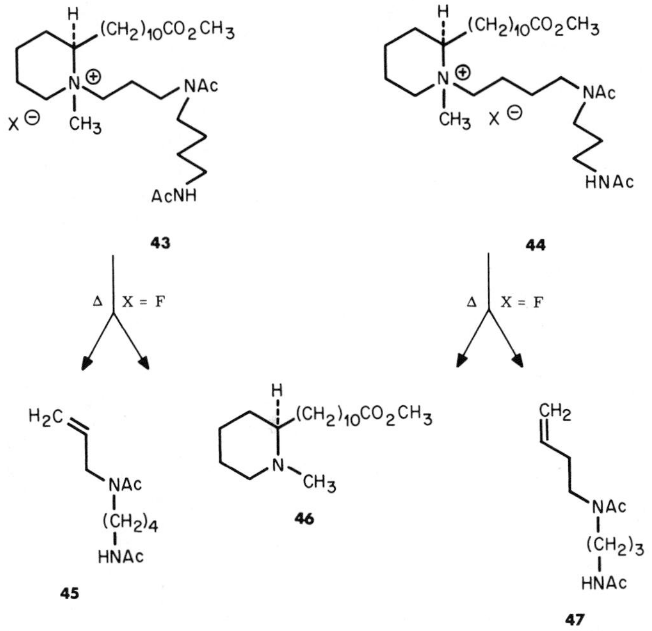

SCHEME 6. Hofmann degradation reaction of quaternary ester mixture.

notine (**51**) was achieved by high-dilution technique using $SOCl_2$ and $(C_2H_5)_3N$.

Using variations of this synthesis (e.g., 1,3-diaminopropane instead of putrescine, different deprotection reactions), the alkaloids (±)-isooncino-

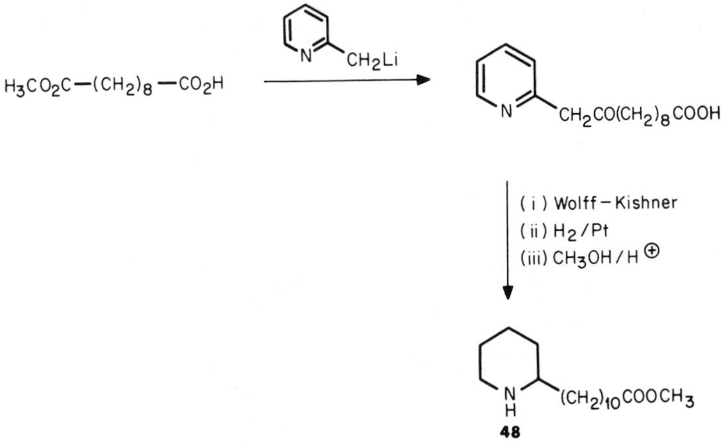

SCHEME 7. Synthesis of piperidine intermediate **48**.

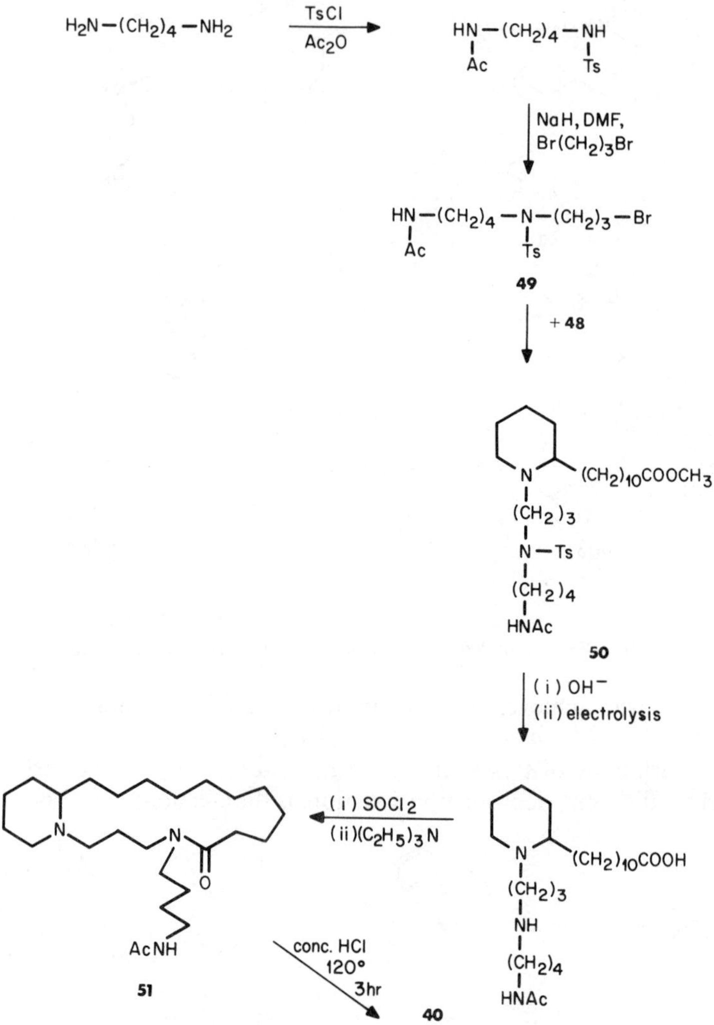

SCHEME 8. Synthesis of oncinotine (40).

tine (42), (±)-neooncinotine (41), and (±)-pseudo-oncinotine (52) as well as (±)-oncinotine (40) were prepared (66). An alternative synthetic pathway has been developed [see Ref. (65)].

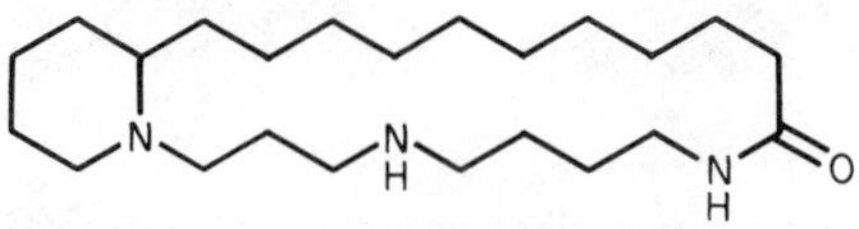

52   Pseudo-oncinotine

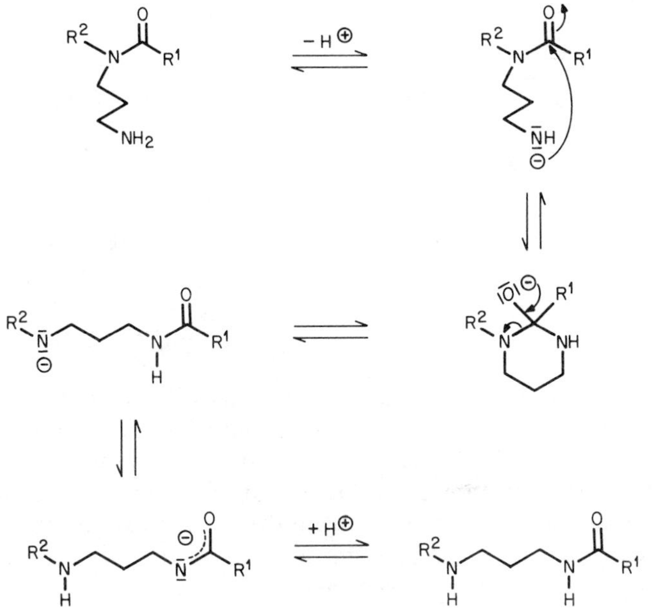

SCHEME 9. The principle of the base-catalyzed transamidation reaction (R^1 and R^2 are alkyl substituents).

As mentioned before, neooncinotine (**41**) is transformed to isooncinotine (**42**) by treatment with potassium *tert*-butoxide in boiling toluene. This transamidation reaction takes place via a six-membered intermediate, as shown in Scheme 9 (*67*). The driving force for this reaction sequence is the higher stability (resonance stabilization) of the product anion **53** compared to the anion **54** of the starting material. A corresponding isomerization of oncinotine (*40*), which would pass through a seven-membered intermediate has not been observed [compare (*68*)]. If two or more trimethyleneamino units are attached to the amide nitrogen, this ring-expansion reaction takes place two or more times to form a ring enlarged by 4 or $n \times 4$ ring members. This repetitive ring-enlargement reaction is called the "Zip-reaction" (*67,69,70*). The same reaction principle was used to enlarge carbocyclic rings (*71*) and to transform carbocyclic rings into enlarged lactones (*72*).

5. The Inandenines

The inandenines are macrocyclic spermidine derivatives that have been isolated as ketones from *Oncinotis* species. The main alkaloids of the leaves of *O. inandensis* inandenin-12-one (**53**) and inandenin-13-one (**54**), exist as a mixture that has not yet been separated. Only by chemical degradation of the mixture to the ethylene acetal derivatives **55** and **56** it has been estab-

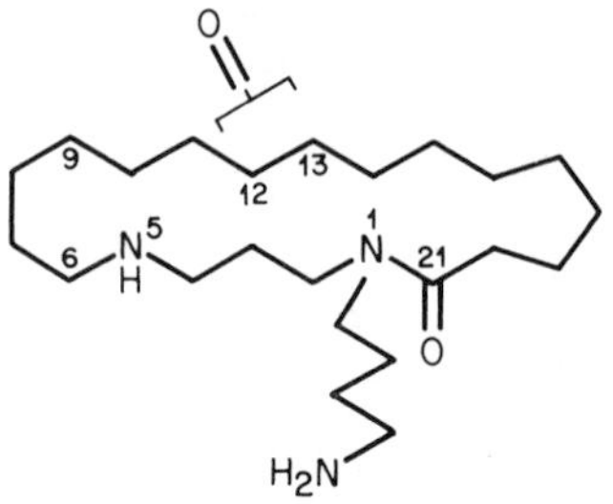

53 Inandenin-12-one

54 Inandenin-13-one

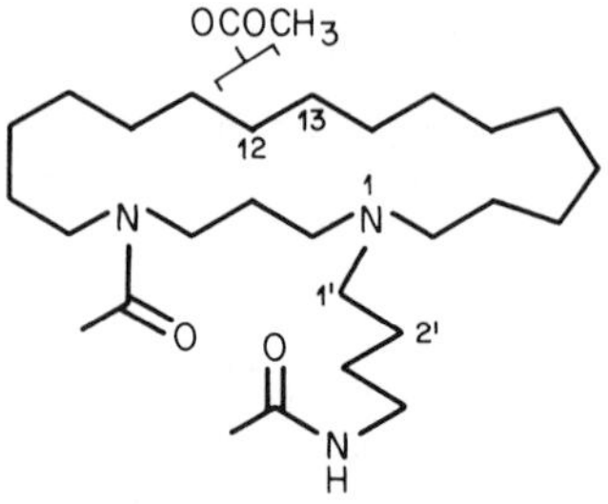

61/62

lished that **53** and **54** are isomeric compounds, differing in the position of the ketone group (73). The products **55** and **56** were obtained from the inandenines **53** and **54** by acid-catalyzed hydrolysis, ethylene glycol/p-toluenesulfonic acid treatment, LAH reduction and acetylation. Besides other fragment ion signals, the mass spectrum shows two peak groups at m/z 440 ($C_{23}H_{42}N_3O_5$), 426 ($C_{22}H_{40}N_3O_5$), and m/z 257 ($C_{14}H_{25}O_4$), 243 ($C_{13}H_{23}O_4$). These fragment ions were formed by α cleavage with respect to the acetal oxygen atoms (see Scheme 10). The difference in the signals within the two peak groups is due to a CH_2 group; this result corresponds to the given structure. At the same time the ratio of the two components can be estimated as 1 : 1.

A similar result was obtained by a Schmidt degradation reaction (use of HN_3 in $H_2SO_4/CHCl_3$) on the inandeninone mixture (see Scheme 11). A mixture of the four lactams **57–60** was obtained, which then gave eight compounds by hydrolysis: two dicarboxylic acids, two ω-aminocarboxylic acids, two triaminocarboxylic acids and two tetraamines; the dicarboxylic acids identification and their quantitative analysis (as dimethyl ester derivatives) was carried out by gas chromatography. The nitrogen-containing compounds were esterified, acetylated, and separated by TLC. Identification of the substances, each occurring in groups of two, has been made by mass spectrometry.

The presence of spermidine in the inandenines has been proved by KOH fusion: The reaction product has been identified as the triacetyl derivative of spermidine. The mass spectrum of the triacetyldeoxo compounds **61** and **62** shows how spermidine is incorporated in the alkaloids. This mixture of compounds was prepared by reduction of **53/54** with LAH followed by acetylation. The base peak at [M$^+$ $-$ 100] in the mass spectrum is formed by α cleavage with respect to N-1, between C-1′ and C-2′. If the alternative incorporation [e.g., as in the case of neoonicinotine (**41**)] had occurred, a cleavage of 86 atomic mass units from the molecular ion could be expected.

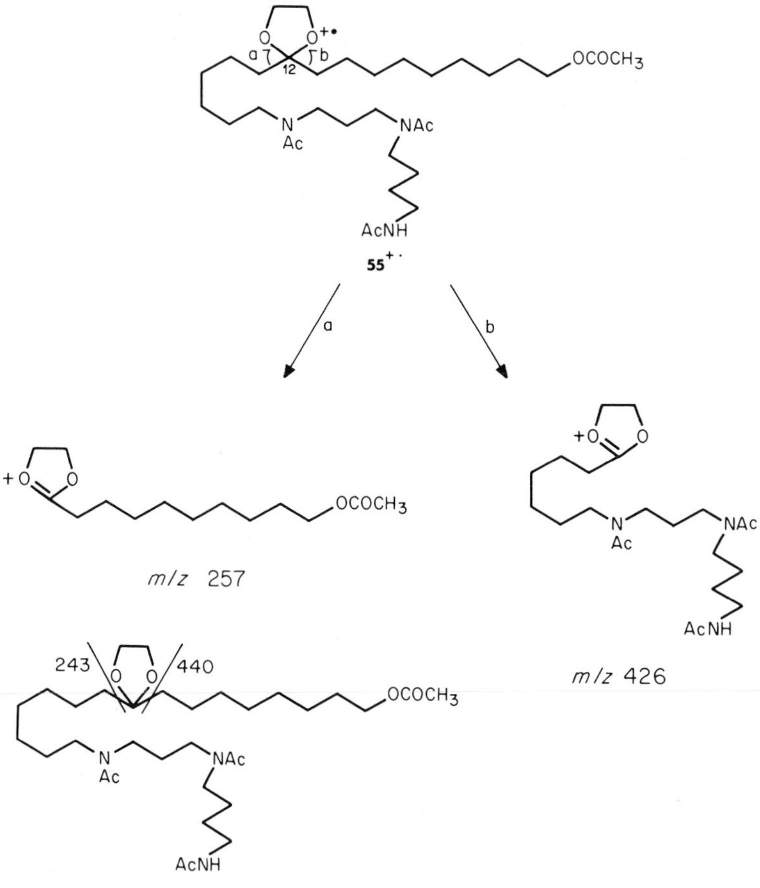

SCHEME 10. Mass spectral fragmentation of inandenine derivatives **55** and **56**.

However, this signal is only observed with an intensity of 5%. Based on this observation, the presence of ~ 5% of these isomers of **61/62** cannot be excluded.

Schmidt degradation of inandenin-10-one (**63**), isolated from the stem bark of *Oncinotis nitida,* followed by hydrolysis and analysis of the resulting mixture in a manner analogous to that described in the case of inandenin-12- and -13-one (see Scheme 11) was the basis of its structure elucidation (*26*).

A minor alkaloid of the stem bark of *O. nitida* is inandenin-10,11-diol (**64**). Its reported structure was based on HCl-catalyzed hydrolysis of its *N,N',O,O'*-tetraacetyl derivative. In this reaction, which proceeds via a

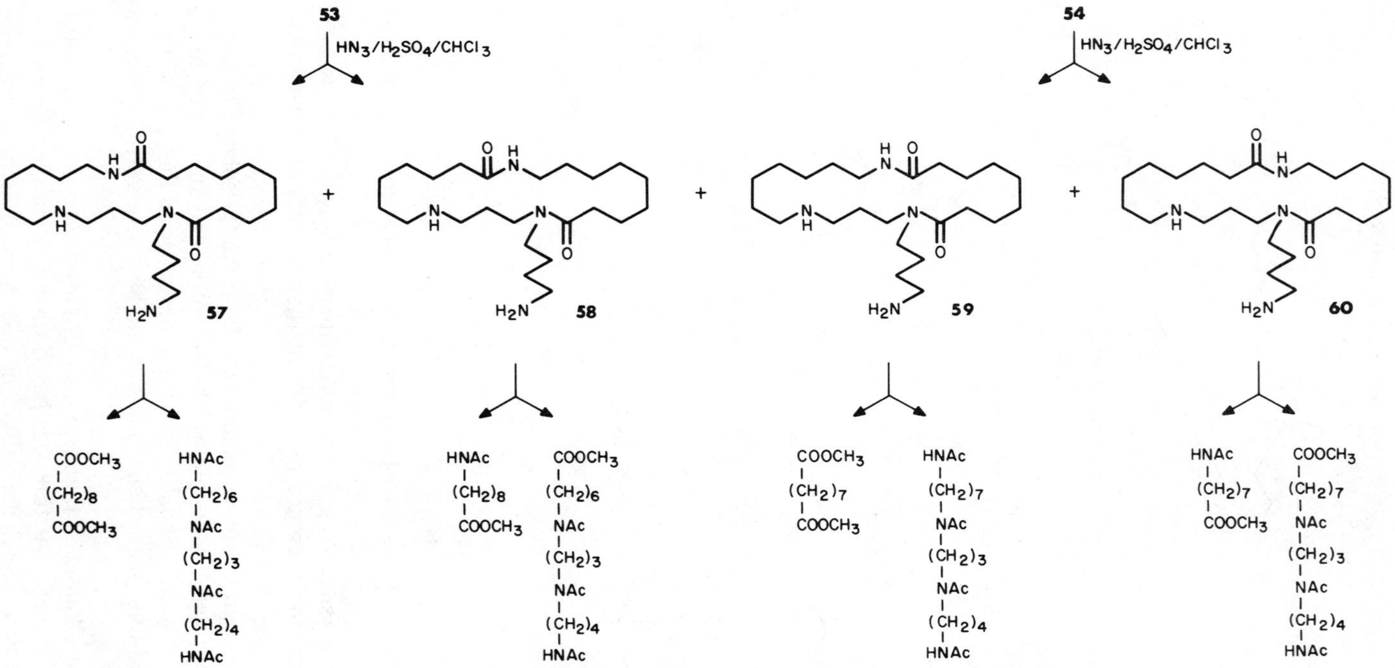

SCHEME 11. Schmidt degradation reaction of the inandenine mixture **53/54**.

hydride shift, a mixture of inandenin-10- and -11-one was formed. Again, Schmidt degradation was performed. Analysis of the reaction mixture led to the structure **64** (*26*).

Preliminary experiments on the alkaloids of *Oncinotis nigra* have afforded a fantastic mixture of related compounds. Besides small amounts of inandeninones, several inandeninols were isolated as an inseparable mixture. To identify the compounds in the mixture, two reactions were performed: a transamidation reaction and Schmidt degradation of the ketones prepared by CrO_3 oxidation of the natural alcohols. The structures **65–72** were proposed on the basis of the two reactions and MS and GC/MS identification of the mixed degradation products. It was not possible to determine the ratios of these alkaloids (*74*).

63	10 = O
64	10-OH, 11-OH
65	10-OH
66	11-OH
67	12-OH
68	13-OH

69	10-OH
70	11-OH
71	12-OH
72	13-OH

6. The Palustrines

Palustrine (**73**) is the main alkaloid of several *Equisetum* species, whereas its *N*-formyl derivative, palustridine (**74**), and its 18-deoxy derivative **75** are minor alkaloids of *E. palustre* (*75 – 77*). These alkaloids are elaborated from spermidine and a substituted C_{10} carboxylic acid. Palustridine afforded palustrine by acid-catalyzed hydrolyses.

Palustrine contains a lactam ring and yields spermidine (*78*) on alkali fusion. Catalytic hydrogenation followed by Eschweiler–Clarke methylation gave first *N*-methyldihydropalustrine. This bis-tertiary base was then treated with methyl iodide and the resulting quaternary product submitted to Hofmann degradation. The de-base mixture obtained was reduced catalytically and the product boiled for several hours with strong HCl. Along with other cleavage products, dihydropalustraminic acid (**76**) was isolated and identified by derivatives, spectroscopic data, and comparison to a synthetic product. The configuration of palustrine follows from this analysis.

The synthesis of (±)-dihydropalustraminic acid (**76**) was performed in the

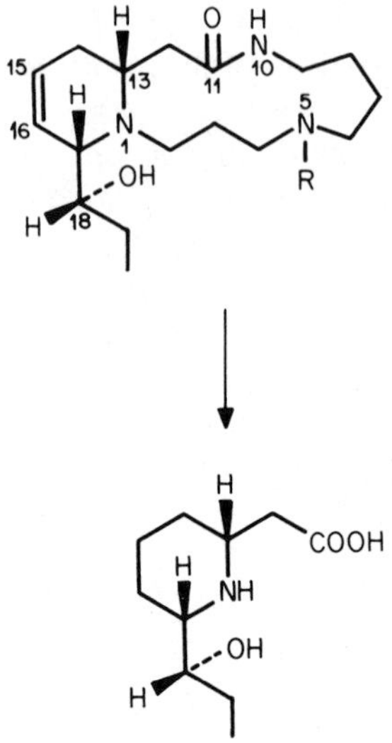

73 R = H   Palustrine
74 R = CHO   Palustridine

76

following manner. Starting from the ethyl ester **77**, compound **78** was prepared by epoxydation of the nonconjugated double bond, which then reacted with benzylamine to give the isomeric piperidine derivatives **79** and **80** (see Scheme 12). Hydrogenolysis and transesterification resulted in **81** and **82**. The (2*S*),(6*R*),(1 '*S*) configuration of (−)-**81** and (−)-*cis*-dihydropalustramine alcohol, combined with the established threo configuration of (±)-(**76**), led to the absolute configuration of palustrine (*79*).

In the mass spectrum of palustrine (**73**), the side chain is the first fragment to be eliminated by α cleavage from the molecular ion ($M^+ = 309$), after which by a McLafferty rearrangement, one of the H atoms attached to C-2 appears at the lactam oxygen fragment ion, m/z 250. From this the position

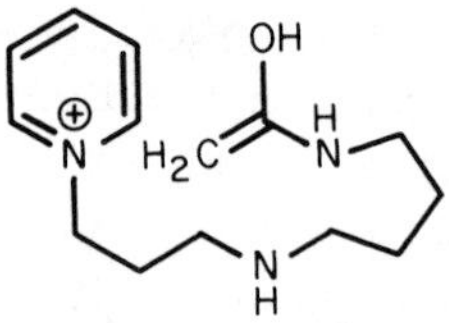

$m/z = 250$

of the double bond, the piperidine ring can be deduced. The structure of the minor base of *Equisetum palustre,* 18-deoxypalustrine (**75**, alkaloid P3), was deduced mainly from spectroscopic evidence and comparison with **73** and its derivatives (*77*).

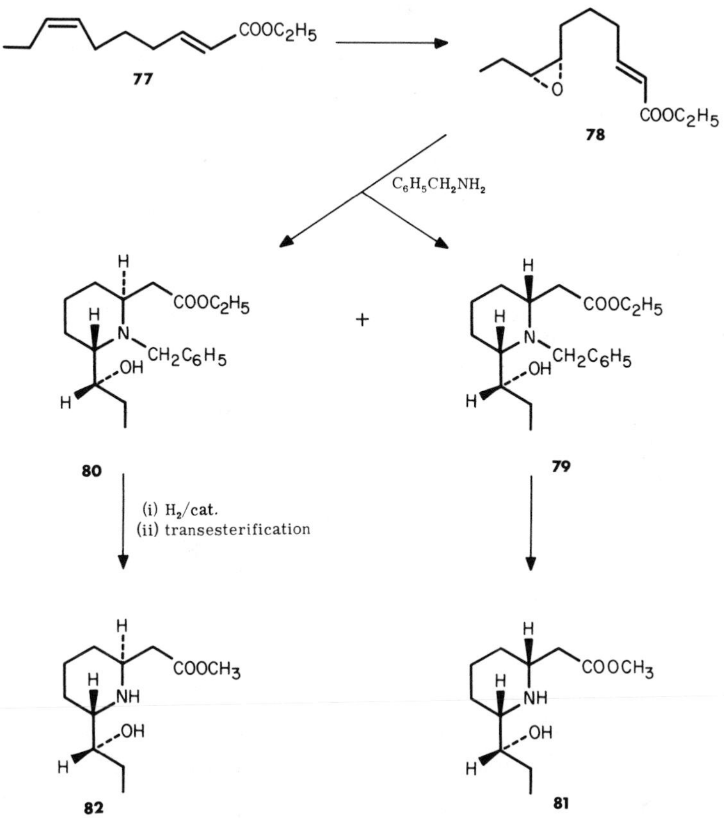

SCHEME 12. Synthesis of (±)-dihydropalustraminic acid methyl ester (**81**).

Recently, the sysnthesis of (±)-18-deoxypalustrine (**75**) has been accomplished (*80*) (see Scheme 13). The starting material was the di-protected piperidine derivative **83**, prepared by photooxygenation of 1,2-dihydro-2-propyl-1-carbobenzoxypyridine, followed by a $SnCl_2$-mediated reaction with ethyl vinyl ether. Treatment of the hydrogenated and tosylated product with 1,8-diazabicyclo[5.4.0]undec-7-ene (DBU) afforded **84**, which after N-deprotection and amidation with $ClCO(CH_2)_2NHTs$, yielded the intermediate **85**. The alcohol **86**, prepared by deprotection and reduction with $NaBH_4$ as well as LAH from **85**, was oxidized with Jones reagent to the corresponding acid, esterified, and alkylated with the partially protected butane substituent to give **87** that contained all carbon and nitrogen atoms of the desired compound. Ring closure to the 13-membered lactam **88** was achieved by preparing the acid chloride and reacting it with the primary amine of **87** under high-dilution conditions. Reductive cleavage of the tosyl

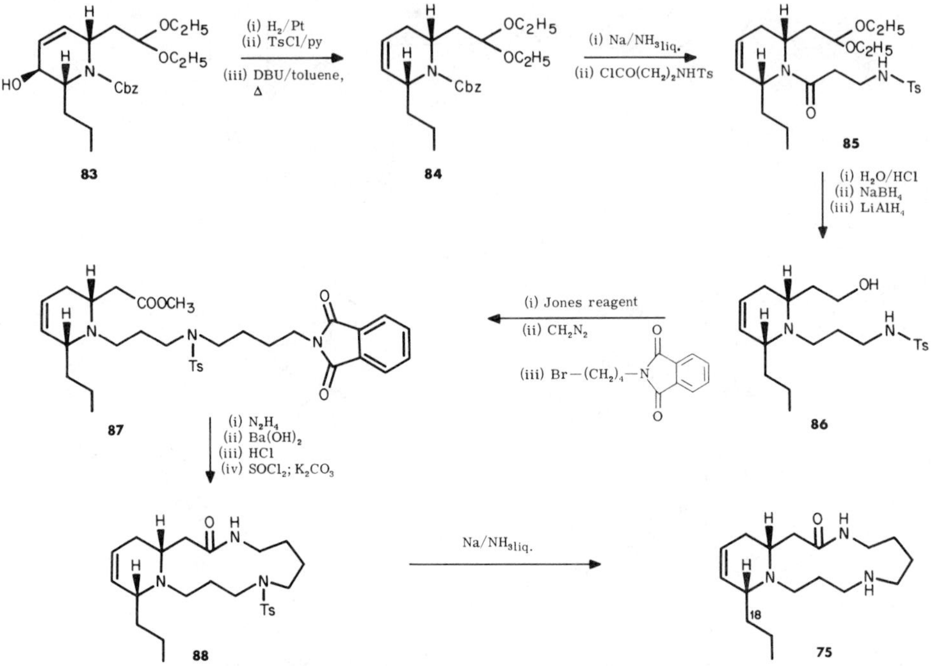

SCHEME 13. Synthesis of (±)-18-deoxypalustrine (**75**).

group completed the total synthesis of **75** in 18% yield (with respect to **83**). The comparison of the synthetic and natural products is not yet completed (*80*).

7. The *Cannabis* Alkaloids

Two alkaloids have been isolated from the roots of the Mexican variant of *Cannabis sativa* L. (marijuana) (*81,82*). The structure of cannabisativine (**89**), together with its relative stereochemistry, was deduced by X-ray

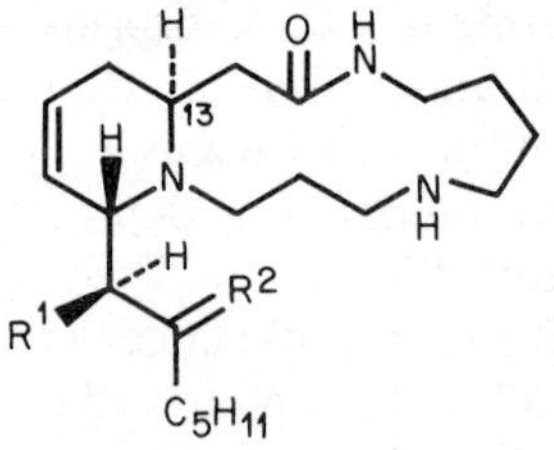

89 R^1 = OH, R^2 = $\overset{OH}{\underset{H}{\diagdown}}$ Cannabisativine

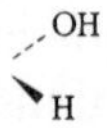

90 R^1 = H, R^2 = O   Anhydrocannabisativine

crystallography (*83*). The transformation of **89** to the second alkaloid anhydrocannabisativine (**90**) was achieved by treatment with oxalic acid at 180–185°C for 2 min.

The similarity between the palustrines (see Section III,B,6) and these *Cannabis* alkaloids is obvious, but it is remarkable that center-13 in both series is epimeric.

8. The Lunarines

The combination of spermidine, as the basic component, and two cinnamic acid units to form an amide is realized in some alkaloids isolated from *Lunaria biennis*. Structure **91** was determined by X-ray crystallography for the main alkaloid lunarine (*84–86*).

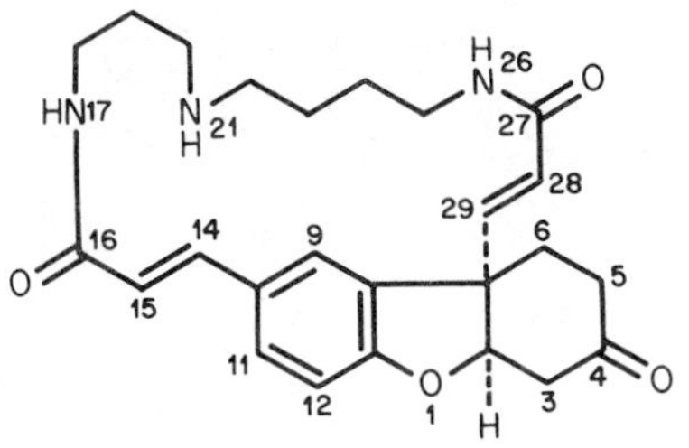

91 Lunarine

The alkali melt of lunarine gives, in addition to spermidine (*87*), the two biphenyl derivatives **92** and **93** (*88*). Although all but two carbon atoms of lunarine are accounted for by these results, a skeletal rearrangement has occurred during the formation of **92** from **91**, which complicated the structural elucidation of this alkaloid [for discussion of this rearrangement, see Warnhof (*89*)].

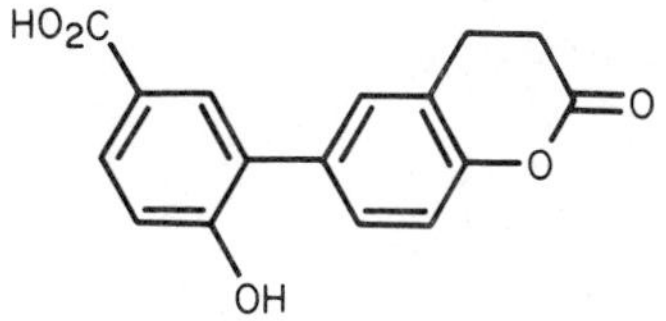

92 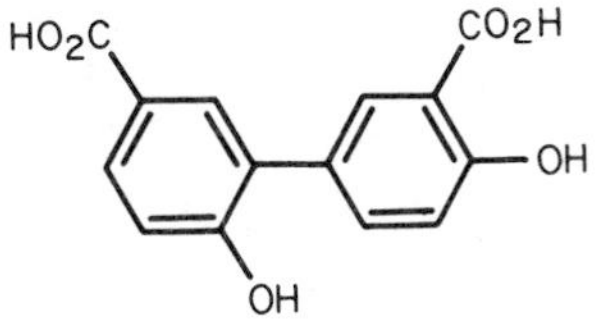 **93**

Two epimeric alcohols are formed by the reduction of lunarine with NaBH$_4$. One of these is identical to the naturally occurring alkaloid LBY. The alkaloids LBX and LBZ possess an additional CH$_2$ group with respect to **91**. Earlier it had been assumed that in these bases, the C-5 atom of lunarine (**91**) and of alkaloid LBY was attached through an additional CH_2 group to the middle N atom of the spermidine part of the molecule. According to other investigations (*90*), however, the CH_2 group must be attached to both N-17 and N-21 of the 1,3-diaminopropane part of the spermidine moiety, as shown in formulas **94** and **95**. Because LBX is readily formed by treatment of lunarine (**91**) with formaldehyde, it is probable that this base was formed during isolation from the plant in the presence of small amounts of formaldehyde as a solvent impurity (*33*).

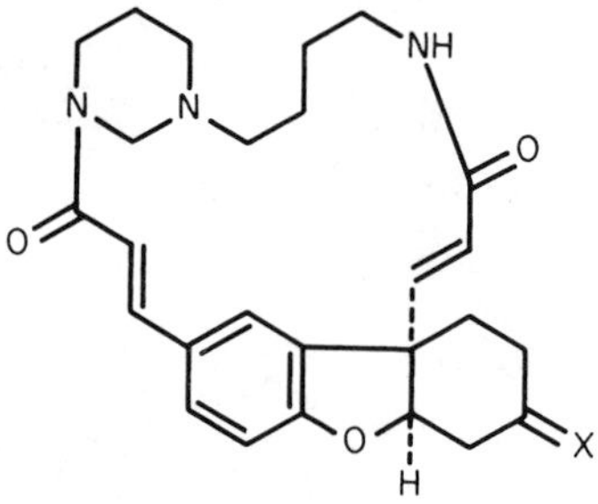

94 Alkaloid LBX   X = O

95 Alkaloid LBZ   X = $\overset{\displaystyle\,OH}{\underset{\displaystyle H}{<}}$

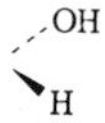

Recently, mention has also been made of lunaridine (**96**), which differs from lunarine (**91**) in that the incorporation of the spermidine part takes place in the opposite way (see Section V). The tetrahydro derivative **97** has been synthesized in the racemic form (cf. Scheme 14). By oxidative coupling of 4-hydroxycinnamic acid the Pummerer ketone-like intermediate **98** is formed, which is then converted to the *N,N'*-diacetylhydroxylamine (**99**) in the manner depicted. Heating the latter with spermidine in boiling THF gives ~ 12% yield of a macrocyclic substance that can be converted to

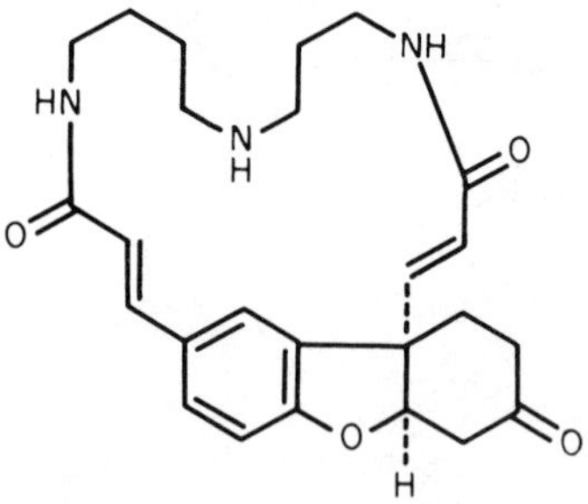

96 Lunaridine

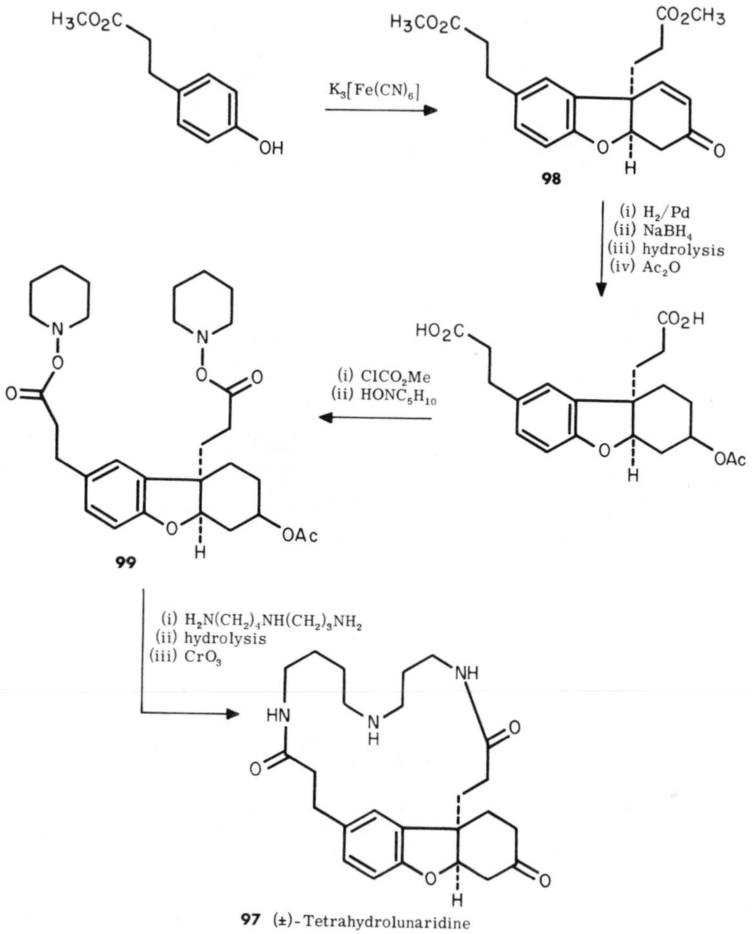

SCHEME 14. Synthesis of tetrahydrolunaridine.

tetrahydrolunaridine (**97**). It is quite remarkable that spermidine is incorporated in a regiospecific manner (*91,92*).

Another satellite alkaloid of *Lunaria biennis* is numismine (**100**). Its structure was deduced by comparing its tetrahydro derivative to hexahydrolunarine and finding both products to be identical (see Section V).

The total syntheses of both lunarine (**91**) and lunaridine (**96**) have been published (*93*). Starting material was the same Pummerer ketone-like compound **98** (see Scheme 14) that was converted by standard procedures to **101** (see Scheme 15). In order to introduce the two double bonds conjugated with the ester, **101** was oxidized to **102** by the reaction sequence depicted in

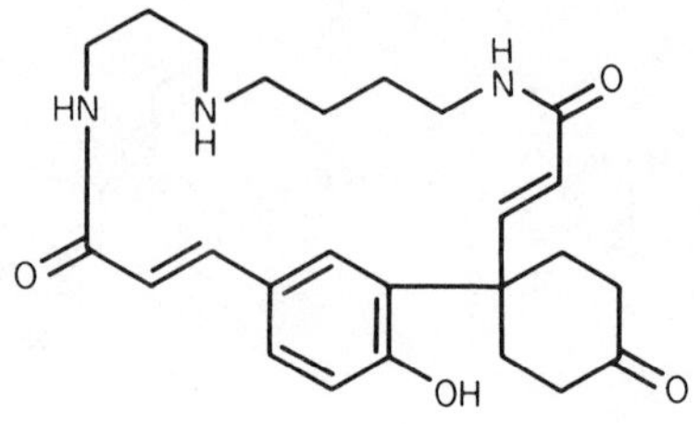

100   Numismine

Scheme 15. Protection of the ketone and transformation of the methyl esters to the activated amides in **102** yielded **103**. The condensation reaction between the two primary amino groups of spermidine and **103** was achieved by using high-dilution techniques. The incorporation of the unsymmetric spermidine took place in both ways so that a nearly 1 : 1 mixture of (±)-lunarine (**91**) and (±)-lunaridine (**96**) was formed.

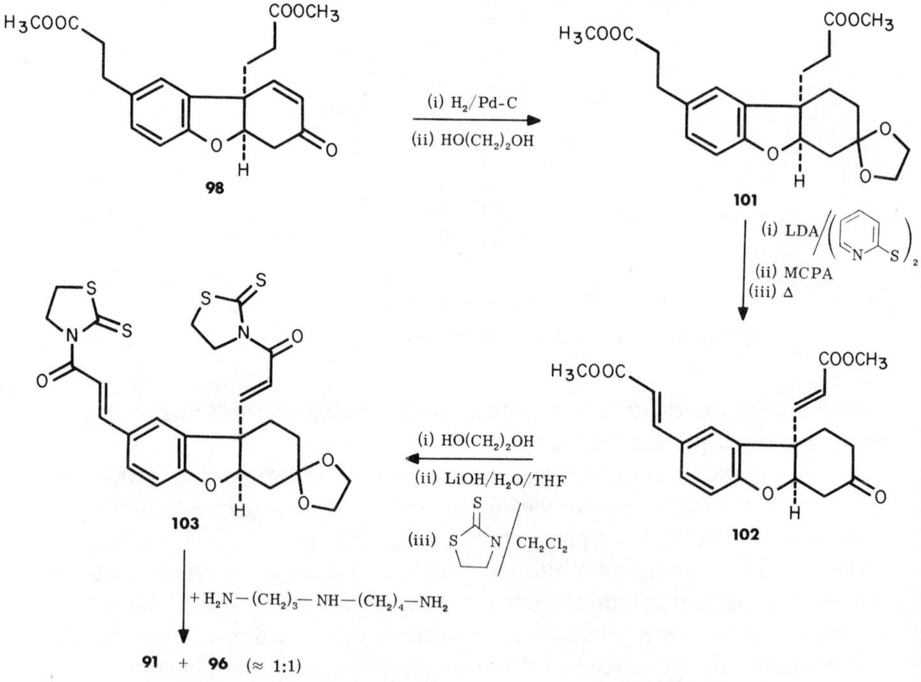

SCHEME 15. Synthesis of lunarine (**91**) and lunaridine (**96**)

9. Codonocarpine and N-Methylcodonocarpine

Other than several alkaloids of unknown structure, the spermidine alkaloid codonocarpine (**104**) and its N-methyl derivative (**105**, alkaloid IV) were isolated from the bark of the Australian *Codonocarpus australis* (*94*).

104 Codonocarpine   **106**

N-Methylation (CH$_2$O followed by reduction with NaBH$_4$) transformed codonocarpine (**104**) to its methyl derivative (**105**). Acid-catalyzed hydrolysis of **104** gave spermidine. The permethylated alkaloid was oxidized with KMnO$_4$ to 2,2'-dicarboxydiphenyl ether (**106**), which was also synthesized.

N-Methylcodonocarpine (**105**) was transformed to derivative **107** by reduction with H$_2$–Raney nickel, followed by O-ethylation with diazoethane (see Scheme 16). Compound **107** did not give the expected Hofmann degradation product, but rather the ring-closed pyrrolidine derivative **108** after transformation to the quaternary hydroxide followed by pyrolysis. After repeating the methylation and pyrolysis of the quaternary hydroxide followed by catalytic hydrogenation, compound **109** was obtained (*95*). The degradation product **109** has been synthesized according to Scheme 17 (*96*).

The central reaction in the total synthesis of tetrahydrocodonocarpine (**110**) was treatment of the activated ester **111** with spermidine (*97*) (see Scheme 18). As expected, the resulting product was a mixture of the two possible isomers **110** (compared to a degradation product of codonocarpine) and **112** in which the incorporation of spermidine is reversed.

A total synthesis of codonocarpine (**104**), similar to the synthesis of the *Lunaria* alkaloids lunarine and lunaridine (see Scheme 15), has been published (*98*). The activated dicarboxylic acid thiazolidine-2-thione diamide (*113*) was treated with spermidine itself. Because of the two possible reaction modes, **104** and the previously unknown isomer with the reversed incorporation of the spermidine moiety were isolated from the reaction products.

In the regiospecific total synthesis of codonocarpine (**104**), a partially protected spermidine derivative and a partially protected diphenyl ether

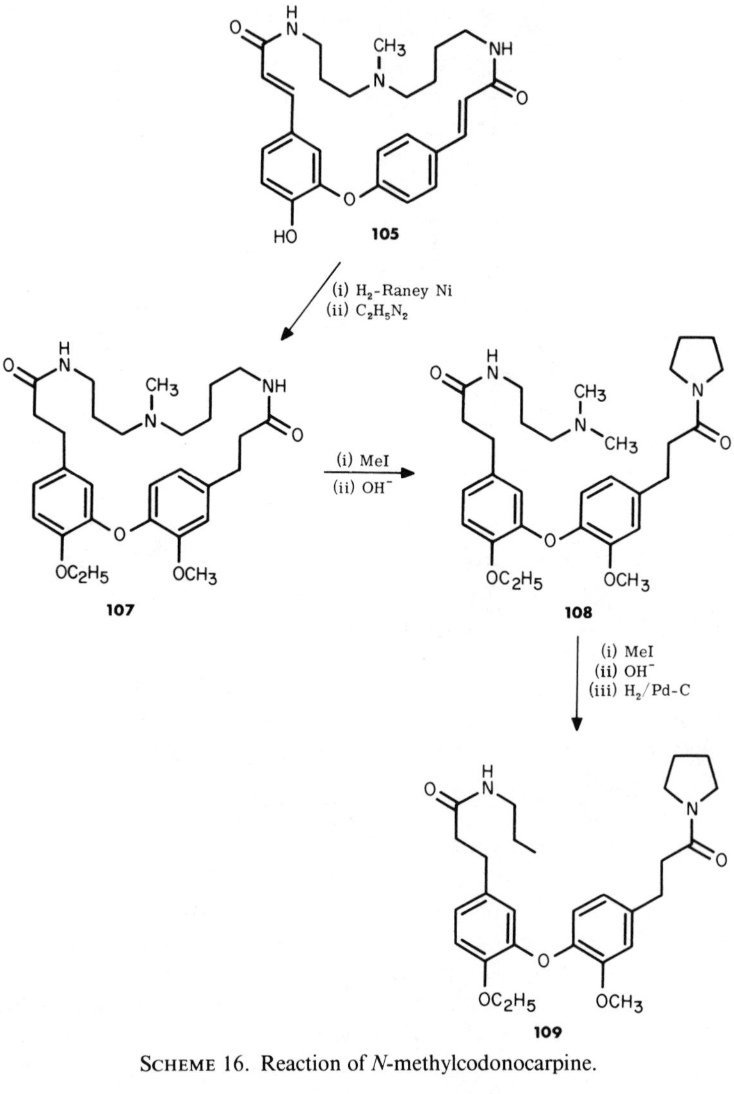

SCHEME 16. Reaction of N-methylcodonocarpine.

derivative were prepared first (99). According to Scheme 19, N,N'-di-tert-butoxycarbonylspermidine (114) was elaborated from 1,4-diaminobutane (100). The nonbasic portion of the codonocarpine molecule (115) was afforded by an Ullmann-type synthesis using the diphenyliodonium bromide prepared from 4-benzyloxybenzaldehyde as one component and methyl ferulate as the other. The resulting diphenylaldehyde was transformed by Knoevenagel condensation to 115. The reaction of 115 with 114 and the conversion of the resulting product to codonocarpine (104) is

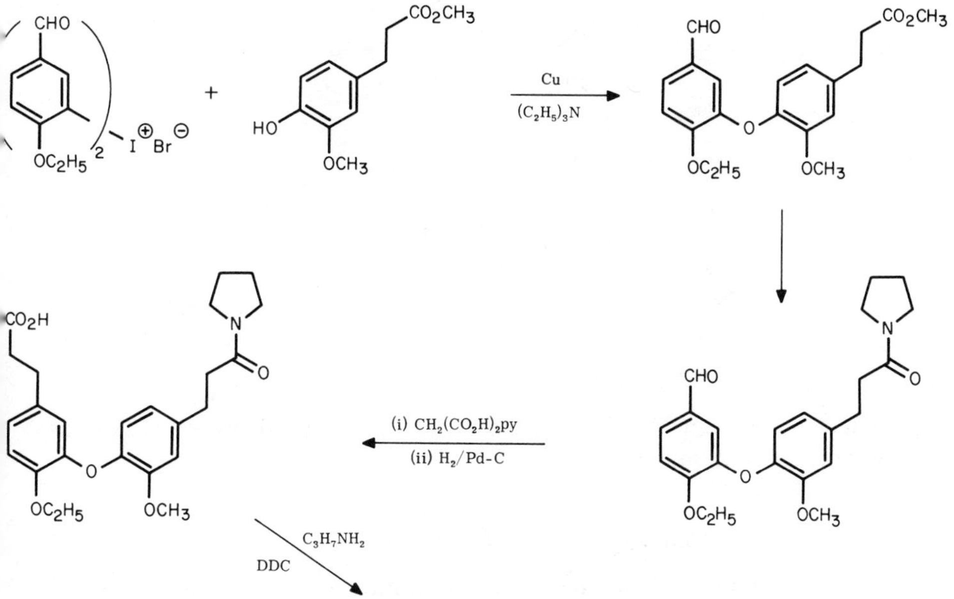

109

SCHEME 17. Synthesis of **109**.

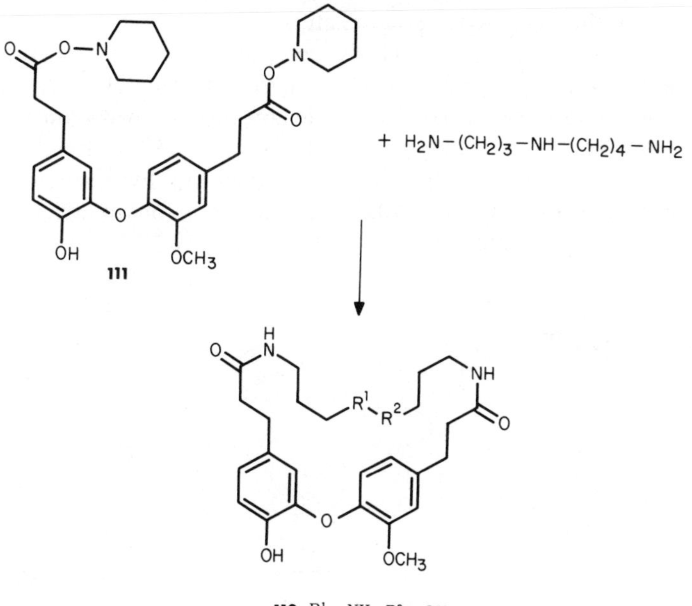

110 $R^1 = NH$, $R^2 = CH_2$
112 $R^1 = CH_2$, $R^2 = NH$

SCHEME 18. Synthesis of tetrahydrocodonocarpine.

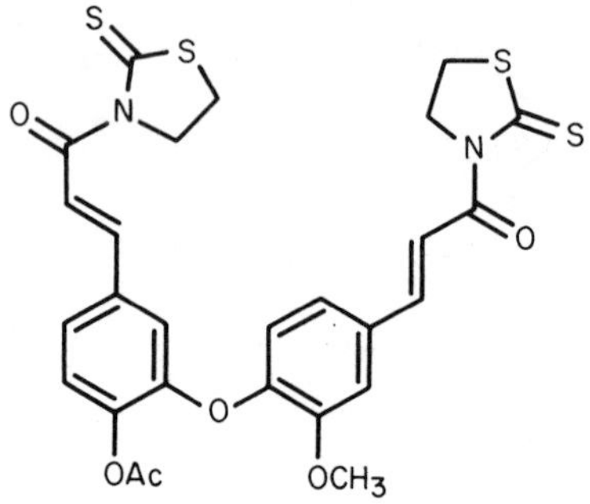

113

depicted in Scheme 20. The ring-closure reaction of the deprotected **116** was performed under high-dilution conditions (*99*).

10. The Celacinnine Group and Mayfolines

The spermidine alkaloids of the Celacinnine group contain a 13-membered ring that is built up from spermidine and one cinnamic acid residue in such a way that the primary amino group of the 1,3-diaminopropane part forms the amide linkage, and the primary amine of the 1,4-diaminobutane part of spermidine is in a benzylic position.

The first alkaloid of this group for which a structure was determined was celacinnine (**117**, $C_{25}H_{31}N_3O_2$) (*101*). On the basis of ^{1}H-NMR and UV spectra, the presence of an (*E*)-cinnamide chromophor was indicated. This proposal was verified by catalytic hydrogenation of **117** to dihydrocelacinnine. This compound was found to be identical to dihydrocelallocinnine, prepared in the same way from celallocinnine (**118**), for which a (*Z*)-cinnamide structure was proposed (*101*). On the basis of ^{1}H-NMR spectral data, it

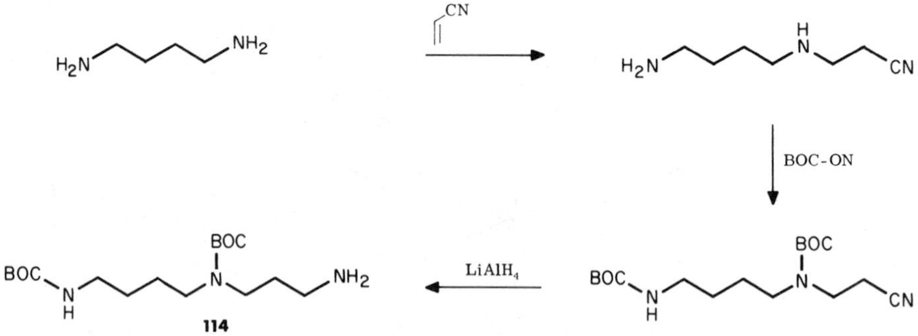

SCHEME 19. Synthesis of *N,N'*-di-*tert*-butoxycarbonylspermidine (**114**).

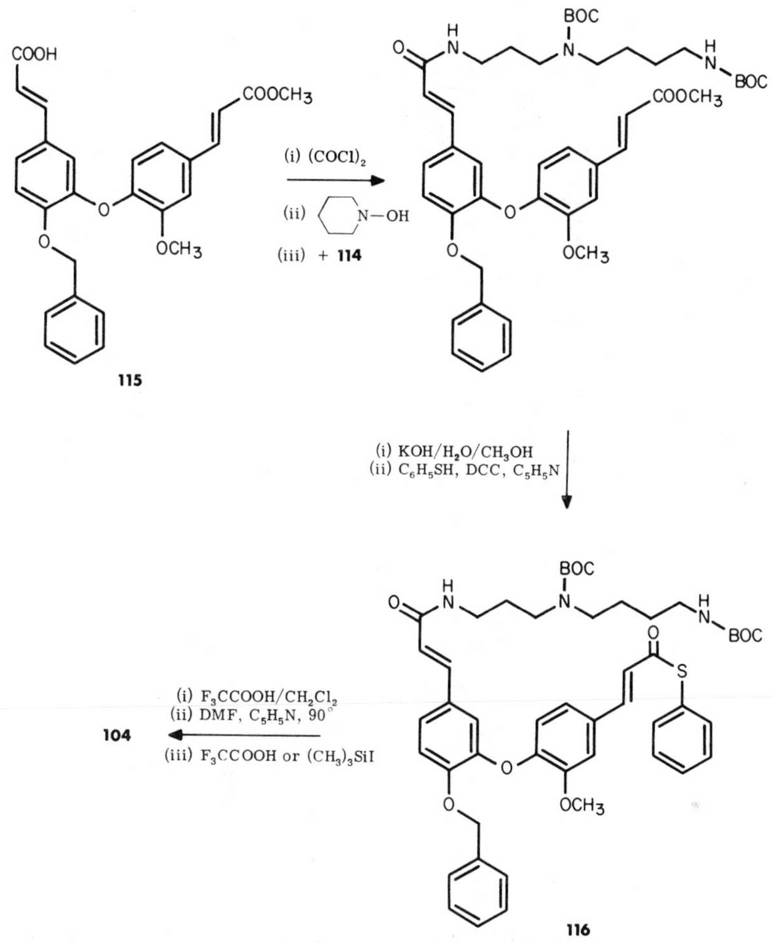

SCHEME 20. Synthesis of codonocarpine (104).

was shown that a secondary amine is in a position benzylic to the second benzene ring (^{1}H-NMR data of 117 and its N-acetyl derivative were prepared in the usual way). Vigorous acid hydrolysis (2 N HCl, 150°C, 18 hr) of 117, followed by acetylation of the reaction mixture, yielded triacetylspermidine. Milder acid hydrolysis (6 N HCl, 100°C, 2 hr) of 117, followed by esterification and acetylation of the reaction products, gave the degradation product 119 in which only the secondary lactam had been hydrolyzed. The structure of 119 was deduced on the basis of extensive mass spectral analysis (peak triad method, see p. 90). The structure of celabenzine (120, $C_{23}H_{29}N_3O_2$) was deduced by a method similar to that employed in the case

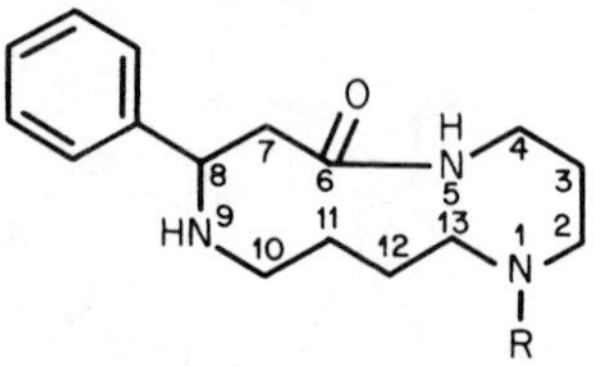

117	R = $C_6H_5CH\overset{E}{=\!\!=}CH$—CO	Celacinnine
118	R = $C_6H_5CH\overset{Z}{=\!\!=}CH$—CO	Celallocinnine
120	R = C_6H_5CO	Celabenzine
121	R =	Celafurine
132	R = OH, S—C(8)	Mayfoline
133	R = $COCH_3$, S—C(8)	N^1-Acetyl-N^1-deoxymayfoline

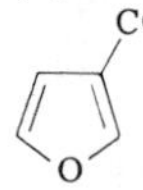

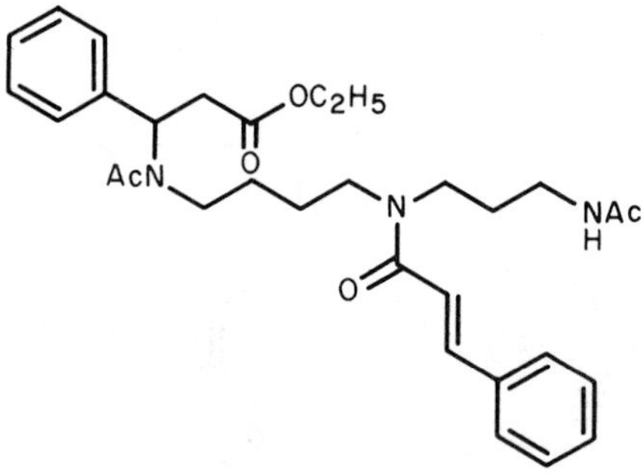

119

of **117**. The related alkaloid celafurine (**121**, $C_{21}H_{27}N_3O_3$) was analyzed using spectral methods (^{1}H-NMR, MS, and UV) (*101,102*). It is interesting to note that further NMR (^{1}H- and ^{13}C-NMR) studies confirmed the structures of celacinnine (**117**) and celallocinnine (**118**), and have shown that **117** exists as rotamers at room temperature in CDCl$_3$ solution (*103*).

Three independent syntheses of alkaloids of the celacinnine group have been published. In the synthesis of celacinnine (**117**) by Wasserman *et al.* (*104*), the central reaction used is the transamidation or "Zip-reaction" (*69,105,106*). In the first part, the preparation of the nine-membered intermediate **122** was achieved by two different routes. Heating of 4-phenylazetidinone (**123**) with 2-methoxypyrroline (**124**) led to the bicyclic 4-oxotetrahydropyrimidine (**125**). The reaction seems to proceed via the tricyclic intermediate **126** and involves an intramolecular ring opening of a β-lactam

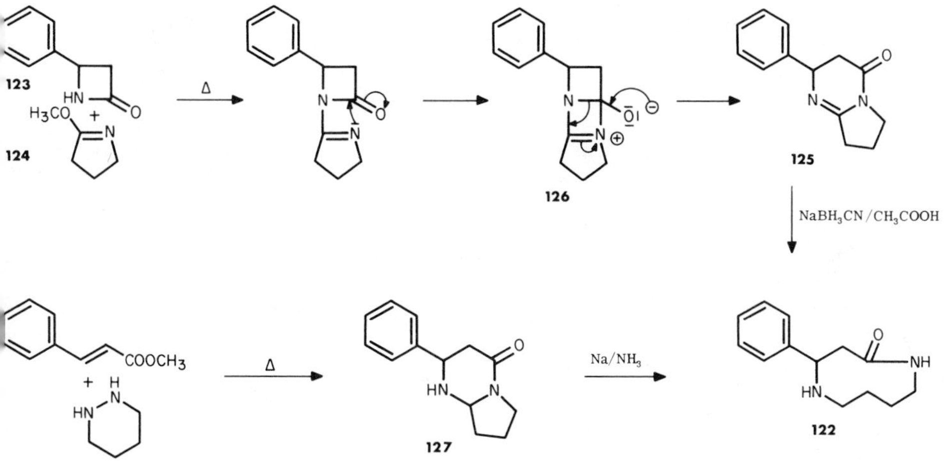

SCHEME 21. Synthesis of 122.

derivative. Reductive cleavage of **125** was accomplished with an excess of NaBH₃CN in the presence of CH₃COOH. The low yield of this reaction (31%) favored an alternative sequence in which piperidazine was heated with ethyl cinnamate to form 2-oxo-4-phenyl-1,5-diazabicyclo[4.3.0]nonane (**127**). The latter was reduced with Na/NH₃ to form **122** in a yield of 80% (see Scheme 21).

The transformation of the intermediate **122** to celacinnine (**117**) is shown in Scheme 22. Alkylation of the anion of **122** with N-(3-iodopropyl)phthalimide, followed by hydrazinolysis afforded the amino amide **128**. It is remarkable that alkylation took place preferentially at the amide nitrogen and not at the benzylic amino function. Treatment of **128** under basic reaction conditions gave, via a ring-enlargement reaction, the celacinnine ring skeleton in 70% yield, which was converted to celacinnine (**117**) itself by acylation with (E)-cinnamoyl chloride. The selective acylation is again remarkable (104).

The following two syntheses of the celacinnine skeleton are very similar to each other. Both employ a boron complex to cyclize an open-chain system

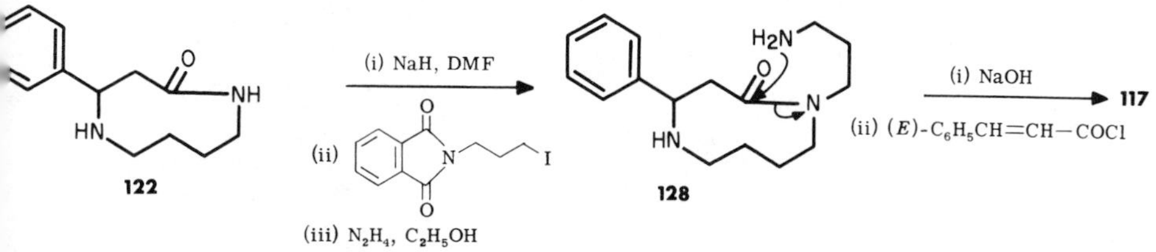

SCHEME 22. Synthesis of celacinnine (**117**).

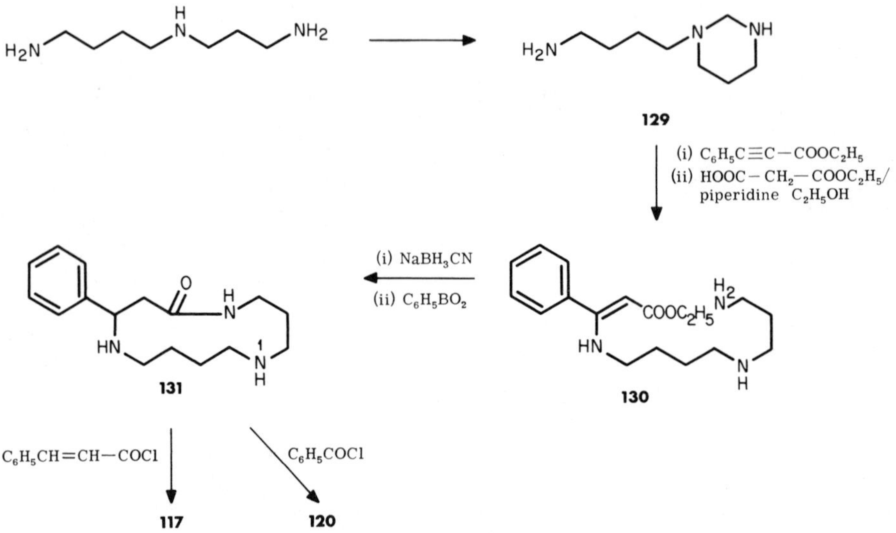

SCHEME 23. Synthesis of celacinnine (**117**).

to a 13-membered lactam. The synthesis of McManis and Ganem (*58*) is presented in Scheme 23. The starting material spermidine reacts with formaldehyde exclusively at the 1,3-diaminopropane moiety to form the hexahydropyrimidine derivative **129**. This protected compound was treated with ethyl phenylpropiolate and then deprotected under Knoevenagel reaction conditions to form **130**. Reduction of the double bond (NaBH$_3$CN), hydrolysis with (C$_4$H$_9$)$_4$NOH, and treatment with catecholborane afforded the same product **131** (Scheme 23). Acylation of **131** at N-1 with cinnamoyl chloride or benzoyl chloride gave celacinnine (**117**) and celabenzine (**120**), respectively. Acylation of **129** with 2 Eq of cinnamoyl chloride, followed by deprotection as before, furnished the spermidine derivative maytenine (**31**).

The third synthesis (*107*) is presented in Scheme 24. Because of argu-

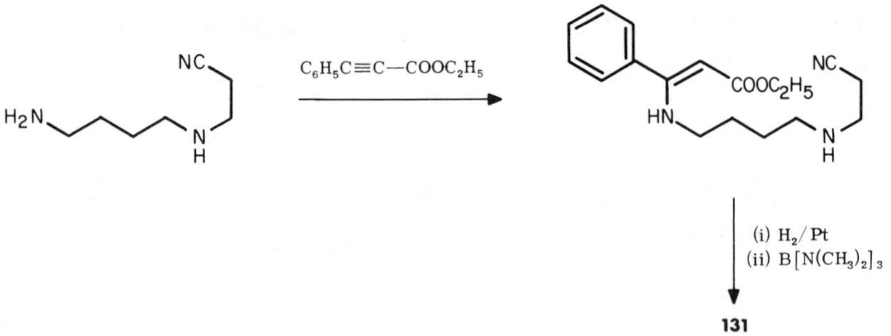

SCHEME 24. Synthesis of **131**.

ments given before, it seems unnecessary to go into details. The authors used compound **131** to prepare, in addition to **117** and **120**, the other two alkaloids celallocinnine (**118**) and celafurine (**121**).

Mayfoline and N^1-acetyl-N^1-deoxymayfoline were isolated from the arial parts of *Maytenus buxifolia*. Mayfoline (**132**) is the main alkaloid of plants grown in the district of Matanzas (Cuba), whereas N^1-acetyl-N^1-desoxymayfoline (**133**) is the main base of those plants grown in the district of Oriente, Santiago de Cuba (*108,109*) (see structure **117**).

The structure elucidation of **132** and **133** (structures, see p. 122) was accomplished mainly on the basis of spectroscopic evidence. Interconversion of the two alkaloids was not reported. Using chiroptic arguments, the absolute configuration at C-8 was deduced as shown on p. 122 (*108–110*).

Some alkaloids were found in nature to be derivatives of the celacinnine alkaloids, having an additional bond between the N^1 substituent and the spermidine moiety. It seems that the precursors of the four alkaloids cyclocelabenzine (**134**), isocyclocelabenzine (**135**), hydroxyisocyclocelabenzine (**136**), and pleurostyline (**137**) (Scheme 25) contain enamine systems that are able to cyclize. The structure elucidation of **134**, **135**, **136**, and **137** was based on extensive spectroscopic analysis, mainly ^{1}H- and ^{13}C-NMR spectra of the alkaloids, their acetyl, and, in case of **137**, their dihydro derivatives

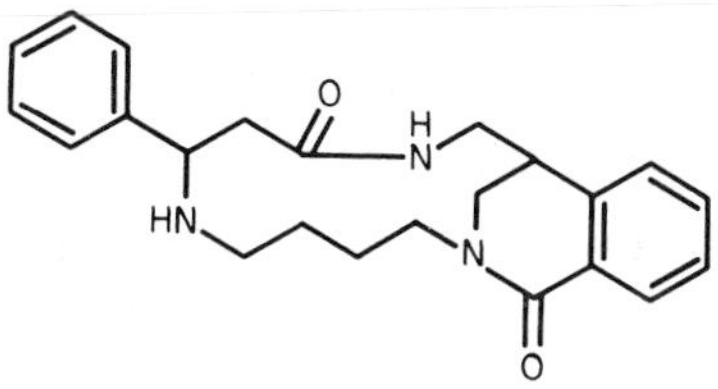

134 Cyclocelabenzine

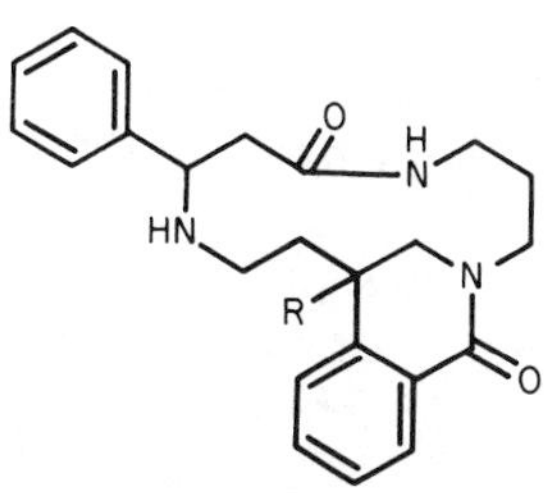

135 R = H Isocyclocelabenzine
136 R = OH Hydroxyisocyclo-
         celabenzine

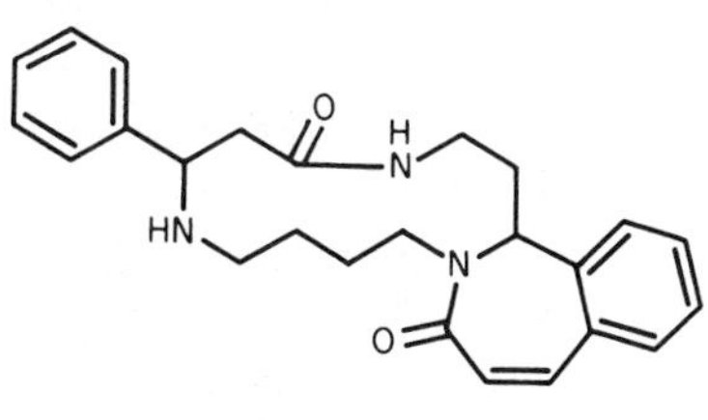

137 Pleurostyline

SCHEME 25. Structures of **134–137**.

(*89,111,112*). Discussion of these interesting spectroscopic arguments is beyond the scope of this chapter.

11. The Dihydroperiphylline Group

The basic skeleton of the dihydroperiphylline group is isomeric to that of the celacinnine group. In contrast to alkaloids of the celacinnine group in which the primary amino group of the 1,3-diaminopropane part forms an amide linkage, the alkaloids of the dihydroperiphylline group have the primary amino group of the 1,4-diaminobutane part involved in an amide-type linkage.

The key alkaloid of the dihydroperiphylline group is periphylline (**138**), on which intensive chemical and spectroscopic analysis was performed. Essential to the structure determination was the detection of spermidine in an alkali melt of tetrahydroperiphylline (**139**), prepared by catalytic hydrogenation of periphylline (**138**). Alkali hydrolysis of **138** yielded (*E*)-cin-

138	R = C$_6$H$_5$— CH$\overset{E}{=}$CH—CO, 11S,	Periphylline
139	R = C$_6$H$_5$—(CH$_2$)$_2$—CO, 2,3-H$_2$	
149	R = C$_6$H$_5$— CH$\overset{Z}{=}$CH—CO, 11S,	Isoperiphylline
150	R = C$_6$H$_5$—CH$\overset{E}{=}$CH—CO, 2,3-H$_2$,	Dihydroperiphylline
152	R = 149 amu	Perimargine
153	R = 151 amu	Dihydroperimargine

namic acid, and similar treatment of **139** yielded dihydrocinnamic acid. The spermidine derivative **140** can be prepared by hydrogenolysis of **139**, followed by acetylation. Identification of **140** was achieved by synthesis of **140** from putrescine (*113,114*).

In the case of periphylline, mass spectral analysis of the hydrolysis product *N*-dedihydrocinnamoyl-*N,N'''*-diacetyltetrahydroperiphylline has also been useful. In this case the peak triad (*m/z* 143, 157, 169) mentioned earlier is in agreement with both 8-membered (**141**) as well as the 13-membered structure **138**. Chemical and spectroscopic arguments do not allow a differentiation between these alternatives (*113,114*); differentiation was only possible by unequivocal total synthesis of a tetrahydroperiphylline deriva-

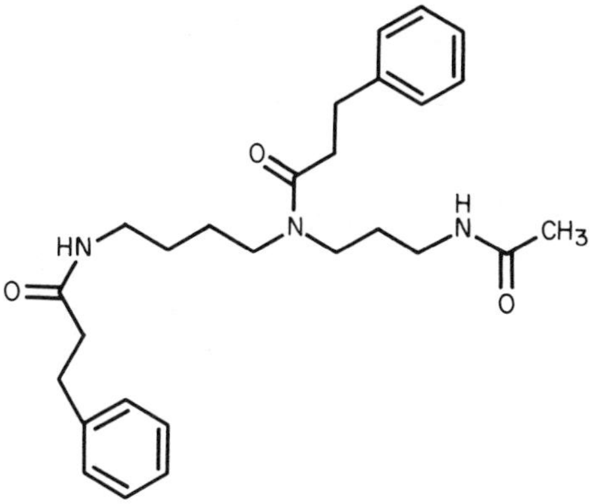

140

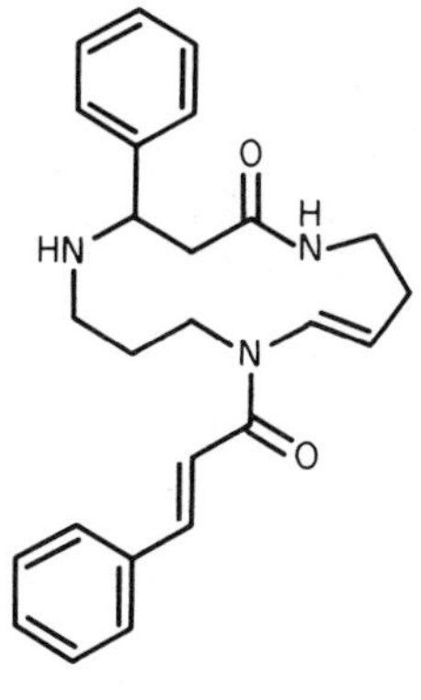

151 Neoperiphylline

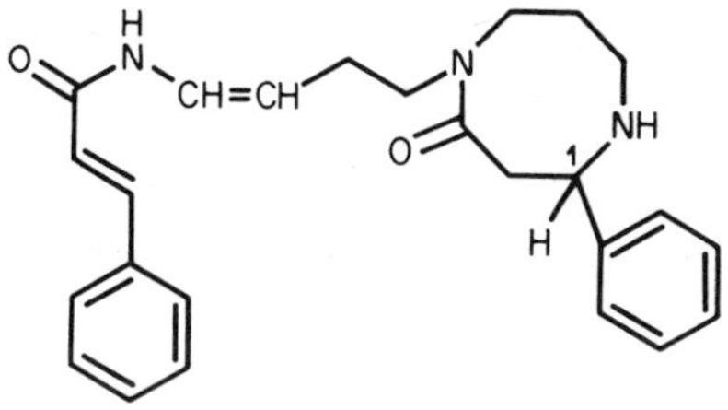

141

tive (*115*). In Scheme 26 the synthesis of the eight-membered periphylline derivative **142** is presented. The 1,3-diaminopropane part was first built up from the amino acid **143** via **144**. The cyclization to the eight-membered ring was performed starting with **145**, in which the third amino group is protected as a nitrile. The conversion of **146** to **142** was accomplished by hydrogenation and acylation. The final product was not identical to a corresponding derivative prepared from periphylline itself. Because of this discrepancy, the 13-membered isomer was also synthesized.

Starting with intermediate **145** (Scheme 26), **147** was prepared using dihydrocinnamoyl chloride, followed by catalytic hydrogenation and saponification of the ester group. The cyclization step was performed in the same manner as described above (see Scheme 27). The resulting product **148** was identical to the analogous derivative obtained from natural periphylline; therefore, periphylline must possess structure **138** (*115*). The chirality of center-11 has been defined as *S* by ORD measurement in relation to the alkaloids chaenorhine and homaline (*113*).

Tetrahydroperiphylline (**139**) was formed by catalytic hydrogenation of the minor alkaloids isoperiphylline (**149**), dihydroperiphylline (**150**), and neoperiphylline (**151**).

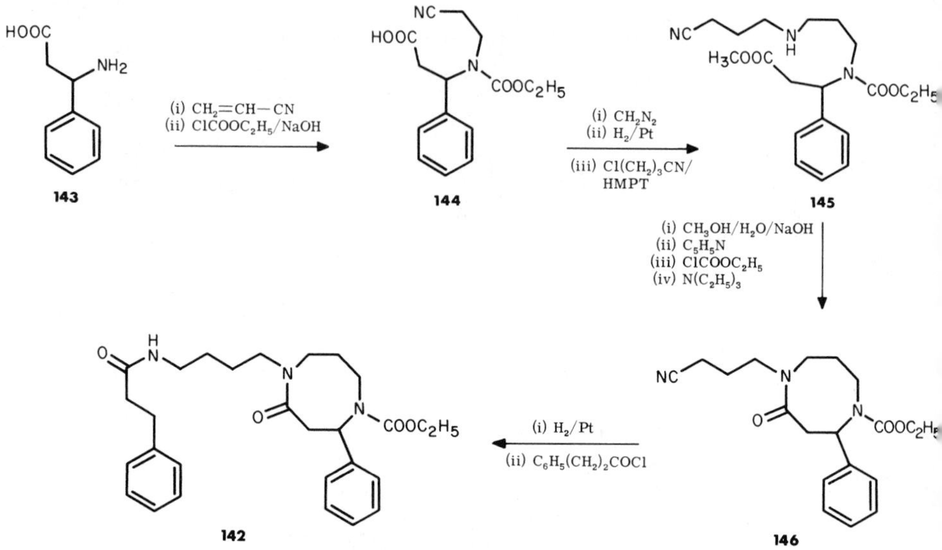

SCHEME 26. Synthesis of the 8-membered periphylline derivative (**142**).

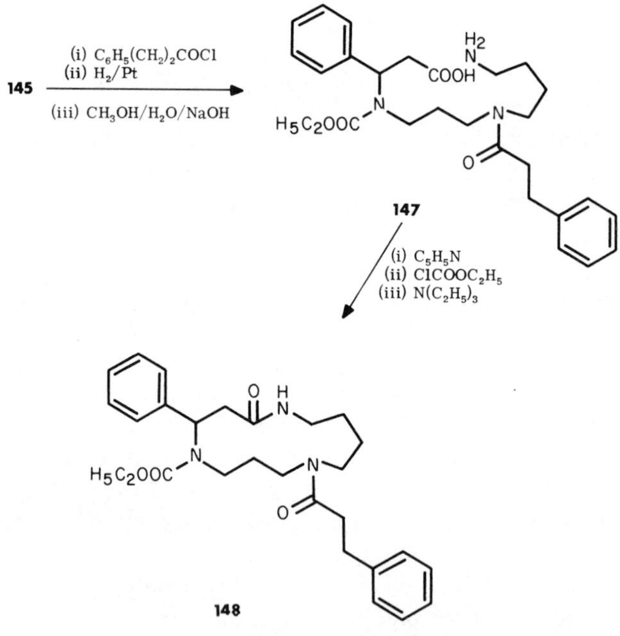

SCHEME 27. Synthesis of **148**.

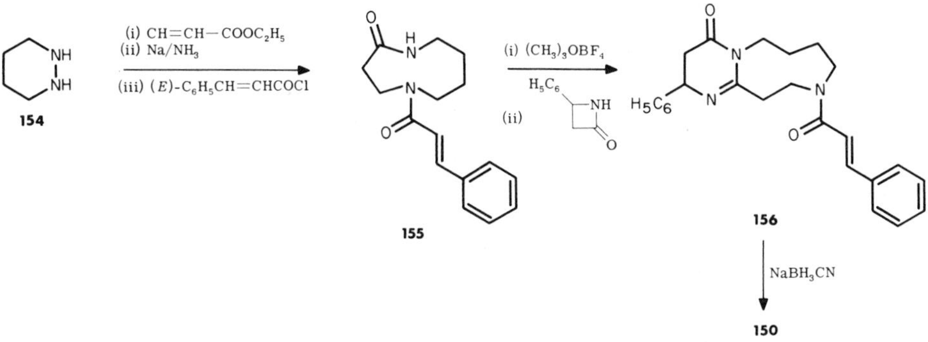

SCHEME 28. Synthesis of dihydroperiphylline (**150**).

Dihydroperiphylline can be prepared from periphylline (**138**) by reduction with NaBH$_3$CN (*116*). The structural differences between the four alkaloids were deduced on the basis of mass spectral molecular weight determination and, of course, ^{1}H-NMR spectral data.

Perimargine (**152**) and dihydroperimargine (**153**) isolated as an inseparable mixture were converted to a simple product by catalytic hydrogenation. Because of the similarity of the spectral data to that of alkaloids of the dihydroperiphylline group, structure **152/153** was proposed. In addition, alkaline hydrolysis gave (*E*)-cinnamic acid and an unstable organic acid that was isolated but not identified. The substituent at N-6 is therefore characterized in terms of atomic mass units (*114*).

A synthesis of dihydroperiphylline (**150**) was published by Wasserman and Matsuyama (*116*) (Scheme 28). The reaction pathway shows great similarity to that of the synthesis of celacinnine by the same authors (*104*). The starting material, piperazine (**154**), was condensed with ethyl acrylate, followed by reductive cleavage and acylation, to give the nine-membered lactam **155**. The third nitrogen atom was introduced as a β-lactam, ring opening formed **156**, which by NaBH$_3$CN treatment, was converted to **150**.

12. Derivatives of Spermidine Lengthened by One Methylene Group (Homospermidine)

The name homospermidine (**157**) is commonly (e.g., *7,11,18*) used for the symmetric triamine but once was used also for its unsymmetric isomer **158** (*117*). Both skeletons occur in derivatives of natural origin.

From plants belonging to the family of Solanaceae, five alkaloids were isolated containing symmetrical homospermidine as the basic backbone: solamine (**159**), solapalmitine (**160**), solapalmitenine (**161**), solacaproine (**162**), and solaurethine (**163**). The characteristic features of all the com-

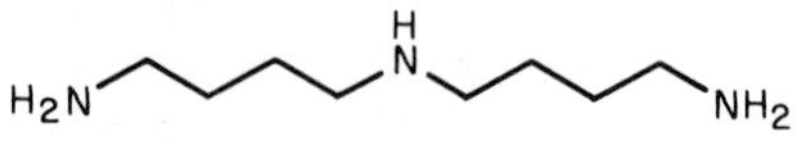

157

158

pounds are the bisdimethylamino group and the fact that the central nitrogen atom is in the form of an amide (except for **159**). The compounds first isolated and elucidated as to structure were solapalmitine and solapalmitenine (the natural mixture of both was called solapartine).

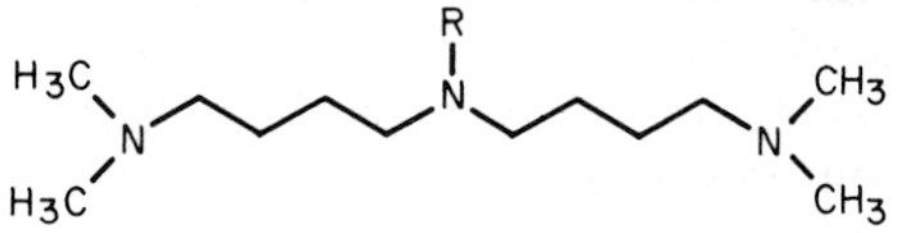

159 R = H   Solamine

160 R = CO—(CH$_2$)$_{14}$—CH$_3$ Solapalmitine

161 R = CO—CH$\overset{E}{=}$CH—(CH$_2$)$_{12}$—CH$_3$ Solapalmitenine

162 R = CO—(CH$_2$)$_4$—CH$_3$ Solacaproine

163 R = COOC$_2$H$_5$ Solaurethine

Hydrogenation of the mixture followed by hydrolysis gave solamine (**159**) and a mixture of palmitic acid (80%) and stearic acid (20%). The alkaloid containing stearic acid was not investigated further. The structure of solamine was confirmed by direct comparison with the synthetic 4,4′-bis(dimethylamino)dibutylamine. Treatment of solamine (**159**) with palmitoyl chloride and (E)-hexadec-2-enoyl chloride produced solapalmitine (**160**) and solapalmitenine (**161**), respectively (118,119).

Solamine (**159**) was found to be the principal component of *Cyphomandra betacea*. By treatment of solamine with *n*-hexanoyl chloride, the second component of the same plant was formed, which was also hydrolyzed to form *n*-hexanoic acid and solamine (120). Solaurethine (**163**), an alkaloidal component of another plant in the Solanaceae, was obtained by treatment of solamine (**159**) with ethyl chloroformate (121).

The so-called acarnidines were isolated from the sponge *Acarnus erithacus* (117) as an inseparable three-component mixture with the structures **164, 165**, and **166**. The mixture of compounds was hydrolyzed and the acid

164   R = CO — (CH$_2$)$_{10}$ — CH$_3$

165   R = CO — (CH$_2$)$_3$ — CH $\overset{Z}{=}$ CH(CH$_2$)$_5$ — CH$_3$

166   R = CO — (CH$_2$)$_3$(CH $\overset{Z}{=}$ CHCH$_2$)$_3$CH$_3$

products methylated with ethereal diazomethane. The methyl esters were identified as (Z)-5-dodecenoic acid, lauric acid, and (5Z,8Z,11Z)-5,8,11-tetradecatrienoic acid. Hydrolysis of the catalytically hydrogenated natural alkaloid mixture gave isovaleric acid in addition to lauric and myristic acids.

The guanidine function of these structures was suggested by a positive Sakaguchi test and confirmed by the formation of the 4,6-dimethylpyrimidine derivatives. The combination of chemical degradation reactions with MS and ^{1}H-NMR studies led to the proposed structures **164–166** (*117*). Additional proof of the structures was obtained by synthesis of the three compounds (*122*).

The synthetic approach is depicted in Scheme 29, as demonstrated with compound **164**. Synthetic pathways for the two other compounds are analogous. The identity of synthetic and natural products was established by their spectral properties.

C. Spermine Type

1. Simple Spermine Derivatives

The structure elucidation of the three simple spermine compounds sinapoylspermine (**167**), disinapoylspermine (**168**), and diferuloylspermine (**169**) was performed on the basis of chromatographic identification of their hydrolysis products (*123*).

Kukoamine A (**170**) was isolated from the crude drug "jikoppi," which is prepared from *Lycium chinense.* Hydrolysis with HCl afforded two products: spermine and dihydrocaffeic acid. Both were identified with authentic samples. The ^{13}C-NMR spectrum shows the presence of 14 signals for the 28 carbon atoms. It follows from this observation that kukoamine A must have a symmetric structure; hence, there are only two structure possibilities. The ^{1}H-NMR spectrum shows the presence of only four hydrogen atoms adja-

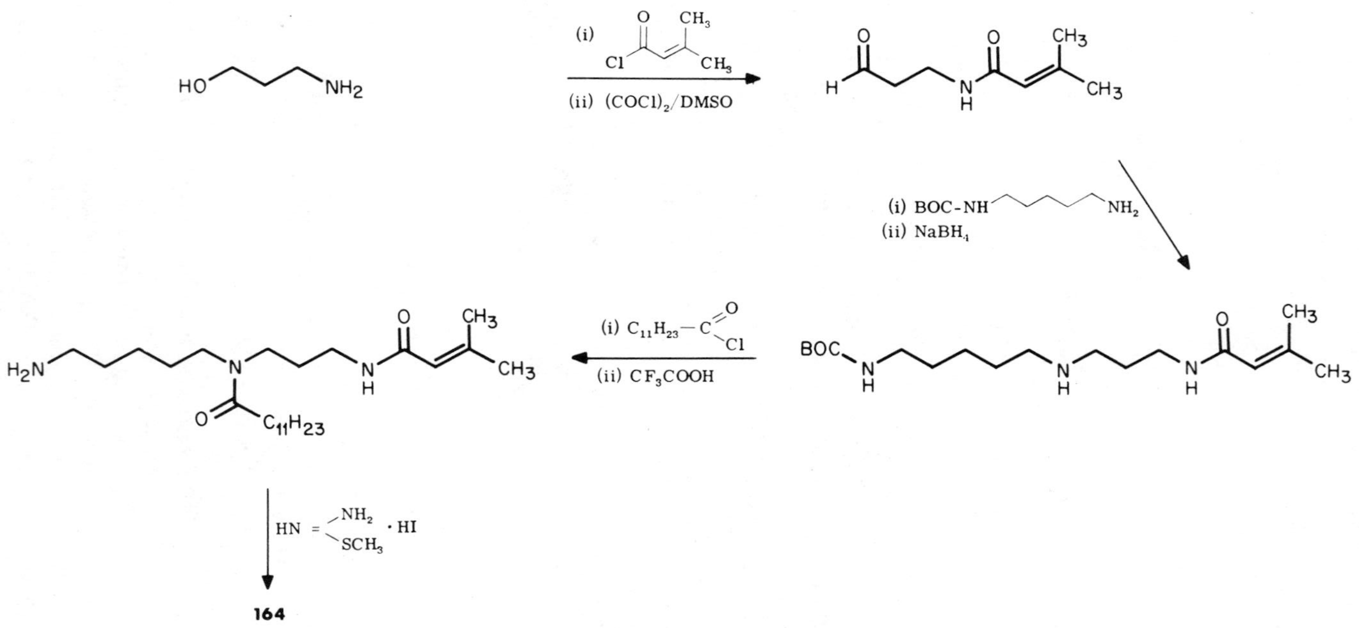

SCHEME 29. Synthesis of **164**.

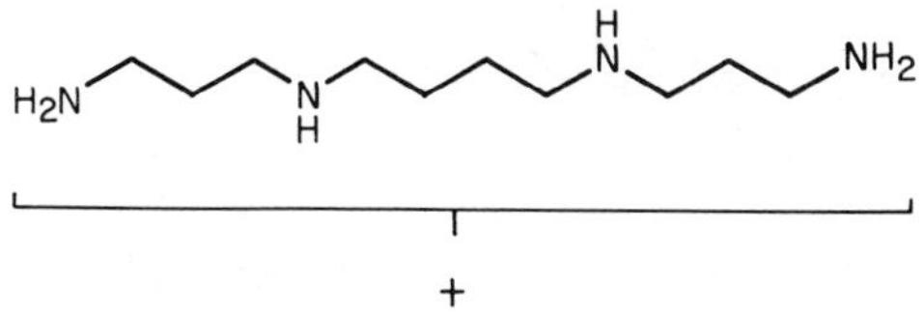

167 1

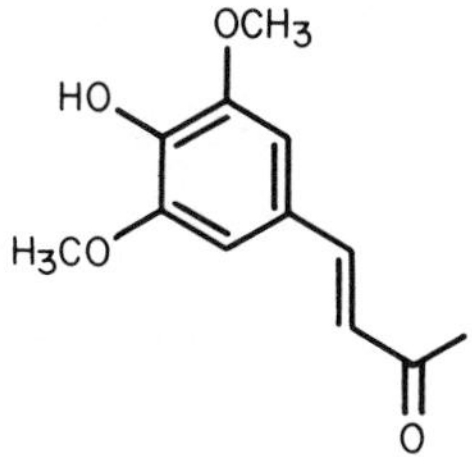

168 2

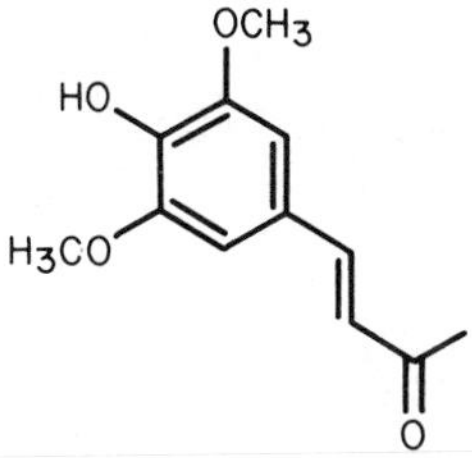

169 2

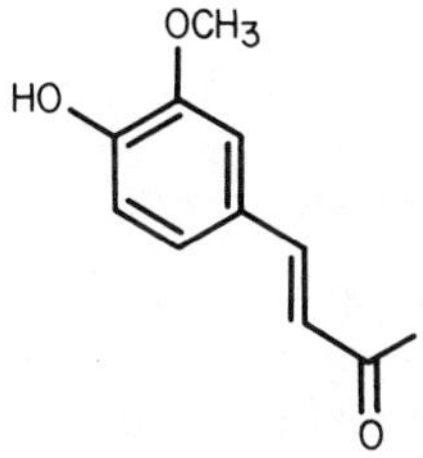

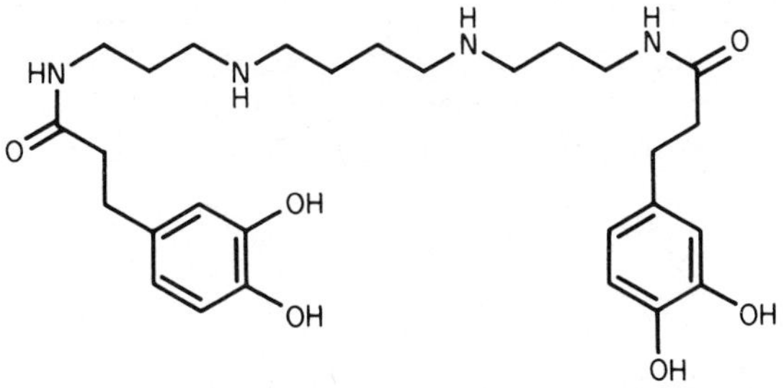

170 Kukoamine A

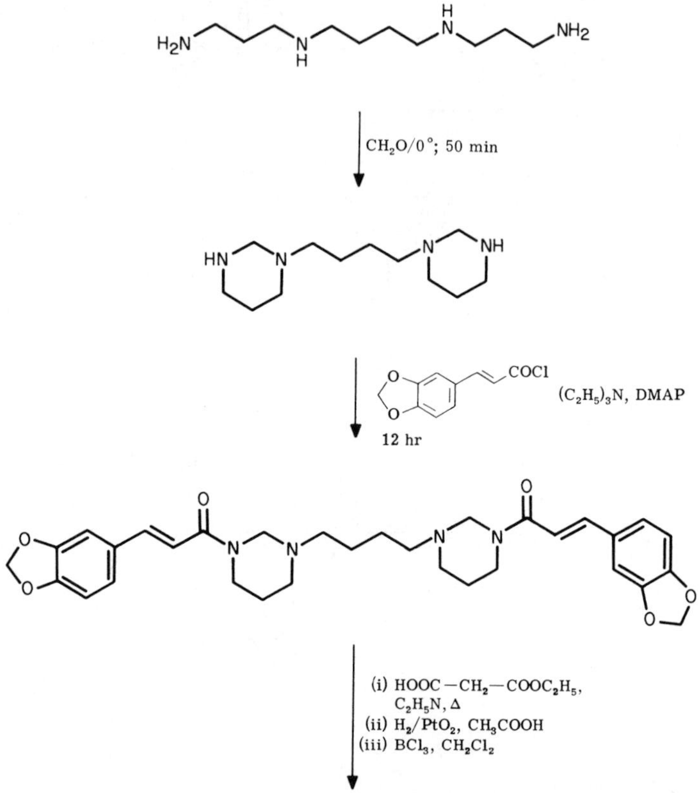

SCHEME 30. Synthesis of kukoamine (**170**).

cent to nitrogen atoms forming amide linkages. From this argument structure, **170** is the only possible structure for kukoamine A (*124*).

In the kukoamine synthesis, formaldehyde was used to form an aminal
with the 1,3-diaminopropane parts of spermine. The 1,4-diaminobutane
portion of the base does not react under these conditions; the central amino
groups will, therefore, be protected by formaldehyde. The two other amino
groups can now react with 3,4-methylenedioxycinnamoyl chloride. After
deprotection and catalytic hydrogenation, **170** was isolated in 62% overall
yield (see Scheme 30) (*125*).

2. Chaenorhine

Chaenorhine (**171**) has been isolated from several *Chaenorhinum* species
(Scrophulariaceae). The principal basic unit of this alkaloid is spermine,

171 Chaenorhine

172

which was obtained by alkali melt and identified as the tetraacetyl derivative. The isolate was different from the isomeric 1,12-diacetamido-4,8-diacetyl-4,8-diazadodecane (24). Potassium permanganate oxidation gave 3,4′-dicarboxy-6-methoxydiphenyl ether (172) in 30% yield.

Two derivatives, namely 13,14,21,22-tetrahydro-21,22-secochaenorhine (173) and the Hofmann base 174, have been particularly important for the structure elucidation of these natural products (see Schemes 31 and 32).

By treating chaenorhine (171) with sodium in liquid ammonia, the diphenyl ether function between atoms 21 and 22 is cleaved and the 13,14-double bond is hydrogenated simultaneously. Acetylation of the thus-formed product 173 yields an N,O-diacetyl derivative. Instead of the three $COCH_3$ groups that are present in the molecule, at least five methyl signals are observed in the $COCH_3$ region of the ^{1}H-NMR spectrum. An exact integration of these signals was not possible because of superposition with other resonances. The presence of several conformers is presumably responsible for the increase in the number of signals. This supposition has been supported by specific deuterioacetylation.

In addition to 4-hydroxydihydrocinnamic acid, spermine is formed in the acid-catalyzed hydrolysis of 173 (2 N HCl, 150°C, sealed tube). The expected 4-methoxycinnamic acid has not been found. If the phenolic hydroxyl group in 173 is ethylated and the product oxidized with $KMnO_4$, 4-methoxy- and 4-ethoxybenzoic acid are the products and they must have been formed from the differently substituted benzene rings in 173 (see Scheme 31).

The first clue to the mode of connection of the two cinnamic acid residues with the spermine moiety was given by analyzing the MS of 175 resulting from acetylation of a partial hydrolysis product of 173. Because of the formation of the peak triades mentioned at the beginning of Section III, the cinnamic acid residue that carries a free phenolic hydroxyl group (after

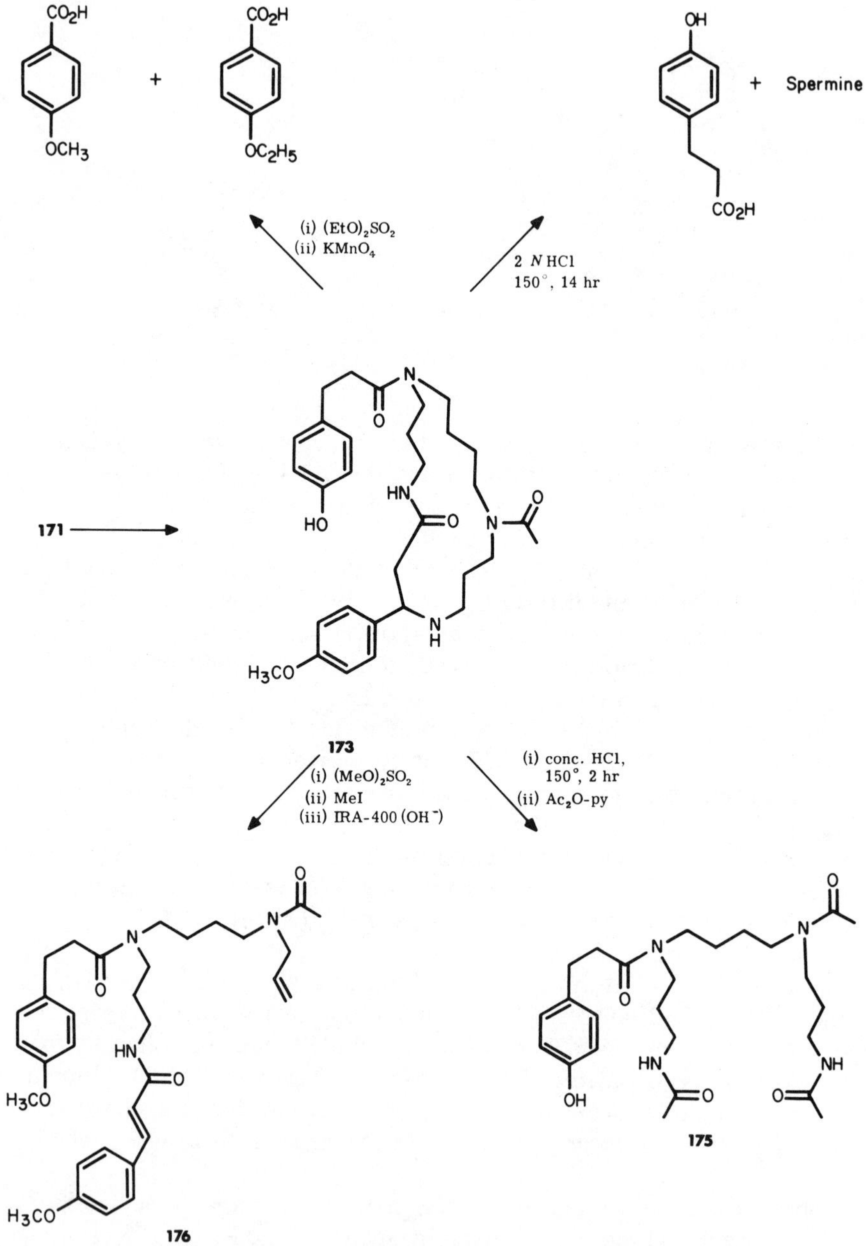

SCHEME 31. Reactions of chaenorhine (171).

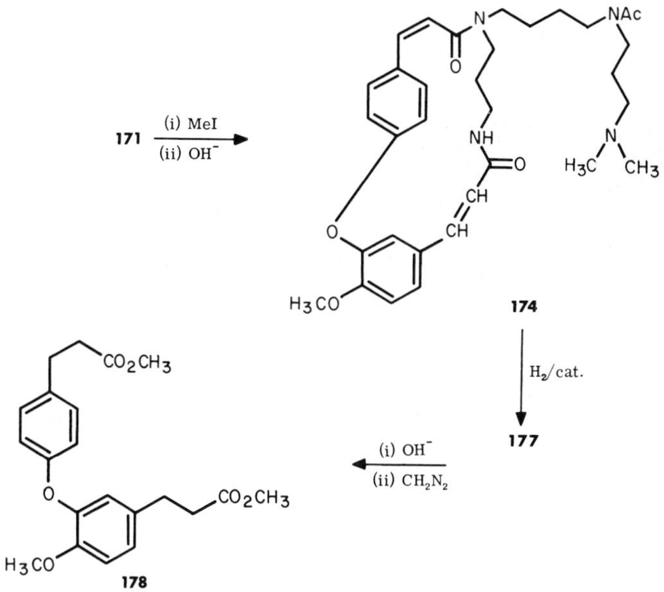

SCHEME 32. Reactions of chaenorhine (171).

reduction of **171** to **173**) must be attached to one of the two secondary nitrogen atoms of the spermidine molecule. The spermine derivative **176** results from Hofmann degradation of the reduction product **173**. The assignment of its structure depends mainly on mass spectral arguments.

The second compound important for the structural determination of chaenorhine (**171**) was obtained from N^2,N^2-dimethylchaenorhine by Hofmann degradation. The resulting Hofmann base **174** was hydrogenated catalytically to the tetrahydro compound **177**. Next, it was hydrolyzed in the presence of base and the product esterified with diazomethane. 6-Methoxy-diphenyl ether (3,4'-β,β'-dipropionic acid dimethyl ester) (**178**), which could be identified by comparison with a synthetic product, was obtained in a yield of $\sim 40\%$ (see Scheme 32).

Structural problems not yet discussed [e.g., connection of C-1 with N-2; substitution of N-6 with an acetyl group, and the configuration of the 13,14-double bond in chaenorhine] have been solved mainly by NMR spectroscopic analysis. The absolute configuration at C-1 given in **171** has been derived from chiroptic measurements of chaenorhine, some of its derivatives, and model compounds. The two benzene rings of the diphenyl ether are twisted around one another in chaenorhine (**171**) and also in its 13,14-dihydro derivative. Therefore, UV spectra of model substances show little agreement with those of the chaenorhine derivative in which ring A exists (*126*).

3. Aphelandrine, Orantine, O-Methylorantine, Ephedradines B, C, and D

Aphelandrine (**179**) is the principal alkaloid of several *Aphelandra* species. Its structure was deduced on the basis of many degradation reactions and extensive analysis of ^{1}H- and ^{13}C-NMR, and MS spectroscopy (*127,128*). The principal reactions of this compound are discussed below.

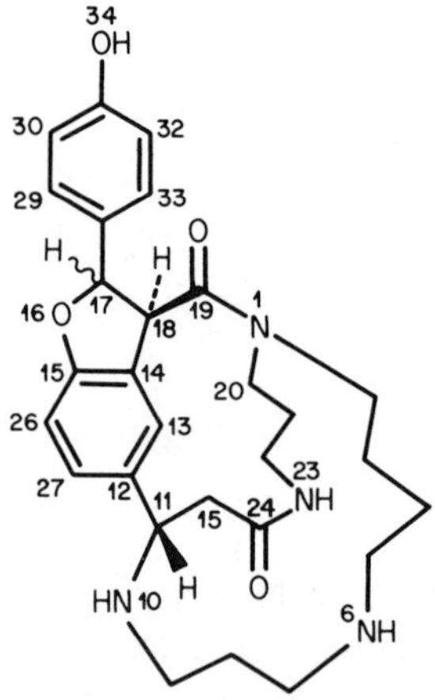

179 Aphelandrine

Acetylation of **179**, followed by partial hydrolysis (phenolic acetyl group) and etherification (CH$_2$N$_2$), gave derivative **180**. By hydrogenolysis, the C-16—C-17 bond in **180** was opened; after etherification the resulting phenol was transformed to **181** (see Scheme 33). Under modified hydrogenation conditions that N-10—C-11 bond can be hydrogenolyzed as well.

In addition to spermidine, spermine was isolated among the products of a KOH melt reaction and characterized as its tetraacetyl derivative [compare with (*24*)]. By an additional degradation of spermine, spermidine is formed.

The nature of the nonbasic portion of aphelandrine was determined in two different ways: KMnO$_4$ oxidation of **181**, followed by CH$_2$N$_2$ treatment afforded two esters methyl 4-methoxybenzoate (**182**) and dimethyl 4-methoxyisophthalate (**183**) (Scheme 33). Both compounds answer the question of the substitution pattern of the cinnamic acid units of aphelandrine. The mode of connection of the two cinnamic acid residues can be deduced from products of the hydrolysis of **180**, which forms lactone **184**. The structure of

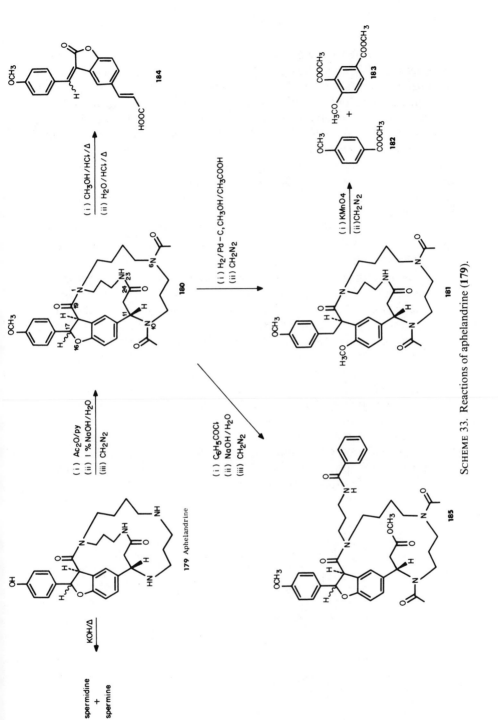

SCHEME 33. Reactions of aphelandrine (**179**).

184 was confirmed by synthesis along with chemical and spectroscopic evidence (*42,127*). Spectroscopic data suggested that **184** as such is not part of aphelandrine. It is formed in a rearrangement analogous to a reaction involving the alkaloid hordatine A (*42*). The chromophore incorporated in **179** is a benzofuran derivative, as depicted in the alkaloid structure.

To resolve the question of the connection of spermine with the benzofuran system, the compounds **185** (Scheme 33) and **186** (Scheme 34) were prepared and analyzed. In **179** the units are connected by three bonds, two of which are amide bonds. One of the amides is N-monosubstituted and must therefore represent one end of the spermine unit. By treatment of **180** with benzoyl chloride this N-substituted amide is transformed to an N-substituted secondary amid (imide), which under mild hydrolysis conditions [preferred cleavage of the N-23—C-24 bond] followed by esterification yielded **185**. Deuteration experiments, transformation reactions with **185**, as well as extensive spectroscopic investigations, have shown that in **179** N-23 is connected to C-24 and not to C-19.

Evidence for the N-10—C-11 bond was the Hofmann degradation reaction summarized in Scheme 34. Methylation of aphelandrine (**179**) afforded a monoquaternized product [dimethylation at N-6], which under mild basic conditions yielded the Hofmann base **186**, and under strong basic conditions (retro-Michael reaction) followed by benzoylation yielded the 1,3-diaminopropane derivative **187**. The latter was identified by synthesis. A combination of the arguments given above and further results not described here led to proposal of the structure of **179**. The relative and absolute configurations of aphelandrine were determined in connection with the structure elucidation of O-methylorantine (**190**) (see below).

Treatment of aphelandrine (**179**) with base (e.g., $NaOCH_3/CH_3OH$, 110°C, 1 hr, sealed tube) gave the isomerization product orantine (**189**).

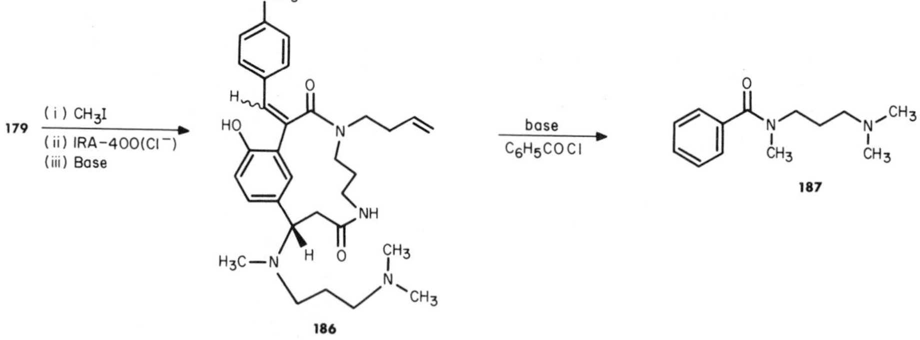

SCHEME 34. Reactions of aphelandrine (**179**).

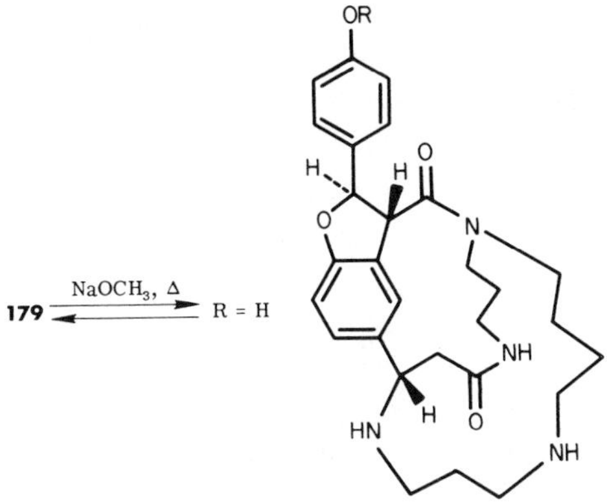

189 R = H   Orantine (Ephedradine A)
190 R = CH₃ O-Methylorantine

SCHEME 35. Equilibrium reaction of **179**.

Under these reaction conditions **179** and **189** are in equilibrium (see Scheme 35). Using similar reaction conditions, several aphelandrine derivatives have been isomerized to the corresponding orantine derivatives (*128*).

Orantine (ephedradine A) was isolated as a natural alkaloid from the crude drug *maō*, which itself is prepared from the aerial part of *Ephedra* plants (*129*). The structure, including the absolute configuration, was deduced by X-ray crystallography after preliminary chemical and spectroscopic work. The natural product and the isomerization product of aphelandrine (**179**) are also identical in their chiroptical properties (*128*).

O-Methylorantine (**190**) was isolated from *Aphelandra* and *Chaenorhinum* species (*130,131*). Its structure was deduced independently of the work on aphelandrine (**179**) but by similar methods. Its correlation with aphelandrine was established using the N^6,N^{10}-diacetyl-O^{34}-methyl derivatives: The aphelandrine derivative **180** was isomerized to the corresponding O-methylorantine derivative with base.

From chiroptic investigations of **179**, **189**, **190**, and some derivatives, the $11S,18S$ configuration for **179** was deduced. It was not possible to determine on the basis of ^{1}H-NMR data the configuration of center-17 relative to center-18 in **179**. The mechanism of the isomerization reaction **179** ⇄ **189** has not yet been investigated (*128*).

The alkaloid ephedradine B (**191**) was isolated from the crude drug *maō-kon*, prepared from the underground parts of *Ephedra* plants. Its structure, including absolute configuration, was deduced from spectroscopic analyses and chiroptic comparison of **189** and **191**. Furthermore, spermine was isolated by hydrolysis of **191** (see Section V).

Two other alkaloids, ephedradine C (**192**) and ephedradine D (**193**), were isolated from the same source (*maō-kon*) (*132,133*). Their structure elucidations were based mainly on spectroscopic evidence. Additionally, **192** was correlated with **191** by treatment with diazomethane. The figures below representing all three alkaloids show their absolute configuration.

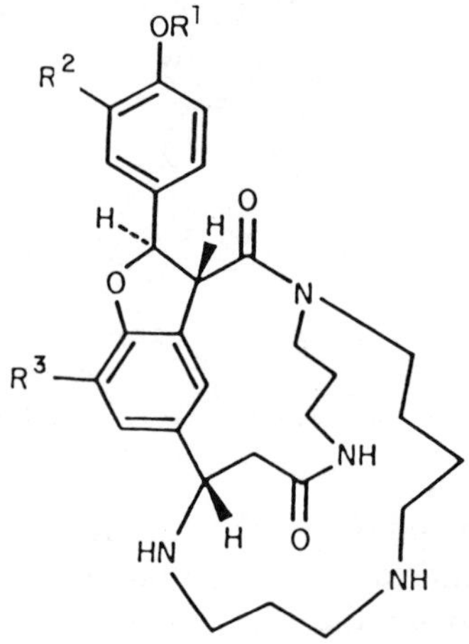

191	R^1 = R^3 H, R^2 = OCH_3	Ephedradine B
192	R^1 = CH_3, R^2 = OCH_3, R^3 = H	Ephedradine C
193	R^1 = R^2 H, R^3 = OCH_3	Ephedradine D

4. The Homalines

Homaline (**194**) is the main alkaloid of the leaves of *Homalium pronyense* (*134–137*). The bases hopromine (**195**), hoprominol (**196**), and hopromalinol (**197**) have been found as minor alkaloids in the same plant (*136*). Homaline is composed of two cinnamic acid units and spermine.

The structure elucidation of the main alkaloid **194** was based primarily on the products of a Hofmann degradation reaction resulting in the base **198** (see Scheme 36). The observation was made that homaline, which contains no formal C=C double bonds apart from the two benzene rings, can be transformed (H_2/Pd-C) to an open-chained spermine derivative. This derivative can be converted to **199** by methylation (H_2CO/H_2/Pd-C). The structure of homaline as **194** follows from these properties. Compound **199** was also obtained by hydrogenation of the Hofmann base **198**. A structural alternative to **194** could be excluded on the basis of spectroscopic arguments. Similar investigations led to the formulas **195**, **196**, and **197** for the three minor alkaloids.

Bisdeoxohomaline (obtained from homaline by treatment with LAH/

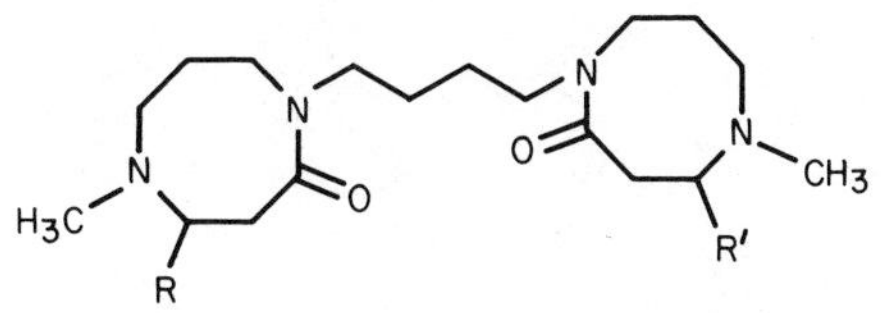

194 R = R′ = Ph, Homaline (config. S,S)
195 R = C_5H_{11}, R′ = C_7H_{15}, Hopromine
196 R = C_5H_{11}, R′ = $CH_2CH(OH)(CH_2)_4CH_3$, Hoprominol
197 R = Ph, R′ = $CH_2CH(OH)(CH_2)_4CH_3$, Hopromalinol

(i) CH_3I
(ii) IRA−400(OH⁻)

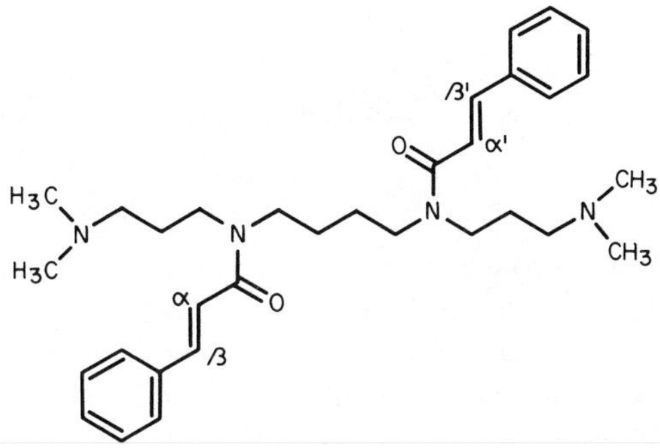

198
199 $\alpha, \alpha', \beta, \beta'$-Tetrahydro

SCHEME 36. Hofmann degradation of homaline (**194**).

THF) has been synthesized starting with (S)-β-phenyl-β-alanine. From this, the structure of homaline (**194**) has been confirmed and the absolute configuration determined (*138*).

The total synthesis of (−)-homaline (**194**) has been published (*139*). The key step of this synthesis is an intramolecular transamidation reaction by which a β-lactam is enlarged to a eight-membered amino lactam (see Scheme 37). Starting with putrescine (**200**), the amino alcohol **201** was obtained via Michael addition with acrylonitrile, followed by hydrolysis, esterification, and reduction with LAH. By protection of the amino functions with the BOC group and O-tosylation, **202** was formed. This compound was first treated with the β-lactam **203**, prepared according to Scheme 37, and afterward deprotected using formic acid to form the key intermediate **204**. The transamidation step was accomplished with either quinoline (1 hr, reflux) or with diphenyl ether (saturated with air). The resulting product was methylated by the Eschweiler–Clarke procedure to yield (−)-homaline (**194**), indistinguishable from the natural product.

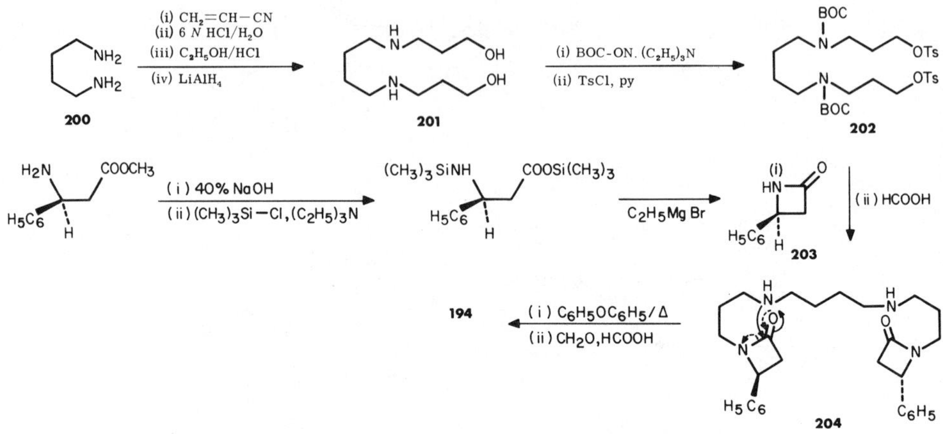

SCHEME 37. Total synthesis of (−)-homaline (**194**).

5. Verbascenine

Verbascenine (**205**) has been isolated from the aerial parts of *Verbascum* species (*140*). Its structure elucidation was based on chemical degradation reactions as well as careful analysis of spectral data (*140*). In addition to spermidine, spermine was isolated from a KOH melt reaction and identified as its tetraacetyl derivative. The two cinnamic acid residues were identified by spectroscopic methods. A C=C double bond was found in only one of these units. The other cinnamic acid residue is part of a β-aminodihydrocin-namide. Therefore, verbascenine must be cyclic. The number of possible structures was reduced to two (**205** and **206**) on the basis of the formation of certain derivatives and their spectroscopic properties. Structure **205** is preferred for verbascenine because of the mass spectral behavior of its

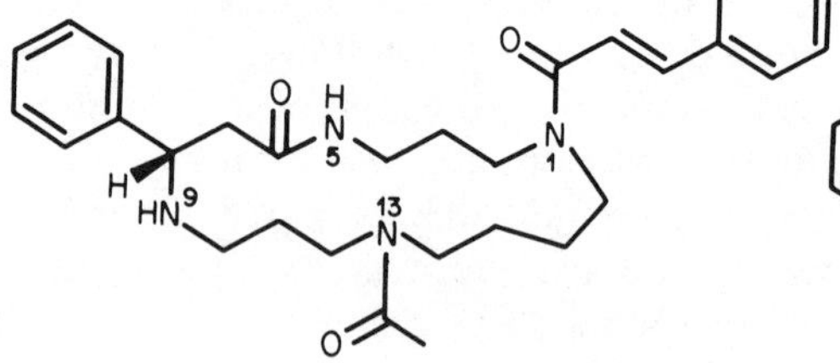

205 Verbascenine

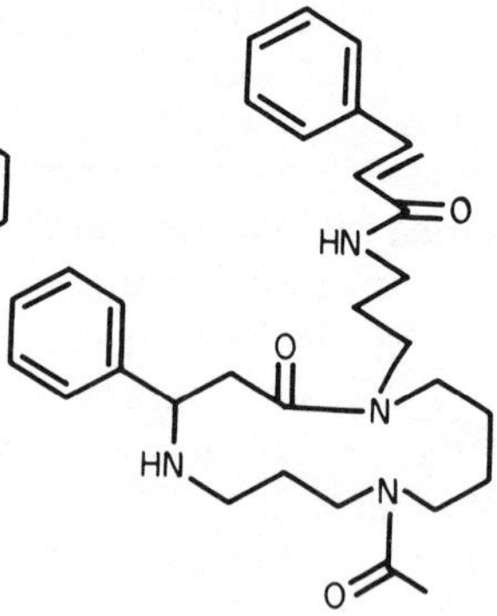

206

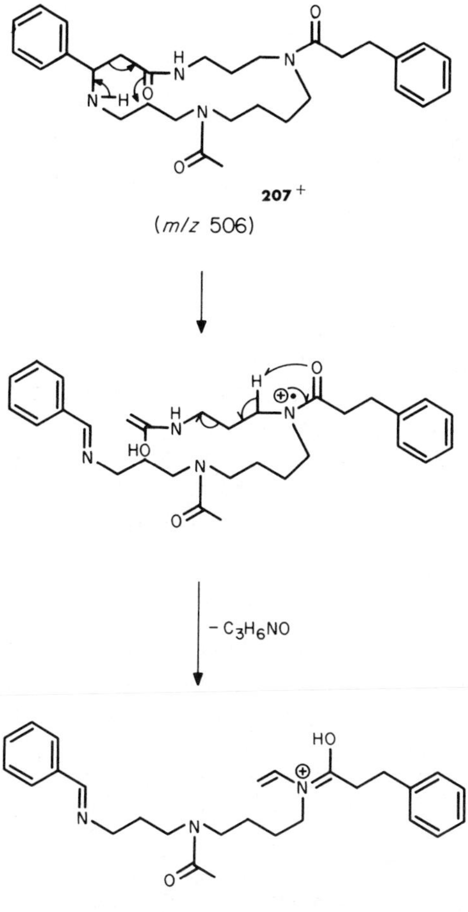

207 $^{+}$

(*m/z* 506)

$- C_3H_6NO$

m/z 434

SCHEME 38. Mass spectral behavior of verbascenine derivatives.

dihydro derivative **207** (Scheme 38). If **206** were the correct structure for verbascenine, fundamental rearrangement reactions should proceed during the mass spectral degradation reaction of the dihydro derivative to form the ion *m/z* 434, which *a priori* was not considered.

6. The Pithecolobines

Structural determination of pithecolobine isolated from *Samanea saman* (*Pithecolobium saman*) (*141*) demonstrated it to be an inseparable mixture of compounds, primarily of the general formula **208**.

$$H_3C-(CH_2)_n$$
$$\backslash CH-(CH_2)_m-C \diagdown \overset{O}{}$$

(structure)

208a	$m = 3$, $n = 6$,	24-30%
208b	$m = 1$, $n = 8$	40-49%
208c	$m = 1$, $n = 10$	13.5-16.5%

$H_3C-(CH_2)_6$ **209**

Hofmann degradation of pithecolobine (**208**) afforded the lactone **209** as well as unsaturated amides that yielded the following carboxylic acids after hydrogenation and hydrolysis: dodecanoic, tridecanoic, and tetradecanoic acids (ratio 65:4:30 by GC analysis of the methyl esters). In order to define the position of the double bonds introduced by the Hofmann degradation reaction, the above-mentioned unsaturated amides were oxidized to the corresponding carboxylic acids with $KMnO_4/KIO_4$ in *tert*-butanol. The composition of pithecolobines is the result of the proportion of the various individual carboxylic acids in the mixture. Lactone **209** is formed by neighboring-group participation of the amide group in **208a** during this reaction. Reduction of **208** with LAH yields the oxygen-free deoxopithecolobine, containing four basic, secondary amine nitrogen atoms (established by formation of a neutral tetraacetyl derivative). The two deoxopithecolobines derived from **208a** and **208b** have been synthesized (see Scheme 39) and compared as tetratosylates with each other and with the corresponding derivative from **208**. The three different preparations were indistinguishable by IR spectroscopy and by TLC (*91*).

Besides synthesis, Hofmann degradation of **208** has established that spermine is part of the pithecolobines. By this degradation reaction, a number of different alkylamines were formed and identified. The question of whether all pithecolobines contain spermine or whether compounds also exist that contain the 1,12-diamino-4,8-diazadodecane (thermospermine) has not yet been answered completely.

IV. Aspects of Biosynthesis and Chemotaxonomy of Polyamine Alkaloids

The alkaloids discussed in Section III contain five basic backbone components: putrescine, spermidine, two homospermidines, and spermine. To date, no other variations are known. The nonbasic part of the alkaloids are in most cases cinnamic acid and fatty acid derivatives. The cinnamic acid

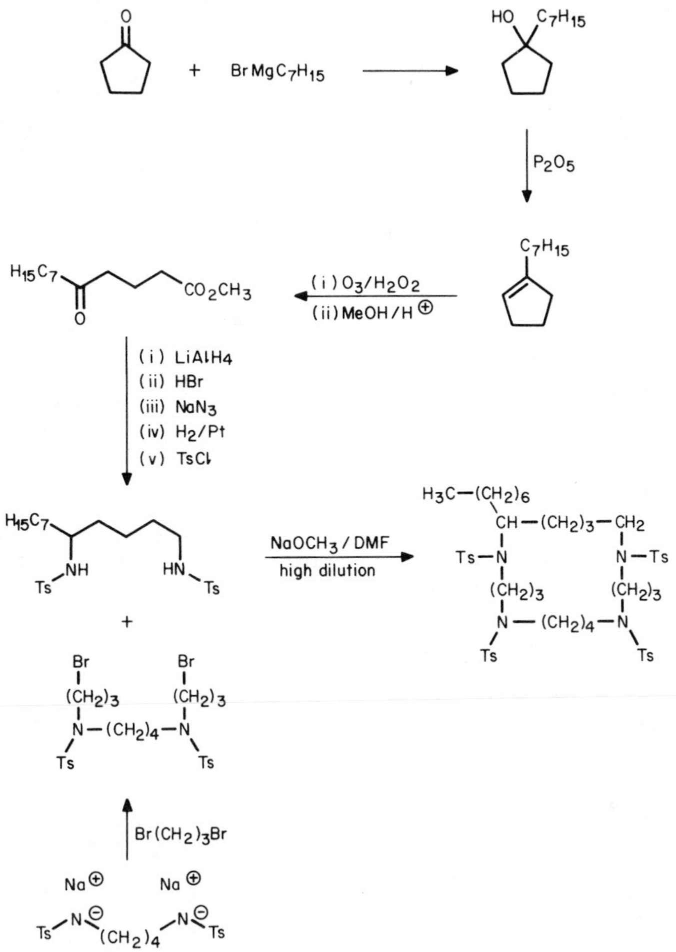

SCHEME 39. Synthesis of N,N',N'',N'''-tetratosyldeoxopithecolobine.

dervatives (one, two, or three units) may contain additional aromatic hydroxyl or methoxyl groups, or two cinnamic acid residues may be condensed with each other (via phenolic coupling, see Scheme 40), or one of the cinnamic acid residues may be connected with a second nitrogen of the polyamine backbone (Michael reaction product) in addition to the amide linkage. In most cases, only a few chemical reactions involved in the formation of polyamine alkaloids from their building blocks in nature are known: amide formation, Michael reaction, and phenolic coupling.

The proposed schematic pathway for some selected alkaloids containing two condensed cinnamic acid units is shown in Scheme 40. In principle, the

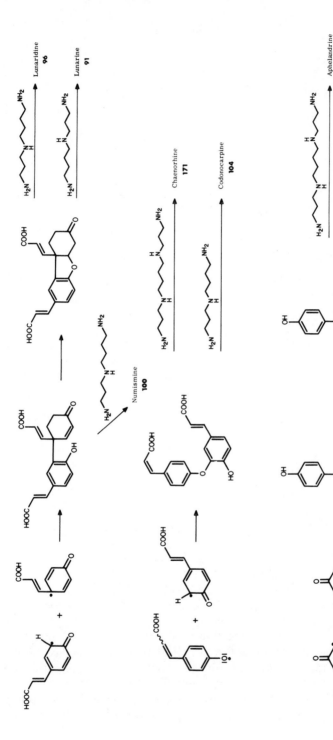

SCHEME 40. Schematic proposed biosynthetic pathway of di- and polyamine alkaloids containing two condensed cinnamic acid units.

oxidative coupling can take place at two different stages. In the first, the two cinnamic acid units are coupled together followed by amide linkage formation with a di- or polyamino compound. This proposal is illustrated in Scheme 40. The second mechanism involves an initial amide bond formation between the di- or polyamine and cinnamic acid. The cinnamic amides of the thus-formed precursors can, in a later step, be coupled in an oxidation reaction. Such a mechanism has been proposed for formation of the hordatines from coumaroylagmatines (45). To date, no experimental biosynthetic investigations in higher plants have been published.

Concerning the nature of the polyamine alkaloids, characteristic differences exist. In the case of putrescine alkaloids the base is always a chain; in spermidine and spermine alkaloids the cyclic form is predominant. The 13-membered and 17-membered lactams are the preferred rings in the spermidine and spermine alkaloids, respectively. As shown in Scheme 41

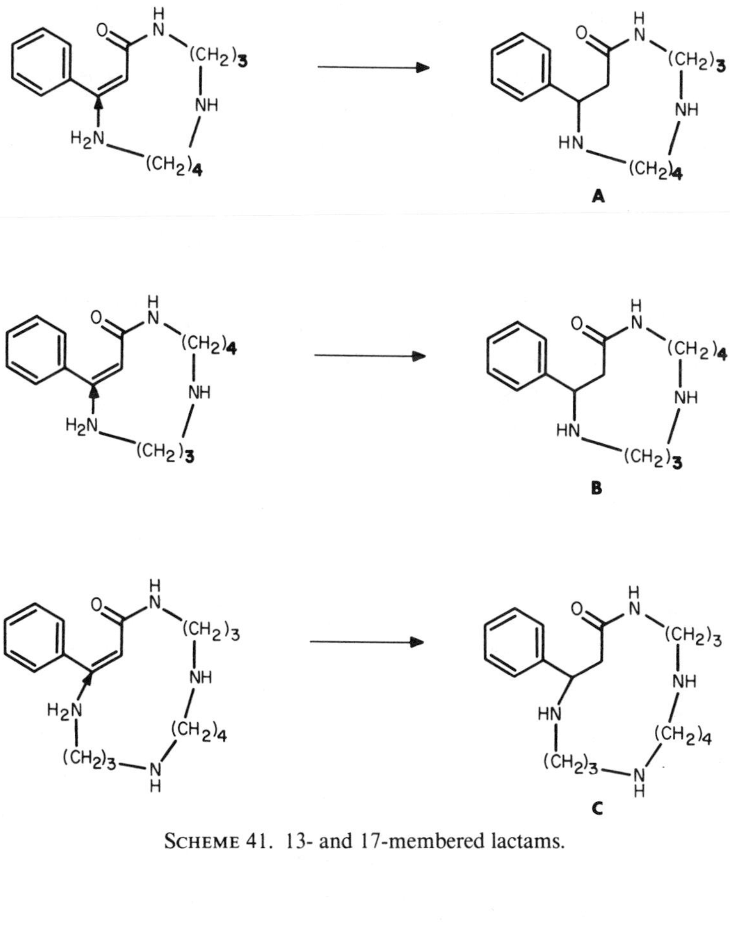

SCHEME 41. 13- and 17-membered lactams.

the result of the attack of the primary amino group in an N-monosubstituted cinnamoylspermidine is a 13-membered ring, whereas the analogous reaction in spermine derivatives is a 17-membered lactam. Because of the symmetry in spermine, only one 17-membered isomer (type C) is possible; but with spermidine two possible ring isomers can be formed representing two alkaloid groups, namely celacinnine (type A) and dihydroperiphylline (type B), which were the subject of a chemotaxonomic study of Celastraceae (*142*).

The naturally occurring di- and polyamine alkaloids are listed in the tables of Section V. At present, nearly 100 compounds have been reported. They were isolated from 56 genera in 30 plant and animal families. Some of the plants contain all three kinds of alkaloids (putrescine, spermidine, and spermine), e.g., *Ananas comosus* Merill. (Bromeliaceae). Intensive investigations of alkaloids were performed in some genera, e.g., *Aphelandra, Chaenorhinum, Equisetum, Maytenus.* The reasons were commercial availability and pharmacological interest (*Maytenus*).

Some of the plant species contain many different alkaloids, e.g., *Lunaria biennis* Moench. (Cruciferaceae). The investigation of this field of plant metabolites seems to have been done randomly and not systematically. Furthermore, the occurrence of di- and polyamine derivatives in many cultivated plants is apparent (*123*). Their structural differences from the "more complicated alkaloids" are not very large. It appears very probable that many more plants and animals contain di- and polyamine alkaloids than are known today.

Chemotaxonomic investigations are useful in cases where natural compounds occur in a very limited number of plant families [e.g., ~ 1200 indole alkaloids in three plant families (*143*)]. The number of species investigated belonging to one plant family should be very high. In the case of di- and polyamine alkaloids, this ratio is only 1 (plant family) : 2 (species) compared with, for example, indole alkaloids of Apocynaceae with a ratio of 1 : 249. Therefore, we think that at present, chemotaxonomic investigations of di- and polyamine alkaloids may not be useful.

V. Catalog of Naturally Occurring Diamine and Polyamine Alkaloids

Table II summarizes the isolation of structurally identified di- and polyamine alkaloids.

TABLE II
NATURALLY OCCURRING DI- AND POLYAMINE ALKALOIDS

Alkaloid	Structure[a]	Occurrence	mp (°C) (Ref.)	$(\alpha)_D$ (solvent) (Ref.)	UV λ_{max} (log ε) (Ref.)	IR	NMR	MS	Structure	Synthesis
Putrescine Alkaloids										
Aerothionin $C_{24}H_{26}Br_4N_4O_8$ (MW = 814)	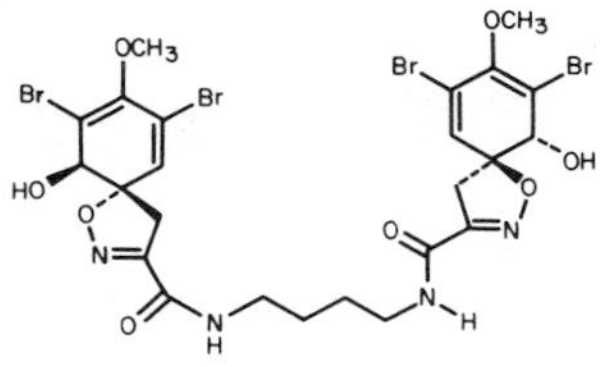 Absolute configuration (47)	*Aplysina fistularis* (47) *A. aerophoba* (sponge) (46) *Verongia thiona* (46)	134–137° (dec.) (47)	+210° (MeOH) (47) +252° (acetone) (46)	282 (47) (MeOH) 284 (4.10) (46)	(46)	(46, 47)	(46)	(46,47)	
N-Carbamoyl-putrescine $C_5H_{13}N_3O$ (MW = 131)	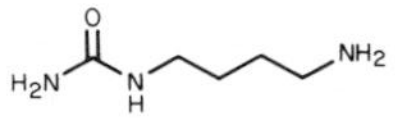	*Hordeum vulgare* (Gramineae) (39) *Sesamum indicum* (Pedaliaceae) (K-deficient plants) (144)	185–186° (39)						(39)	(144, 145)
4-Coumaroylagmatine $C_{14}H_{20}N_4O_4$ (MW = 276)	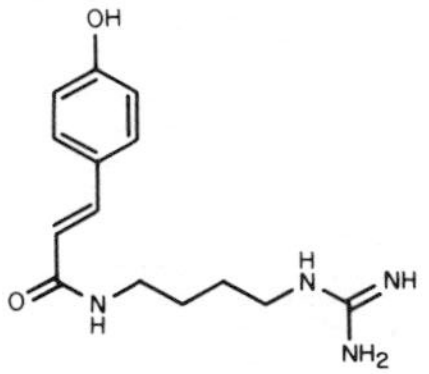	Barley seedlings (9) *Hordeum bulbosum* (Gramineae) (146) *H. distichon* (146) *H. jubatum* (146) *H. murinum* (146) *H. spontaneum* (146)	215–217° (picrate) (9)		·HCl 220 (4.26) 292 (4.34) 304 (4.32) (9)				(9)	(9,45)

(Continued)

TABLE II (*Continued*)

Alkaloid	Structure[a]	Occurrence	mp (°C) (Ref.)	(α)ᴅ (solvent) (Ref.)	UV λ_max (log ε) (Ref.)	IR	NMR	MS	Structure	Synthesis
4-Coumaroyl-2-hydroxyputrescine $C_{13}H_{18}N_2O_3$ (MW = 250)	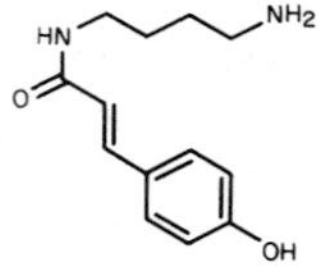	*Triticum aestivum* (Gramineae) (147)							(147)	
4-Coumaroylputrescine $C_{13}H_{18}N_2O_2$ (MW = 234)	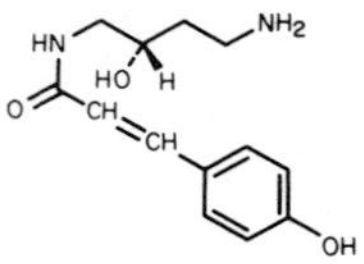	*Lycopersicum esculentum* (Solanaceae) (123) *Nicotiana tabacum* (Solanaceae) (93,148) *Pennisetum americanum* (Gramineae) (123) *Persea gratissima* (Lauraceae) (123) *Petunia* hybrid (Solanaceae) (123) *Salix* sp. (Salicineae) (123) *Triticum vulgare* (Gramineae) (123) *Zea mays* (Gramineae) (123)	182–183° (93)		235 285 sh 304 (148)				(148)	(93)

152

Compound	Structure	Source	UV	Ref.
Dicaffeoylputrescine $C_{22}H_{24}N_2O_6$ (MW = 412)	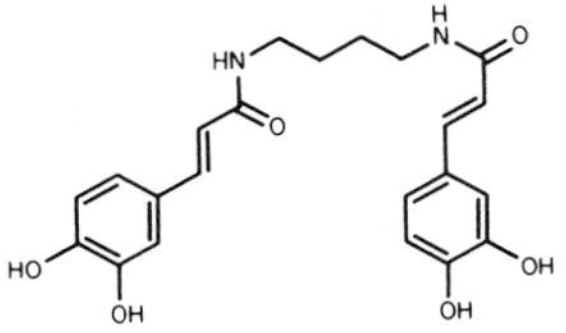	*Helianthus annuus* (Compositae) (123) *Nicotiana tabacum* (Solanaceae) (149) *Pyrus communis* (Rosaceae) (123) *Salix* sp. (Salicineae) (123)	242 299 sh 320 (149)	(149)
Di-4-coumaroylputrescine $C_{22}H_{24}N_2O_4$ (MW = 380)	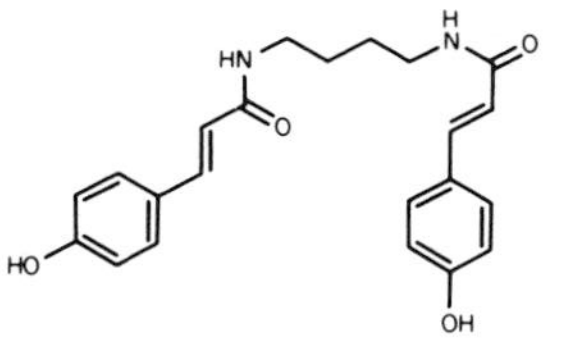	*Dianthus caryophyllus* (Caryophyllaceae) (123) *Helianthus annuus* (Compositae) (123) *Nicotiana tabacum* (Solanaceae) (148) *Pyrus communis* (Rosaceae) (123) *Rubus idaeus* (Rosaceae) (123) *Vicia faba* (Leguminosae) (123)	235 285 sh 310 (148)	(148)

(Continued)

153

TABLE II (*Continued*)

Alkaloid	Structure[a]	Occurrence	mp (°C) (Ref.)	$(\alpha)_D$ (solvent) (Ref.)	UV λ_{max} (log ε) (Ref.)	References				
						IR	NMR	MS	Structure	Synthesis
Diferuloylputres- cine $C_{24}H_{28}N_2O_6$ (MW = 440)	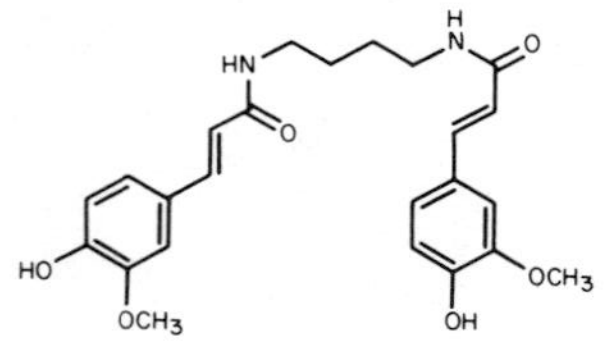	*Ananas comosus* (Bromeliaceae) (123) *Dianthus caryo- phyllus* (Caryophylla- ceae) (123) *Gomphrena glo- bosa* (Amarantha- ceae) (123) *Lycopersicum esculentum* (Solanaceae) (123) *Nicotiana taba- cum* (Solanaceae) (150) *Petunia* hybrid (Solanaceae) (123) *Triticum vulgare* (Gramineae) (123) *Vicia faba* (Leguminosae) (123)			240 285 sh 318 (150)				(150)	(150)
Disinapoylputres- cine $C_{26}H_{32}N_2O_8$ (MW = 500)	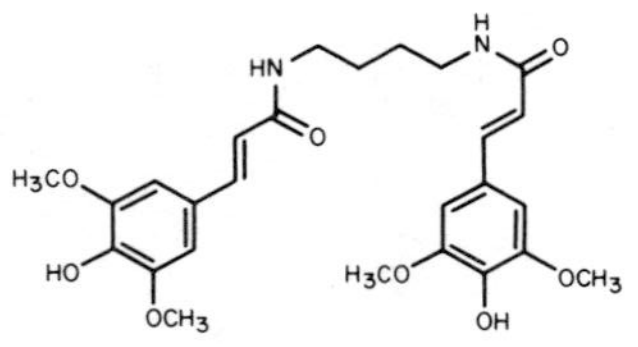	*Lilium* sp. (Liliaceae) (123)							(123)	

154

Compound	Structure	Source	mp / [α]	UV	References
Feruloyl-2-hydroxyputrescine $C_{14}H_{20}N_2O_4$ (MW = 280)	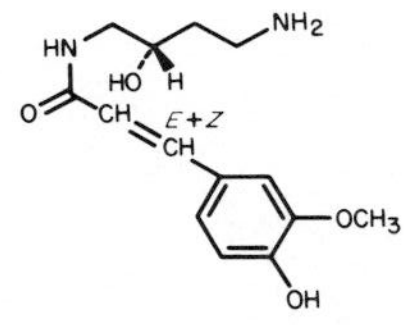 Absolute configuration (146)	Triticum aestivum (Gramineae) (147)	N,O,O'-triacetyl derivative −20° (CHCl₃) (147)	293 (3.85) 316 (3.92) (147)	(147) (147) (147) (147)
Hordatine A $C_{28}H_{38}N_8O_4$ (MW = 550)	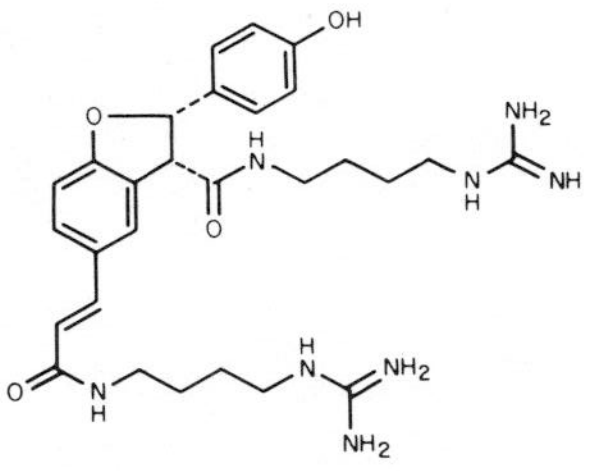 Relative configuration (44)	Hordeum bulbosum (Gramineae) (146) H. distichon (146) H. jubatum (146) H. murinum (146) H. spontaneum (146) H. vulgare (42)	127–128° (dipicrate) (42) +69° (H₂O) (42)	(EtOH) 229 (4.32) 298 (4.26) 307 (4.26) (42)	(44) (42) (42–44) (42,44)
Hordatine B $C_{29}H_{40}N_8O_5$ (MW = 580)	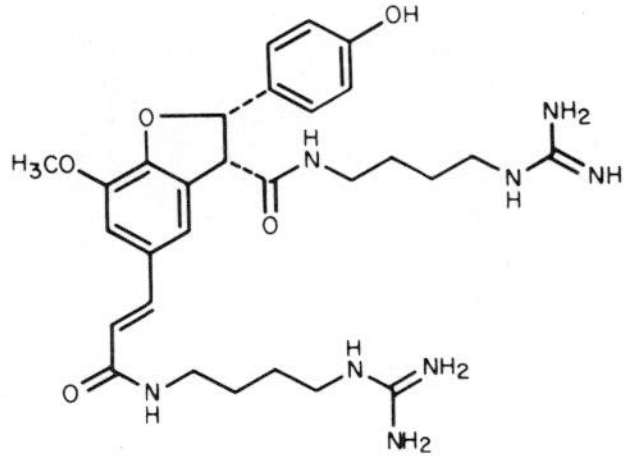 Relative configuration (44)	Hordeum bulbosum (Gramineae) (146) H. distichon (146) H. jubatum (146) H. murinum (146) H. spontaneum (146) H. vulgares (42)	132–135° (dipicrate) (42) +54° (H₂O) (42)	(EtOH) 224 (4.34) 301 (4.19) 316 (4.21)	(42) (42)

(Continued)

155

TABLE II (*Continued*)

Alkaloid	Structure[a]	Occurrence	mp (°C) (Ref.)	$(\alpha)_D$ (solvent) (Ref.)	UV λ_{max}(log ϵ) (Ref.)	References				
						IR	NMR	MS	Structure	Synthesis
Hordatine M C$_{34}$H$_{48}$N$_8$O$_9$ (MW = 712)	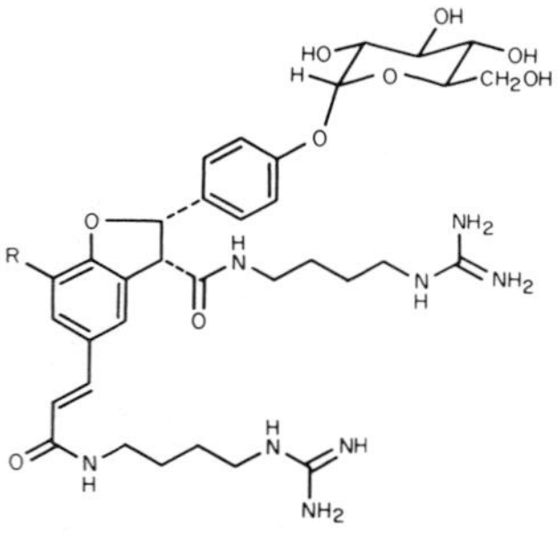 Inseparable mixture of R = H and R = OCH$_3$	*Hordeum vulgare* (Gramineae) (42)		+60° (H$_2$O) (42)	(EtOH) 224 (4.69) 320 (4.49) 357 (4.47) (42)		(42)		(42)	
Paucine (=Caffeoyl-putrescine) C$_{13}$H$_{18}$N$_2$O$_3$ (MW = 250)	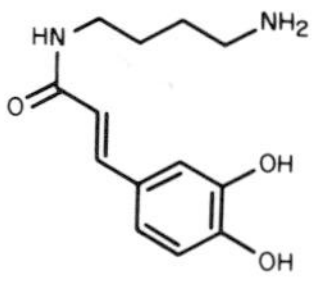	*Nicotiana tabacum* (Solanaceae) (93, 148,149,151, 152) *Pentaclethra macrophylla* (Leguminosae) (37,153) *Persea gratissima* (Lauraceae) (123) *Petunia* hybrid (Solanaceae) (123) *Salix* sp. (Salicineae) (123)	245–50° (HCl) 247–250° (HCl) (93)		245 295 sh 318 (149) 218 (4.11) 322 (4.10) (37)	(37,39, 151, 152)	(37, 93, 151, 152)	(37)	(37,148, 151)	(93)

156

Compound	Source	mp	UV λmax (nm)	Ref.
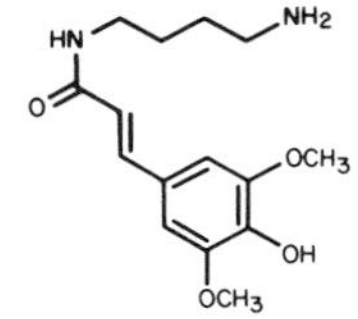 Sinapoylputrescine $C_{13}H_{22}N_2O_4$ (MW = 294) Structure: NH₂ ... OCH₃, OH, OCH₃, HN, O	*Ananas comosus* (Bromeliaceae) (123) *Lilium* sp. (Liliaceae) (123)			(123)
Subaphylline (= Feruloylputrescine) $C_{14}H_{20}N_2O_3$ (MW = 264) Structure: NH₂ ... OCH₃, OH, HN, O 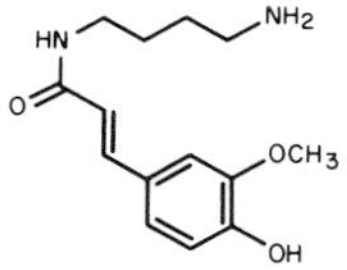	*Ananas comosus* (Bromeliaceae) (123)	151–152° (HCl) (93)	240 285 sh 315 (148,154)	(154) (154) (148, 154, 155) (93, 150, 156)
	Citrus paradisi (Rutaceae) (154)			
	Gomphrena globosa (Amaranthaceae) (123)	225–226° (dec.) (picrate) (154)		
	Lycopersicum esculentum (Solanaceae) (123)			
	Nicotiana tabacum (Solanaceae) (93,148)			
	Pennisetum americanum (Gramineae) (123)	185° (picrate) (154)		
	Persea gratissima (Lauraceae) (123)	230° (HCl) (38)		
	Petunia hybrid (Solanaceae) (123)	139° (HCl) (156)		

(Continued)

TABLE II (*Continued*)

Alkaloid	Structure[a]	Occurrence	mp (°C) (Ref.)	(α)_D (solvent) (Ref.)	UV λ_max(log ε) (Ref.)	References IR	References NMR	References MS	References Structure	References Synthesis
		Salix sp. (Salicineae) (*123*)								
		Salsola subaphylla (Chenopodiaceae) (*38*)								
		Triticum vulgare (Gramineae) (*123*)								
		Zea mays (Gramineae) (*123*)								
Tetramethylputrescine C₈H₂₀N₂ (MW = 144)		*Duboisia leichhardtii* (Solanaceae) (*157*)	oil (*40*) 197–200° (picrate) (*160*)			(*159*)	(*160, 161*)	(*161*)	(*40, 160*)	
		D. myoporoides (*158*)	280° (HCl) (*159*)							
		Hyoscyamus muticus (Solanaceae) (*40*)	273° (HCl) (*40*)							
		H. niger (*162*)								
		H. reticulatus (*163*)								
		Oldenlandia affinis (Rubiaceae) (*160*)								
		Ruellia rosea (Acanthaceae) (*159*)								

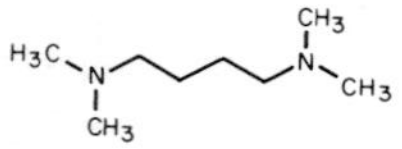

Structure: $H_3C-N(CH_3)-CH_2CH_2CH_2CH_2-N(CH_3)-CH_3$

Compound	Source	mp	$[\alpha]_D$	UV λ_{max} nm (log ε)	Refs.
N,N,N'-Trimethyl-N'-(4-hydroxy-Z-cinnamoyl)putrescine $C_{16}H_{24}N_2O_2$ (MW = 276)	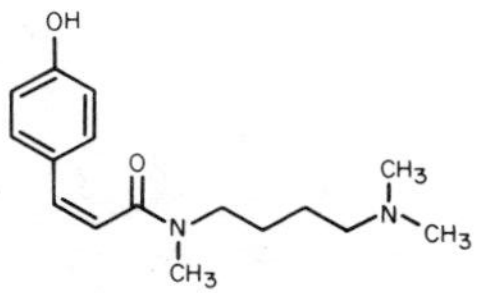 Kniphofia flavovirens (Liliaceae) (41) K. foliosa (41) K. tuckii (41)	60–65° (41)		(MeOH) 285 (4.13) (41)	(41) (41) (41) (41)
N,N,N'-Trimethyl-N'-(4-methoxy-Z-cinnamoyl)putrescine $C_{17}H_{26}N_2O_2$ (MW = 290)	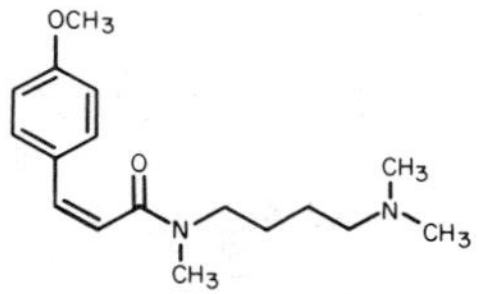 Kniphofia flavovirens (Liliaceae) (41) K. foliosa (41) K. tuckii (41)	oil (41)		(MeOH) 270 (4.26) (41)	(41) (41) (41) (41)

Spermidine Alkaloids

Compound	Source	mp	$[\alpha]_D$	UV λ_{max} nm (log ε)	Refs.
N^1-Acetyl-N^1-deoxymayfoline $C_{18}H_{27}N_3O_2$ (MW = 317)	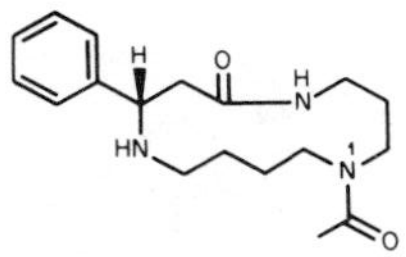 Maytenus buxifolia (Celastraceae) (109)	177–178° (109)	−8° (CHCl₃) (109)	252 (2.48) 259 (2.49) 264 (2.37) 268 (2.23) (109)	(109) (109) (109) (109, 110)

Absolute configuration (109). The structure given in (109) has to be revised (110).

(Continued)

159

TABLE II (Continued)

Alkaloid	Structure[a]	Occurrence	mp (°C) (Ref.)	$(\alpha)_D$ (solvent) (Ref.)	UV λ_{max}(log ϵ) (Ref.)	References IR	NMR	MS	Structure	Synthesis
Agrobactine $C_{32}H_{36}N_4O_{10}$ (MW = 636)	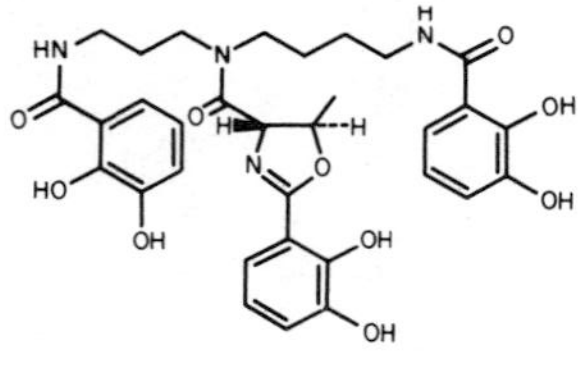 Relative configuration (60)	Agrobacterium tumefaciens (49)	108–112° (dec.) (49)		(EtOH) 252 (4.45) 316 (3.98) (49)	(60)	(49, 60)		(49,60, 163a)	
Alkaloid LBX $C_{24}H_{31}N_3O_3$ (MW = 449)	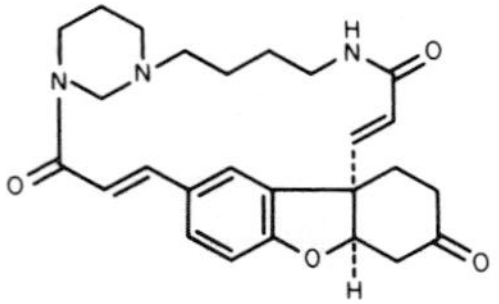 Artifact ? (167) Absolute configuration (166)	Lunaria biennis (Cruciferae) (164,165)	250° (dec.) (164)	+201° (CHCl$_3$) (164)	(EtOH) 225 (4.42) 303 (4.34) 325 (4.30) (164)	(164)	(164)		(164–166)	
Alkaloid LBY (=Lunarinol II) $C_{24}H_{33}N_3O_4$ (MW = 439)	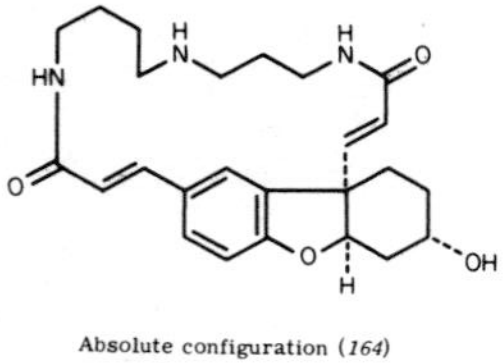 Absolute configuration (164)	Lunaria biennis (Cruciferae) (164,165)	268–273° (165) 270–272° (164)	+108° (EtOH) (165) +152° (EtOH) (164)	(EtOH) 226 (4.40) 298 (4.33) 320 (4.29) (164)	(164)	(164)		(164, 165)	

160

Compound	Source						
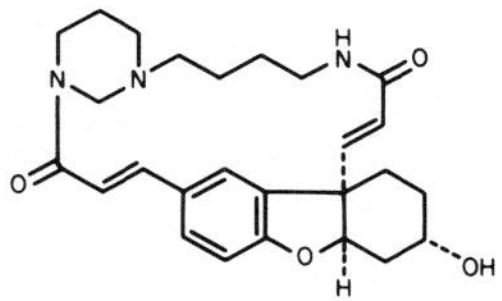 Alkaloid LBZ $C_{26}H_{33}N_3O_4$ (MW = 451) Absolute configuration (166)	*Lunaria biennis* (Cruciferae) (164,165)	amorph (164)		(164)	(164)	(164–166)	
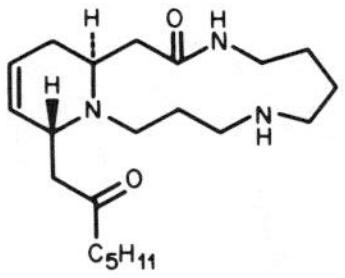 Anhydrocannabis- ativine $C_{21}H_{37}N_3O_2$ (MW = 363) Absolute configuration (82)	*Cannabis sativa* (Moraceae) (82)	+19° (MeOH) (82)	(82)	(82)	(82)	(82)	
1,8-Bis(2,3-di- hydroxybenz- amido)-4-azaoc- tane (compound II) $C_{21}H_{27}N_3O_6$ (MW = 417) 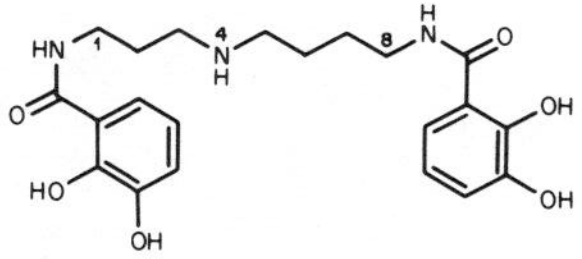	*Paracoccus* (= *Micro- coccus*) *denitrificans* (50)	205–207° (52)		(51, 52, 54, 168)	(50)	(51–54)	
Caffeoylspermidine $C_{16}H_{25}N_3O_3$ (MW = 307)	*Nicotiana tabacum* (So- lanaceae) (148,151)		(151)		(151)		
R=H, R=	*Pennisetum americanum* (Gramineae) (123)						

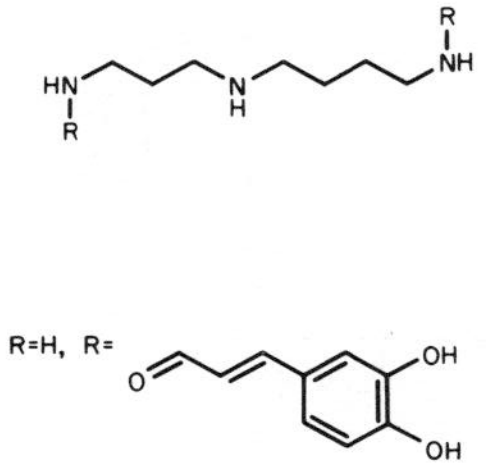

(Continued)

161

TABLE II (*Continued*)

Alkaloid	Structure[a]	Occurrence	mp (°C) (Ref.)	$(\alpha)_D$ (solvent) (Ref.)	UV λ_{max}(log ε) (Ref.)	References				
						IR	NMR	MS	Structure	Synthesis
Cannabisativine C$_{21}$H$_{39}$N$_3$O$_3$ (MW = 381)	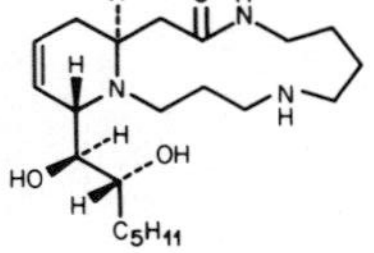	*Cannabis sativa* (Moraceae) (83)	167–168° (83)	+55° (CHCl$_3$) (83)	(81)	(81,83)	(81, 83)	(81,83)	(81,83)	
Celabenzine (Celabenzene) C$_{23}$H$_{29}$N$_3$O$_2$ (MW = 379)	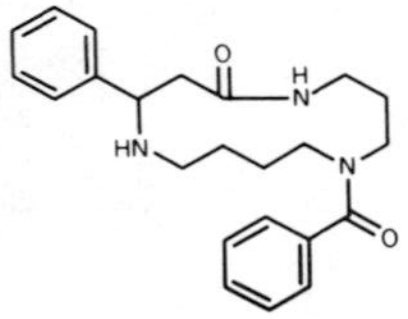	*Maytenus mossambicensis* (Celastraceae) (111) *Tripterygium wilfordii* (Celastraceae) (102)	156–158° (101) 163–167° (111)	0° (CHCl$_3$) (102)	(EtOH) 258(3.14)Inf. 264(2.97)Inf. 268(2.84)Inf. (102)	(102, 112)	(102, 111, 112)	(102, 111, 112)	(101, 102)	(107)
Celacinnine C$_{25}$H$_{31}$N$_3$O$_2$ (MW = 405)	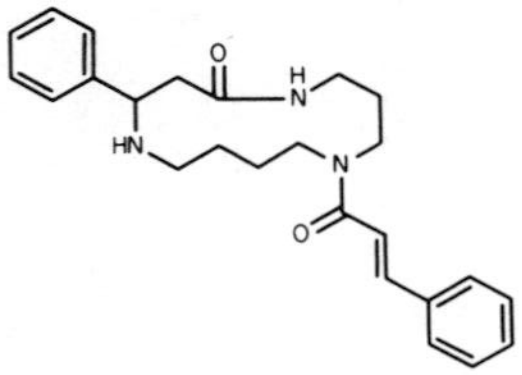	*Maytenus heterophylla* (Celastraceae) (169) *M. serrata* (*M. arbutifolia*) (101,102) *Pleurostylia africana* (Celastraceae) (103,169) *Tripterygium wilfordii* (Celastraceae) (101, 102)	203–204° (102) 201–203° (103)	−19° (CHCl$_3$) (102) −21° (CHCl$_3$) (103)	(MeOH) 223(4.20)Inf. 277(4.36) (101,102) (MeOH) 218 (4.33) 224 (4.25) 278 (4.40) (103)	(101– 103)	(101– 103)	(101– 103)	(101, 102)	(58,104, 107)

162

Name / Formula / MW	Structure	Plant source	mp	$[\alpha]$	UV					Refs
Celafurine $C_{21}H_{27}N_3O_3$ (MW = 369)	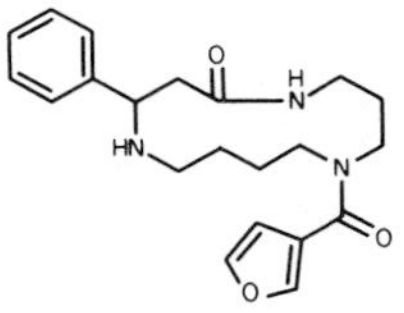	*Tripterygium wilfordii* (Celastraceae) *(102)*	154–155° *(101)*	−11° (CHCl$_3$) *(101)*	(EtOH) 222 (4.00)Inf. 232 (3.84)Inf. 285 (3.09) *(102)*	*(102)*	*(102)*	*(102)*	*(101, 102)*	*(107)*
Celallocinnine $C_{25}H_{31}N_3O_2$ (MW = 405)	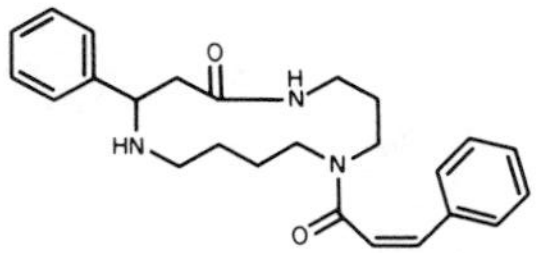	*Maytenus serrata* (*M. arbutifolia*) (Celastraceae) *(101,102)* *Pleurostylia africana* (Celastraceae) *(103,170)*	172–173° *(101)*	−24° (CHCl$_3$) *(101,102)* −22° (CHCl$_3$) *(103)*	(MeOH) 255 (4.07) 264 (3.98)Inf. *(103)*	*(102, 103)*	*(102, 103)*	*(103)*	*(101, 102)*	*(107)*
Codonocarpine $C_{26}H_{31}N_3O_5$ (MW = 465)	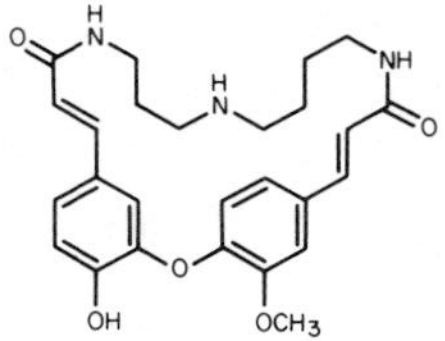	*Codonocarpus australis* (Phytolaccaceae) *(94)*	187° (dec.) *(171)* 212–214° ·HCl *(94)*		(MeOH) 218 (4.42) 283 (4.44) 312 (4.33) *(171)*	*(95, 171)*	*(95, 171)*	*(95, 171)*	*(95,171)*	*(96–99)*
4-Coumaroylspermidine $C_{16}H_{25}N_3O_2$ (MW = 291)	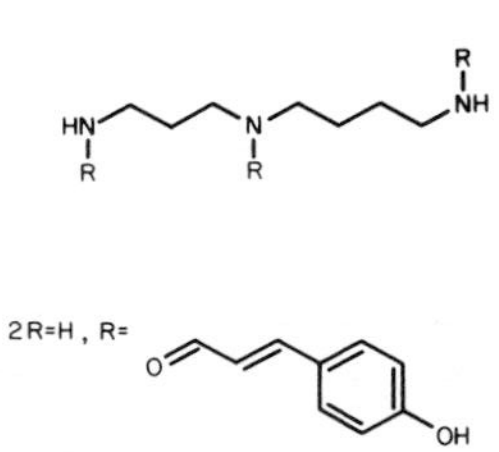	*Salix* sp. (Salicineae) *(123)*							*(123)*	

(Continued)

163

TABLE II (*Continued*)

Alkaloid	Structure[a]	Occurrence	mp (°C) (Ref.)	$(\alpha)_D$ (solvent) (Ref.)	UV λ_{max} (log ε) (Ref.)	References				
						IR	NMR	MS	Structure	Synthesis
Cyclocelabenzine $C_{23}H_{27}N_3O_2$ (MW = 377)	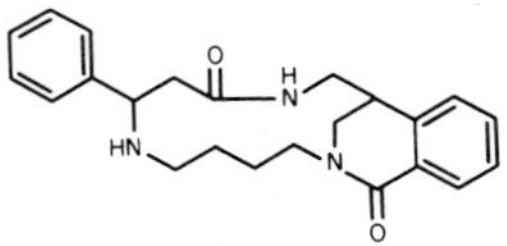	*Maytenus mossambicensis* (Celastraceae) (*112*)	180–183° (*112*)	+30° (CHCl₃) (*112*)	(MeOH) 253 (3.75) 263 (3.65) (*112*)	(*112*)	(*112*)	(*112*)	(*111*, *112*)	
18-Deoxypalustrine $C_{17}H_{31}N_3O$ (MW = 293)	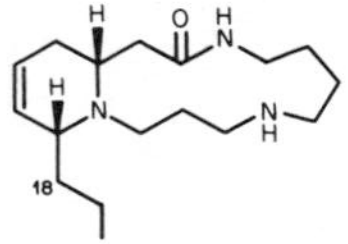	*Equisetum palustre* (Equisetaceae) (*77*)						(*77*)	(*77*)	(*80*)
Dicaffeoylspermidine $C_{25}H_{31}N_3O_6$ (MW = 469)	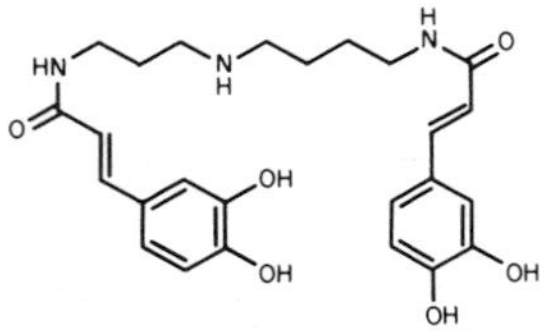	*Aesculus hippocastanum* (Hippocastanaceae) (*123*) *Nicotiana tabacum* (Solanaceae) (*151*) *Salix* sp. (Salicineae) (*123*)			(*151*)		(*151*)	(*151*)	(*151*)	
Di-4-coumaroyl-spermidine $C_{25}H_{31}N_3O_4$ (MW = 437)	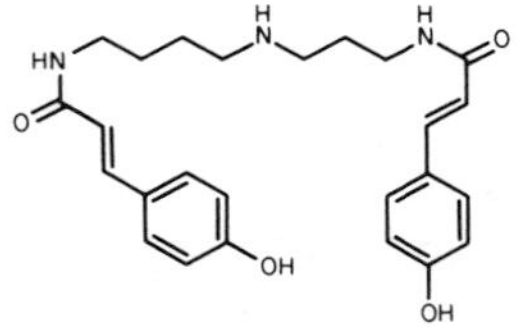	*Aesculus hippocastanum* (Hippocastanaceae) (*123*) *Dianthus caryophyllus* (Caryophyllaceae) (*123*)			235 285 sh 310 (*148*)				(*148*)	

164

Diferuloylspermidine
$C_{27}H_{35}N_3O_6$
(MW = 497)

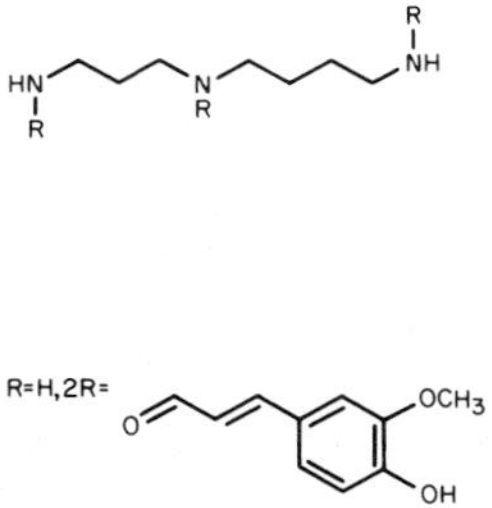

R = H, 2R =

Helianthus annuus (Compositae) (*123*)
Lycopersicum esculentum (Solanaceae) (*123*)
Nicotiana tabacum (Solanaceae) (*148*)
Pyrus communis (Rosaceae) (*123*)
Rubus idaeus (Rosaceae) (*123*)
Vicia faba (Leguminosae) (*123*)

(*123*)

Ananas comosus (Bromeliaceae) (*123*)
Dianthus caryophyllus (Caryophyllaceae) (*123*)
Gomphrena globosa (Amaranthaceae) (*123*)
Lycopersicum esculentum (Solanaceae) (*123*)
Vicia faba (Leguminosae) (*123*)

(Continued)

165

TABLE II (*Continued*)

Alkaloid	Structure[a]	Occurrence	mp (°C) (Ref.)	$[\alpha]_D$ (solvent) (Ref.)	UV λ_{max}(log ϵ) (Ref.)	References				
						IR	NMR	MS	Structure	Synthesis
Dihydroperiphylline $C_{25}H_{31}N_3O_2$ (MW = 405)	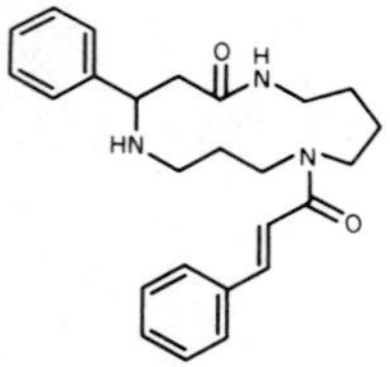	*Peripterygia marginata* (Celastraceae) (*114*)		−21° (CHCl₃) (*114*)	(EtOH) 219 225 280 300 sh. (*114*)	(*114*)	(*114*)	(*114*)	(*114*)	(*116*)
E-5,12-Dimethyl-1-dimethyl-amino-5,9-diazaheneicos-11-en-10-one [Cytotoxic spermidine derivative **1**, (*61*)] $C_{23}H_{47}N_3O$ (MW = 381)	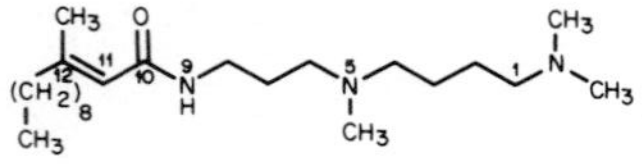	*Sinularia brongersmai* (soft coral) (*61*)	oil (*61*)			(*61*)	(*61*)	(*61*)	(*61*)	(*62*)
5,12-Dimethyl-1-dimethylamino-5,9-diazahenei-cosan-10-one [Cytotoxic spermidine derivative **2**, (*61*)] $C_{23}H_{49}N_3O$ (MW = 383)	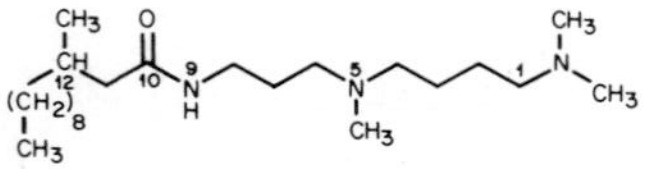	*Sinularia brongersmai* (soft coral) (*61*)	oil (*61*)			(*61*)	(*61*)	(*61*)	(*61*)	(*62*)

Disinapoylspermidine
$C_{29}H_{39}N_3O_8$
(MW = 557)

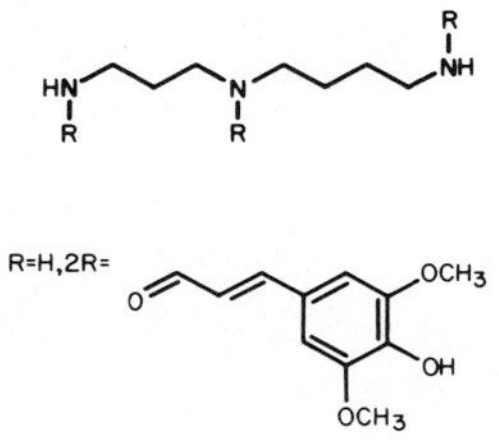

R=H, 2R=

Lilium sp.
(Liliaceae)
(123)

(123)

Feruloylspermidine
$C_{17}H_{27}N_3O_3$
(MW = 321)

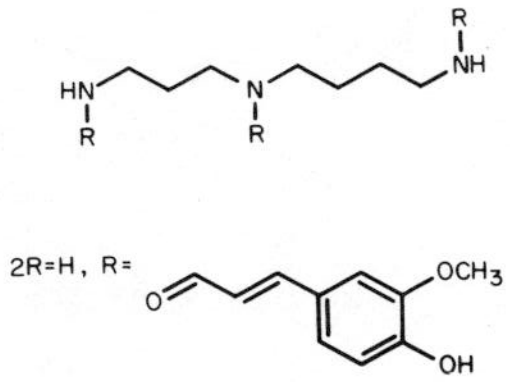

2R=H, R=

Ananas comosus
(Bromeliaceae)
(123)
Lycopersicum
esculentum
(Solanaceae)
(123)

(123)

Hydroxyisocyclocelabenzine
$C_{23}H_{27}N_3O_3$
(MW = 393)

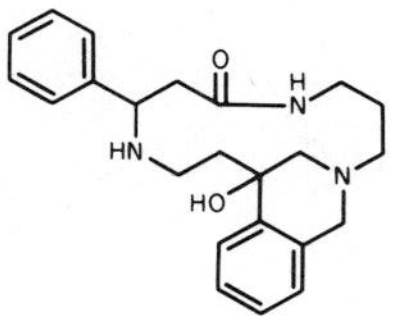

Maytenus
mossambicensis
(Celastraceae)
(112)

135° (112)   + 113° (CHCl₃) (112)   (MeOH)
256 (3.69)sh
265 (3.62)sh (112)

(112)   (112)   (112)

(112)

(Continued)

167

TABLE II (Continued)

Alkaloid	Structure[a]	Occurrence	mp (°C) (Ref.)	$(\alpha)_D$ (solvent) (Ref.)	UV λ_{max}(log ϵ) (Ref.)	References				
						IR	NMR	MS	Structure	Synthesis
Inandenin-10,11-diol $C_{23}H_{47}N_3O_3$ (MW = 413)	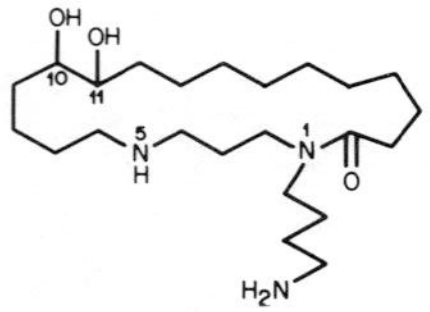	*Oncinotis nitida* (Apocynaceae) (26)	oil (26)					(26)	(26)	
Inandenin-10-one $C_{23}H_{45}N_3O_2$ (MW = 395)	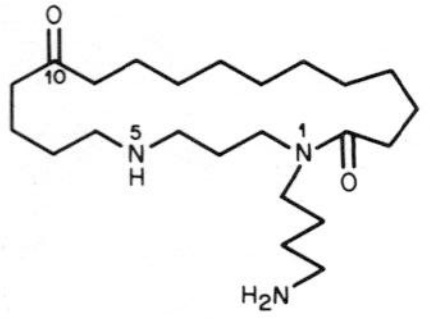	*Oncinotis nitida* (Apocynaceae) (26)	oil (26)					(26)	(26)	
Inandenin-12-one	Until now the mixture of the two alkaloids could not be separated.									
Inandenin-13-one $C_{23}H_{45}N_3O_2$ (MW = 395)	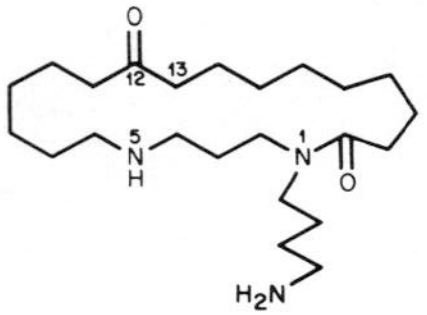	*Oncinotis inandensis* (Apocynaceae) (73) *O. nitida* (172)	150–151° ·HCl (73)			(73)	(73)	(73, 172)	(73, 173)	

	Source	mp	$[\alpha]$	UV λ_{max} nm (log ε)				
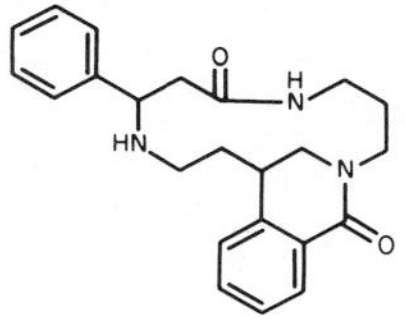 Isocyclocelabenzine C₂₃H₂₇N₃O₂ (MW = 377)	*Maytenus mossambicensis* (Celastraceae) (*112*)	227–228° (*112*)	+139° (CHCl₃) (*112*)	(MeOH) 253 (3.77) 264 (3.69) (*112*)	(*112*)	(*112*)	(*112*)	(*111, 112*)
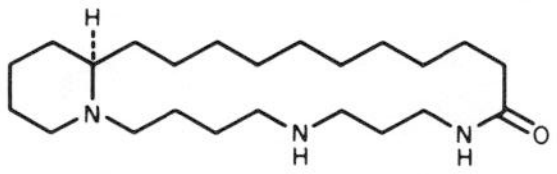 Isooncinotine C₂₃H₄₅N₃O (MW = 379) Absolute configuration (*63*)	*Oncinotis nitida* (Apocynaceae) (*63*)	66–71° (*63*)	−37° (MeOH) (*63*)		(*63*)	(*63*)	(*63*)	(*63,66*)
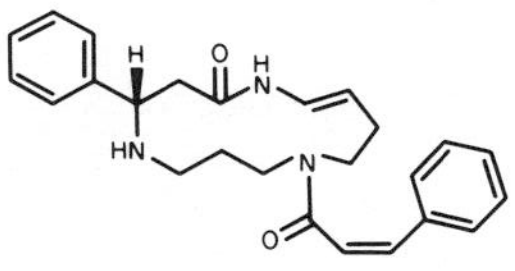 Isoperiphylline C₂₅H₂₉N₃O₂ (MW = 403) Absolute configuration (*114*)	*Peripterygia marginata* (Celastraceae) (*114*)	197° (*114*)	−120° (CHCl₃) (*114*)	(EtOH) 208 (4.64) 245 (4.47) (*114*)	(*114*)	(*114*)	(*114*)	(*114*)
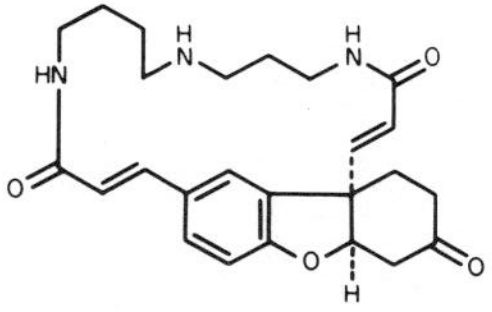 Lunaridine C₂₅H₃₁N₃O₄ (MW = 437) Absolute configuration (*175*)	*Lunaria biennis* (Cruciferae) (*174*)	>265° (dec.) (*175*)	+267° (EtOH) (*175*)	(EtOH) 225 (4.33) 290 (4.22) 297 (4.26) 317 (4.16) (*175*)	(*175*)	(*175*)	(*175*)	(*92,176, 177*)

(Continued)

169

TABLE II (*Continued*)

Alkaloid	Structure[a]	Occurrence	mp (°C) (Ref.)	$(\alpha)_D$ (solvent) (Ref.)	UV λ_{max} (log ε) (Ref.)	References				Synthesis
						IR	NMR	MS	Structure	
Lunarine $C_{25}H_{31}N_3O_4$ (MW = 437)	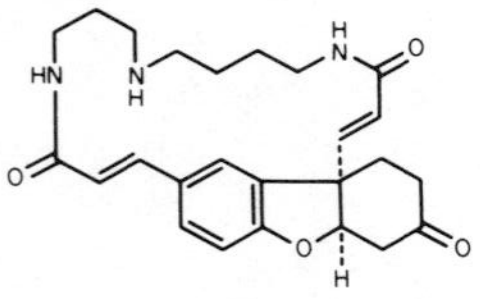 Absolute configuration (*86*)	*Lunaria biennis* (Cruciferae) (*164, 178–182*) *L. rediviva* (*178*	238–240° (*164*)	+344° (EtOH) (*164*)	(EtOH) 226 (4.41) 296 (4.35) 317 (4.25) (*164*)	(*164*)	(*164*)	(*175*)	(*84–86, 164, 175*)	(*92,176*)
Mayfoline $C_{16}H_{25}N_3O_2$ (MW = 291)	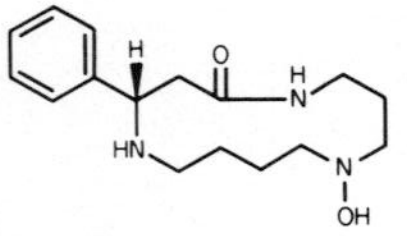 Absolute configuration (*108*)	*Maytenus buxifolia* (Celastraceae) (*108*)	200–204° (*108*)	+11° (CHCl₃) (*108*)	(EtOH) 251 (2.33) 257 (2.39) 261.5[2.33] 264 (2.32)sh 267.5[2.21] (*108*)	(*108*)	(*108*)	(*108*)	(*108*)	
Maytenine $C_{25}H_{31}N_3O_2$ (MW = 405)	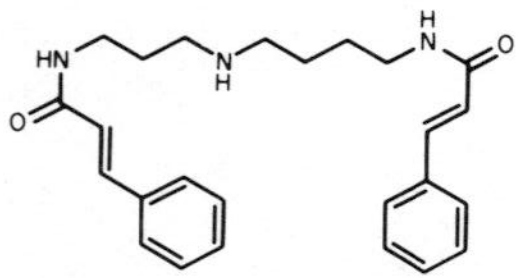	*Maytenus chuchuhuasha* (Celastraceae) (*48*)	158° (*48*)		215 (4.55) 221 (4.48) 273 4.70) (*48*)		(*48*)	(*48*)	(*48*)	(*55–58*)
N-Methylcodonocarpine $C_{27}H_{33}N_3O_5$ (MW = 479)	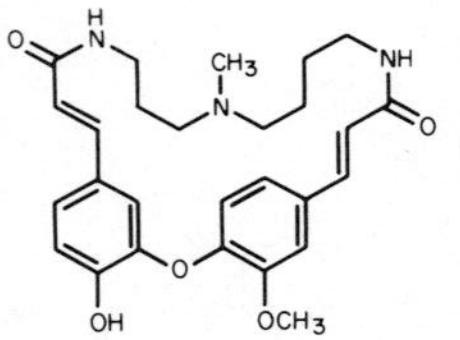	*Codonocarpus australis* (Phytolaccaceae) (*94*)	164–166° (*94*)		(MeOH) 218 283 310 (*94*)	(*94*)	(*94*)	(*95*)	(*94,95*)	(*94,95*)

170

Neooncinotine					

Neooncinotine
$C_{23}H_{45}N_3O$
(MW = 379)

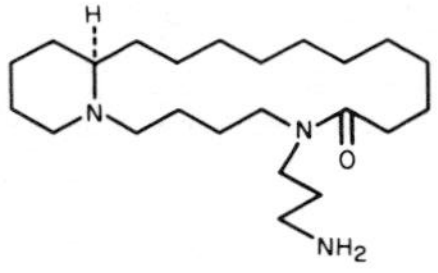

Absolute configuration (63)

Oncinotis nitida
(Apocynaceae)
(63)

(63) (63) (66)

Neoperiphylline
$C_{25}H_{29}N_3O_2$
(MW = 403)

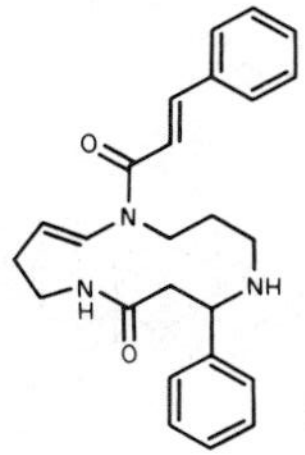

Periptygia marginata (Celastraceae) (114)

amorph (114)

$-34°$ (CHCl$_3$) (114)

(114) (114)

Numismine
$C_{25}H_{33}N_3O_4$
(MW = 439)

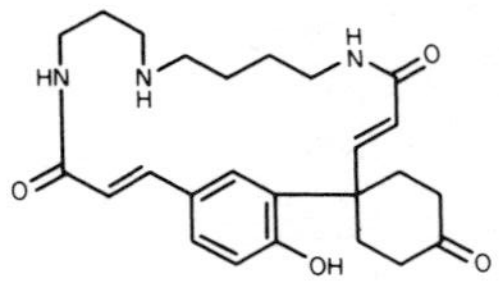

Lunaria biennis
(Cruciferae)
(164)

$>270°$
(dec.)
(164)

0° (164)

(EtOH)
209 (4.44)
232 (4.40)
300 (4.40)
315 (4.32) (164)

(164) (164) (164)

(Continued)

171

TABLE II (Continued)

Alkaloid	Structure[a]	Occurrence	mp (°C) (Ref.)	$[\alpha]_D$ (solvent) (Ref.)	UV λ_{max}(log ϵ) (Ref.)	IR	NMR	MS	Structure	Synthesis
								References		
Oncinotine $C_{23}H_{45}N_3O$ (MW = 379)	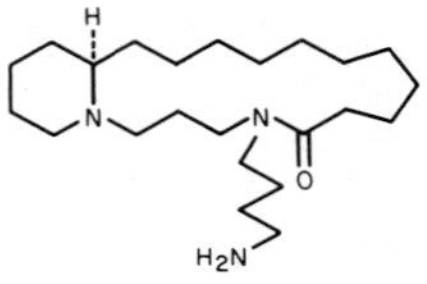 Absolute configuration (63)	*Oncinotis nitida* (Apocynaceae) (183)	oil (63)	−33° (MeOH) (63)	(EtOH) 208 (3.79) (63)	(63)		(63, 172)	(63,183)	(65,66)
Palustridine $C_{18}H_{31}N_3O_3$ (MW = 337)	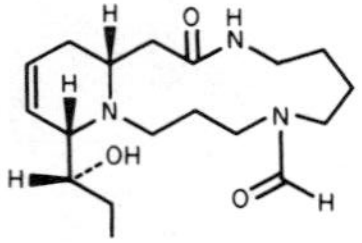 Absolute configuration (186)	*Equisetum palustre* (Equisetaceae) (75)	204° (dec.) ·HCl (184)	·HCl + 50° (H_2O) (184)		(184)	(184)	(184)	(184, 185)	(184)
Palustrine $C_{17}H_{31}N_3O_2$ (MW = 309)	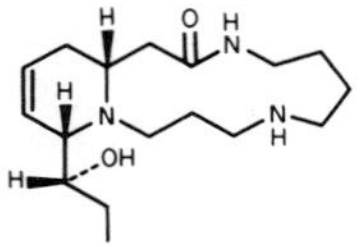 Absolute configuration (186)	*Equisetum arvense* (Equisetaceae) (75) *E. limosum* (75) *E. palustre* (75) *E. ramosissimum* (187) *E. silvaticum* (75)	121° (185) 150–152° ·1 HCl (185) 188–190° (dec.) ·2 HCl (185)	+16° (H_2O) (185) +19° (EtOH) (185)		(185)	(188)	(184)	(77,185, 188a)	(79,189)

172

Compound	Structure	Source	mp / [α]	UV (EtOH)				
Parabactin $C_{32}H_{36}N_4O_9$ (MW = 620)	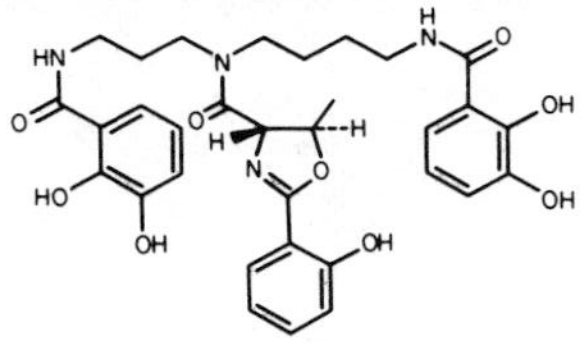 Relative configuration (60)	*Paracoccus denitrificans* (50,60)	114–117° (60)	(EtOH) 250 (4.43) 309 (4.01) (59)	(60)	(60)	(50,59, 60)	
Perimargine Dihydroperimargine (MW = 423 + 425)	Until now the mixture of the two alkaloids could not be separated. Partial structure: 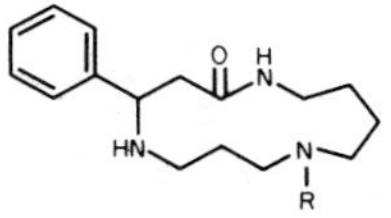 Perimargine R = 149 mass units Dihydroperimargine R = 151 mass units	*Peripterygia marginata* (Celastraceae) (114)		(EtOH) 206 270 (114)	(114)	(114)	(114)	
Periphylline $C_{25}H_{29}N_3O_2$ (MW = 403)	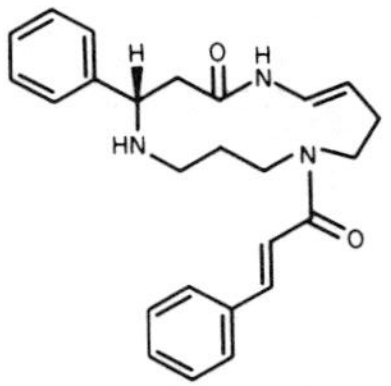 Absolute configuration (114)	*Peripterygia marginata* (Celastraceae) (113,114)	165° (114) −291° (CHCl$_3$) (114)	(EtOH) 219 (5.47) 225 (5.43) 275 (5.75) (114)	(114)	(114)	(113, 114)	(113, 114)

(Continued)

173

TABLE II (*Continued*)

Alkaloid	Structure[a]	Occurrence	mp (°C) (Ref.)	$(\alpha)_D$ (solvent) (Ref.)	UV λ_{max}(log ϵ) (Ref.)	References				
						IR	NMR	MS	Structure	Synthesis
Pleurostyline $C_{25}H_{29}N_3O_2$ (MW = 403)	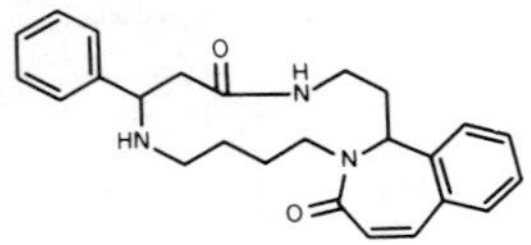	*Pleurostylia africana* (Celastraceae) (*69, 103.170*)	246–247° (*103*)	−194° (CHCl$_3$) (*103*)	(MeOH) 267 (4.07) (*103*)	(*103*)	(*103*)	(*103*)	(*103*)	
Sinapoylspermidine $C_{18}H_{29}N_3O_4$ (MW = 351)	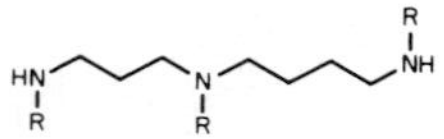 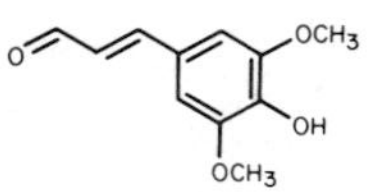	*Lilium* sp. (Liliaceae) (*123*)							(*123*)	
N,N',N''-Tri-4-coumaroylspermidine $C_{34}H_{37}N_3O_6$ (MW = 583)	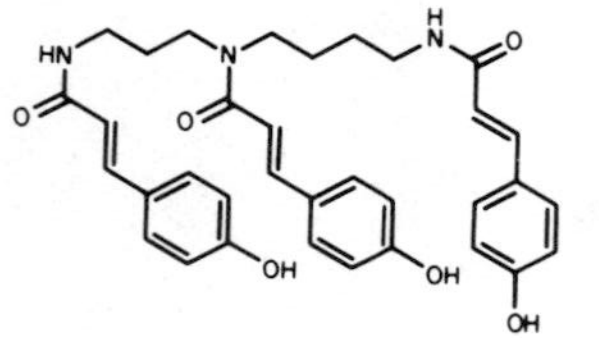	*Crataegus* (Rosaceae) (*190*)		0° (*190*)					(*190*)	(*190*)

Homospermidine Alkaloids

Alkaloid	Structure[a]	Occurrence	mp (°C) (Ref.)				IR	NMR	MS	Structure	Synthesis
Acarmidines (MW = 464) (MW = 466) (MW = 488)	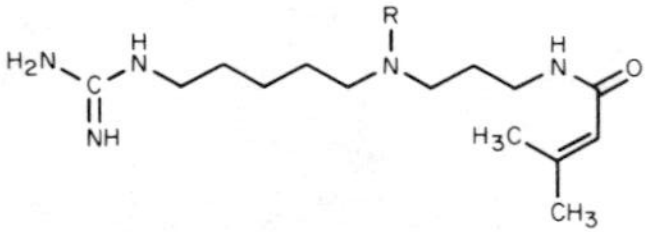 Until now the mixture of the three alkaloids could not be separated.	*Acarnus erithacus* (sponge) (*117*)	oil (*117*)				(*117*)	(*117*)	(*117*)	(*117*)	(*122*)

R = CO—(CH$_2$)$_{10}$—CH$_3$

R = CO—(CH$_2$)$_5$—CH=CH(CH$_2$)$_4$—CH$_3$ (Z)

R = CO—(CH$_2$)$_5$—CH=CHCH$_2$)$_3$—CH$_3$

Name	Structure	Source	Derivative	References
Solacaproine $C_{18}H_{39}N_3O$ (MW = 313)	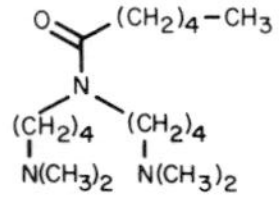 $O{=}C{-}(CH_2)_4{-}CH_3$, N with $(CH_2)_4 N(CH_3)_2$ and $N(CH_3)_2 (CH_2)_4$	Cyphomandra betaceae (Solanaceae) (120)	149–150° picrate (120)	(120) (120) (120)
Solamine $C_{12}H_{29}N_3$ (MW = 215)	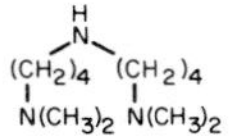 $H{-}N$ with $(CH_2)_4 N(CH_3)_2$ and $N(CH_3)_2 (CH_2)_4$	Cyphomandra betacea (Solanaceae) (120) Solanum carolinense (Solanaceae) (121)	113–114° picrate (120) 151–152° picrate (121)	(118, 120) (118, 120) (118, 120) (118, 120, 121)
Solapalmitenine $C_{28}H_{57}N_3O$ (MW = 451)	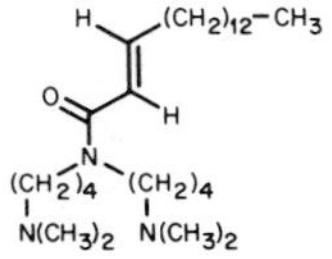 $H{-}C{=}C(H){-}(CH_2)_{12}{-}CH_3$, $O{=}C{-}N$ with $(CH_2)_4 N(CH_3)_2$ and $N(CH_3)_2 (CH_2)_4$	Solanum tripartitum (Solanaceae) (118)		(118, 119) (118, 119) (118, 119) (118)
Solapalmitiine $C_{28}H_{59}N_3O$ (MW = 453)	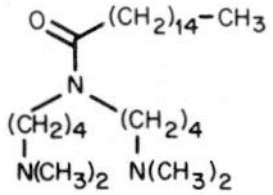 $O{=}C{-}(CH_2)_{14}{-}CH_3$, N with $(CH_2)_4 N(CH_3)_2$ and $N(CH_3)_2 (CH_2)_4$	Solanum tripartitum (Solanaceae) (118)		(118, 119) (118, 119) (118, 119) (118)
Solaurethine $C_{15}H_{33}N_3O_2$ (MW = 287)	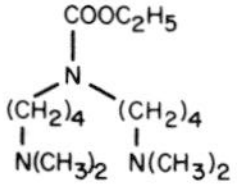 $COOC_2H_5$, N with $(CH_2)_4 N(CH_3)_2$ and $N(CH_3)_2 (CH_2)_4$	Solanum carolinense (Solanaceae) (121)	122° picrate (121)	(121) (121) (121) (121)

(Continued)

TABLE II (*Continued*)

Alkaloid	Structure[a]	Occurrence	mp (°C) (Ref.)	$(\alpha)_D$ (solvent) (Ref.)	UV $\lambda_{max}(\log \epsilon)$ (Ref.)	References				
						IR	NMR	MS	Structure	Synthesis
Spermine Alkaloids										
Aphelandrine $C_{28}H_{36}N_4O_4$ (MW = 492)	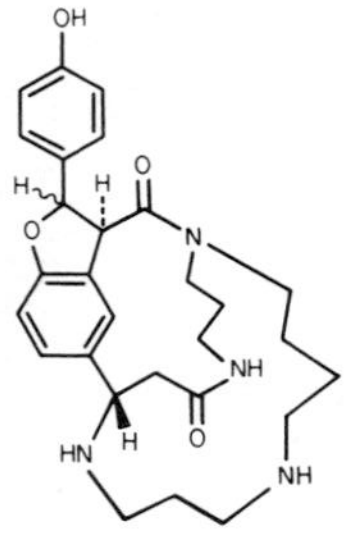	*Aphelandra aurantiaca* (Acanthaceae) (*130*)	263° (dec.) (*127*)	+230° (MeOH) (*127*) +214° (CHCl₃/ MeOH = 9:1) (*127*)	(EtOH) 228 (4.23) 277 (3.46)sh 283 (3.51) 291 (3.38)sh (*127*)	(*127*)	(*127*, *128*)	(*127*)	(*127*, *128*)	
		A. aurantiaca var. *nitens* (*130*)								
		A. chamisson-iana (*130*)								
		A. pepe-parodii (*130*)								
		A. sinclairiana (*130*)								
		A. squarrosa (*127*)								
		A. tetragona (*130*)								
		Encephalo-sphaera lasian-dra (Acanthaceae) (*130*)								
		Premna integri-folia (Verbenaceae) (*191*)								

Absolute configuration (*128*)
Correlation: *O*-Methylorantine (*128*)
            Orantine (*128*)

Compound	Structure	Source	mp	$[\alpha]$	UV λ_{max} nm (log ε)	Ref.
Chaenorhine $C_{31}H_{40}N_4O_5$ (MW = 548)	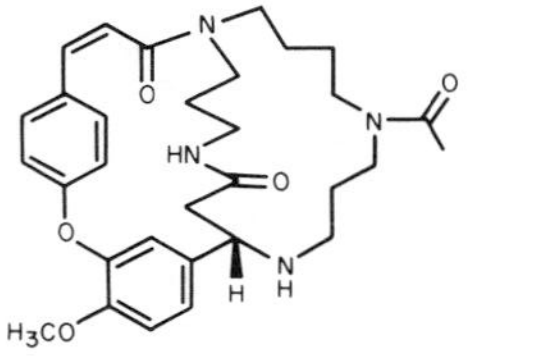 Absolute configuration (126)	Chaenorhinum crassifolium (Scrophulariaceae) (192) C. glareosum (192) C. grandiflorum (192) C. nevadense (192) C. nummulariifolium (192) C. origanifolium (126) C. rubrifolium (131)	263–268° (dec.) (126)	+47° (CHCl₃/MeOH = 9:1) (126)	(MeOH) 264(4.00) 280(3.84)Inf. (126)	(126) (126) (126) (126, 172) (126) (167)
Diferuloylspermine $C_{30}H_{42}N_4O_6$ (MW = 554)	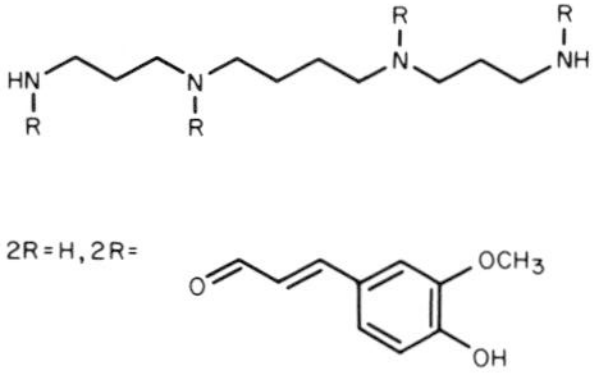 2R=H, 2R=	Ananas comosus (Bromeliaceae) (123) Gomphrena globosa (Amarantha-ceae) (123)				(123)
Disinapoylspermine $C_{32}H_{46}N_4O_8$ (MW = 614)	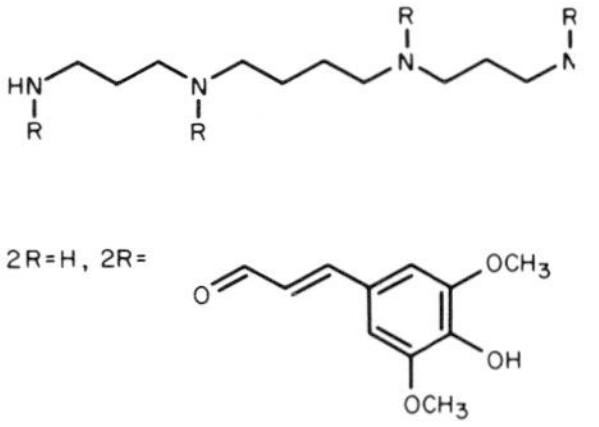 2R=H, 2R=	Lilium sp. (Liliaceae) (123)				(123)

(Continued)

TABLE II (*Continued*)

Alkaloid	Structure[a]	Occurrence	mp (°C) (Ref.)	$(\alpha)_D$ (solvent) (Ref.)	UV λ_{max} (log ε) (Ref.)	References				
						IR	NMR	MS	Structure	Synthesis
Ephedradine B $C_{29}H_{38}N_4O_5$ (MW = 522)	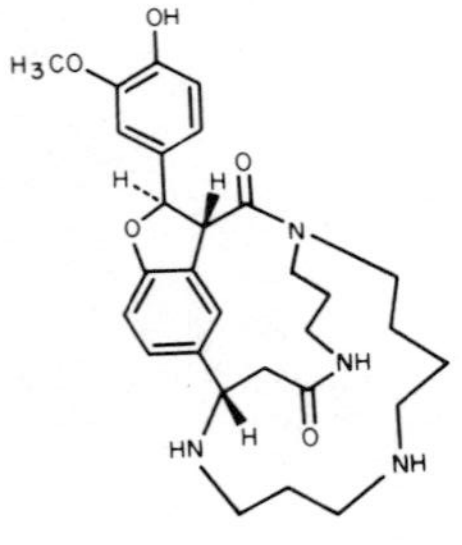	maō-kon (prepared from *Ephedra* roots) (Ephedraceae) (*193*)	219–221° ·2 HBr (*193*)	−102° (H₂O) (*193*)	·2 HBr (MeOH) 232 (4.32) 281 (3.92) (*193*)		(*193*)		(*193*)	
Ephedradine C $C_{30}H_{40}N_4O_5$ (MW = 536)	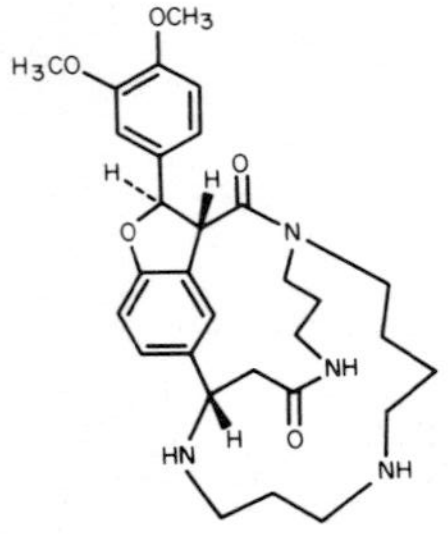	maō-kon (prepared from *Ephedra* roots) (Ephedraceae) (*132*)	224–225° ·2 HBr (*132*)	−101° (H₂O) (*132*)	·2 HBr (MeOH) 231 (4.16) 280 (3.62) (*132*)		(*132*)		(*132*)	(*132*)

Absolute configuration (*193*)

Absolute configuration (*132*)

Compound	Source	mp	$[\alpha]_D$	UV	Ref.			
Ephedradine D $C_{29}H_{38}N_4O_5$ (MW = 522)	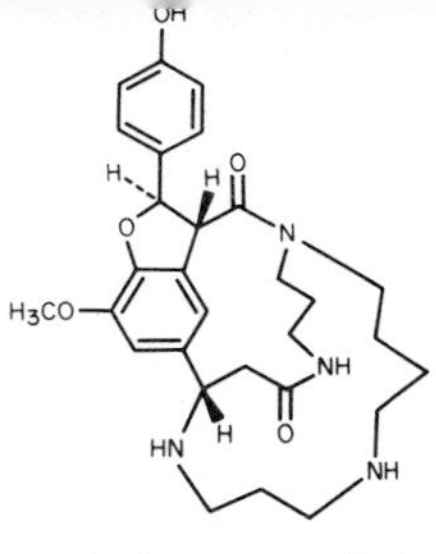 maô-kon (prepared from *Ephedra* roots) (Ephedraceae) *(133)*	219–221° *(133)*	−85.3° (H₂O) *(133)*	(MeOH) 228 283 *(133)*	*(133)*	*(133)*	*(133)*	*(133)*
Homaline $C_{30}H_{42}N_4O_2$ (MW = 490) 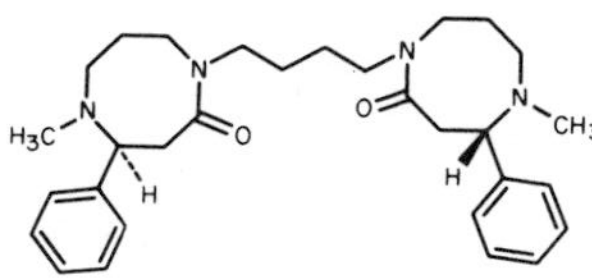	*Homalium pronyense* (Flacourtia- ceae) *(134,136)*	134° *(134)*	−34° (CHCl₃) *(134)*		*(134, 136)*	*(134, 136)*	*(136)*	
Hopromalinol $C_{31}H_{52}N_4O_3$ (MW = 528) 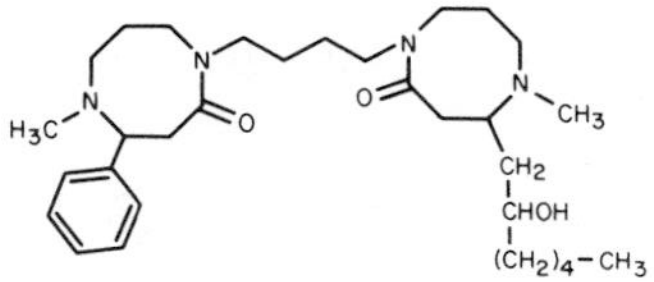	*Homalium pronyense* (Flacourtia- ceae) *(136)*	amorph *(136)*	−17° (CHCl₃) *(136)*		*(136)*	*(136)*	*(136)*	
Hopromine $C_{30}H_{58}N_4O_2$ (MW = 506) 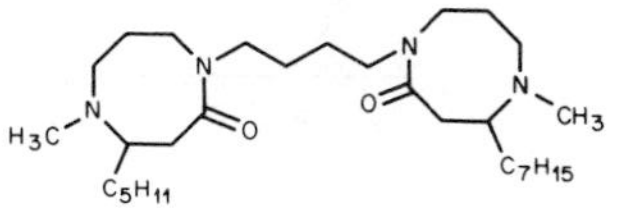	*Homalium pronyense* (Flacourtia- ceae) *(136)*	amorph *(136)*	−10° (CHCl₃) *(136)*		*(136)*	*(136)*	*(136)*	

Absolute configuration *(133)*

Absolute configuration *(136)*

*(135–
137)*

*(137–
139)*

(Continued)

179

TABLE II (*Continued*)

Alkaloid	Structure[a]	Occurrence	mp (°C) (Ref.)	$[\alpha]_D$ (solvent) (Ref.)	UV λ_{max}(log ϵ) (Ref.)	References				
						IR	NMR	MS	Structure	Synthesis
Hoprominol $C_{30}H_{58}N_4O_3$ (MW = 522)	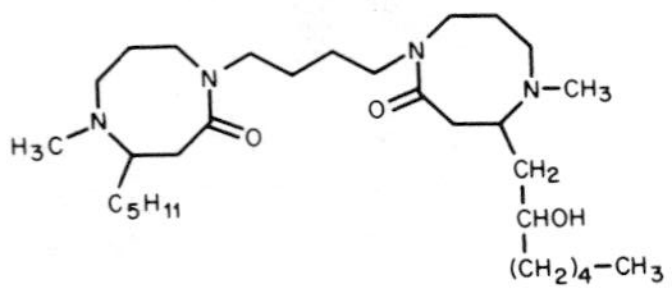	*Homalium pronyense* (Flacourtiaceae) (*136*)	amorph (*136*)	−19° (CHCl₃) (*136*)		(*136*)	(*136*)	(*136*)	(*136*)	
Kukoamine A $C_{28}H_{42}N_4O_6$ (MW = 530)	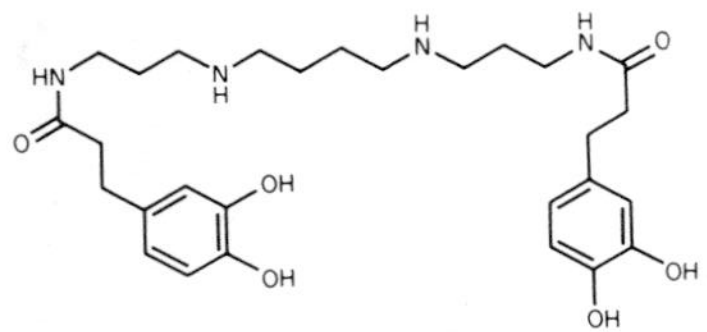	*Lycium chinense* (Solanaceae) (*124*)			284 (*124*)	(*124*)	(*124*)	(*124*)	(*124*)	(*125*)
O-Methylorantine $C_{29}H_{38}N_4O_4$ (MW = 506)	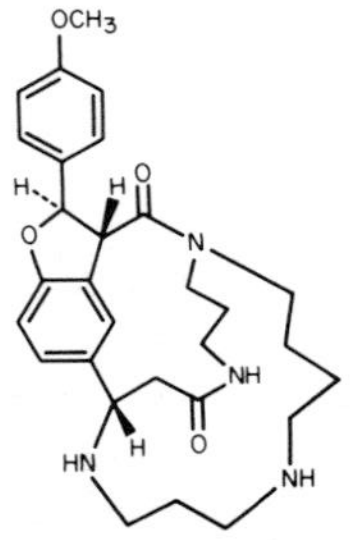	*Aphelandra sinclairiana* (Acanthaceae) (*130*) *Chaenorhinum minus* (Scrophulariaceae) (*131*) *C. villosum* (*128*)	167° (dec.) (*128, 129*)	−84° (CHCl₃/ MeOH = 9 : 1) (*128*)	(EtOH) 229 (4.19) 247 (3.00)sh 253 (2.92) 260 (3.04) sh 275 (3.48) sh 280 (3.54) 288 (3.37) (*128*)	(*128*)	(*128*)	(*128*)	(*128*)	

Absolute configuration (*128*)
Correlation: Aphelandrine (*128*)
Orantine (*128*)

180

Orantine
(=Ephedradine
A) (128)
$C_{28}H_{36}N_4O_4$
(MW = 492)

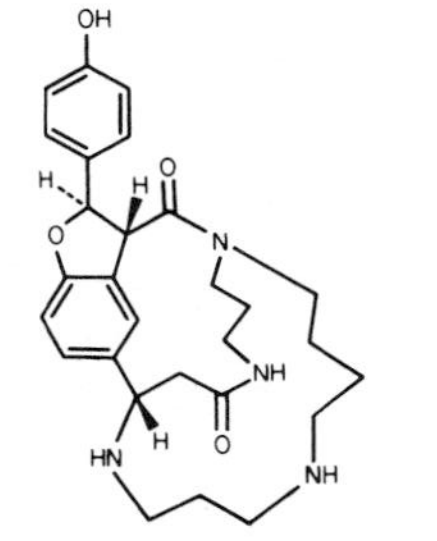

Absolute configuration (129)
Correlation: Aphelandrine (128)
O-Methylorantine (128)

mao (prepared
from aerial
parts of
Ephedra plants
(Ephedraceae)
(129)

166° (dec.)
(129)
222–225°
·2 HCl
(129)
233–235°
·2 HBr
(129)

− 72° (CHCl₃/
MeOH = 9:1) (128)

(EtOH)
228 (4.24)
281 (3.54)
288 (3.40)sh (128)
·2 HCl
(EtOH/H₂O = 5:2)
229 (4.33)
282 (3.62) (128,193)

(128)   (128,   (128)   (128,   (128,
        129,   129)   167)
        193)

Pithecolobines
$C_{22}H_{46}N_4O$
mixture of
(1) m = 3,
    n = 6
(2) m = 1,
    n = 8
(MW = 382)
$C_{24}H_{50}N_4O$
(3) m = 1,
    n = 10
(MW = 410)

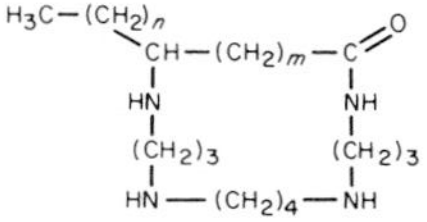

Samanea saman
(Leguminosae)
(141,194)

(90)   (90)   (90)   (91)

(Continued)

TABLE II (*Continued*)

Alkaloid	Structure[a]	Occurrence	mp (°C) (Ref.)	$(\alpha)_D$ (solvent) (Ref.)	UV λ_{max}(log ϵ) (Ref.)	References				
						IR	NMR	MS	Structure	Synthesis
Sinapoylspermine $C_{21}H_{36}N_4O_4$ (MW = 408)	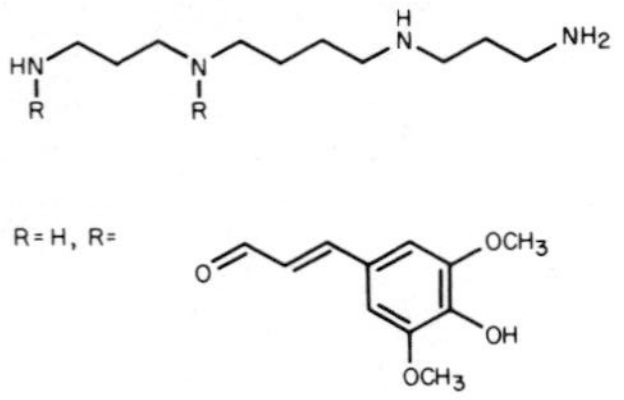	*Brassica oleracea* var. *botrytis* (Cruciferae) (123)								(123)
Verbascenine $C_{30}H_{40}N_4O_3$ (MW = 504)	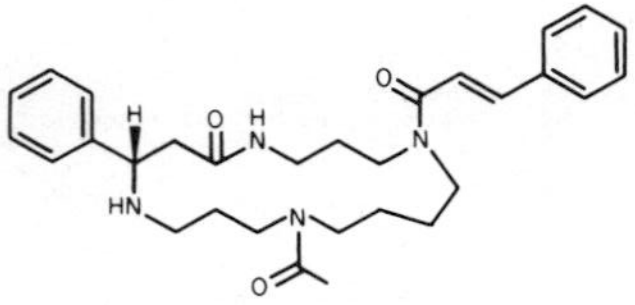	*Verbascum phoeniceum* (Scrophulariaceae) (140) *V. nigrum* (140)	amorph (140)	−15° (MeOH) (140)	(EtOH) 217 (4.33)Inf. 223 (4.38)Inf. 280 (4.27) 299 (4.09)Inf. (140)	(140)	(140)	(140)	(140)	

[a] The number and structure of R groups are defined but their positions in the molecule are interchangeable.

Acknowledgments

We are thankful to Schweizerischer Nationalfonds zur Förderung der wissenschaftlichen Forschung for supporting this work and Dr. Robert P. Borris for linguistic corrections.

REFERENCES

1. D. T. Dubin, *Biochem. Biophys. Res. Commun.* **1**, 262 (1959).
2. H. Tabor and C. W. Tabor, *J. Biol. Chem.* **250**, 2648 (1975).
3. C. W. Tabor and H. Tabor, *Biochem. Biophys. Res. Commun.* **41**, 232 (1966).
4. T. Shiba and T. Kaneko, *J. Biol. Chem.* **244**, 6006 (1969).
5. T. Takita, Y. Muraoka, T. Yoshioka, A. Fujii, K. Maeta, and H. Umezawa, *J. Antibiot.* **26**, 755 (1973).
6. T. Miyaki, K. Numata, Y. Nishiyama, O. Tenmyo, M. Hatori, H. Imanishi, M. Konishi, and H. Kawaguchi, *J. Antibiot.* **34**, 665 (1981).
7. T. A. Smith, *Prog. Phytochem.* **4**, 27 (1977).
8. T. A. Smith, *in* "The Biochemistry of Plants" (E. E. Conn, ed.), Vol. 7, p. 249. Academic Press, New York, 1981.
9. A. Stoessl, *Phytochemistry* **4**, 973 (1965).
10. T. A. Smith, *in* "Polyamines in Biology and Medicine" (D. R. Morris and L. J. Marton, eds.), p. 77. Dekker, New York, 1981.
11. U. Bachrach, "Function of Naturally Occurring Polyamines." Academic Press, New York, 1973.
12. C. W. Tabor and H. Tabor, *Annu. Rev. Biochem.* **45**, 285 (1976).
13. S. S. Cohen, "Introduction to the Polyamines." Prentice-Hall, Englewood Cliffs, New Jersey, 1971.
14. D. H. Russell, "Polyamines in Normal and Neoplastic Growth." Raven Press, New York, 1973.
15. D. H. Russell and B. G. M. Durie, "Polyamines as Biochemical Markers of Normal and Malignant Growth." Raven Press, New York, 1978.
16. R. A. Campbell, D. R. Morris, D. Bartos, G. D. Daves, and F. Bartos, "Advances in Polyamine Research," Vols. 1 and 2. Raven Press, New York, 1978.
17. V. R. Villanueva, R. C. Adlakha, and R. Clavayrac, *Phytochemistry* **19**, 787 (1980).
18. T. A. Smith, *Phytochemistry* **14**, 865 (1975).
19. S. Ramakrishna and P. R. Adiga, *Phytochemistry* **12**, 2691 (1973).
20. Y. Robin, C. Audit, and M. Landon, *Comp. Biochem. Physiol.* **22**, 787 (1967).
21. Y. Robin and N. van Thoai, *C. R. Hebd. Seances Acad. Sci., Ser. C* **252**, 1224 (1961).
22. T. Oshima, *Biochem. Biophys. Res. Commun.* **63**, 1093 (1975).
23. T. Oshima, *J. Biol. Chem.* **254**, 8720 (1979).
24. A. Guggisberg, R. W. Gray, and M. Hesse, *Helv. Chim. Acta* **60**, 112 (1977).
25. K. Chantrapromma, J. S. McManis, and B. Ganem, *Tetrahedron Lett.* 2475 (1980).
26. M. M. Badawi, K. Bernauer, P. van den Broek, D. Gröger, A. Guggisberg, S. Johne, I. Kompiš, F. Schneider, H.-J. Veith, M. Hesse, and H. Schmid, *Pure Appl. Chem.* **33**, 81 (1973).
27. M. Hesse and H. Schmid, *Int. Rev. Sci.: Org. Chem., Ser. Two* **9**, 265 (1976).
28. M. Hesse and H. Schmid, *Bulg. Acad. Sci., Comm. Dep. Chem.* **5**, 279 (1972).
29. E. Fujita, *Yuki Gosei Kagaku Kyokaishi* **38**, 333 (1980).
30. E. Fujita, *Pure Appl. Chem.* **53**, 1141 (1981).
31. H. Matsuyama, *Yuki Gosei Kagaku Kyokaishi* **39**, 1151 (1981).

32. H. H. Wasserman and J. S. Wu, *Heterocycles* **17**, 581 (1982).
33. H. Kühne and M. Hesse, *Helv. Chim. Acta* **65**, 1470 (1982).
34. H. Bosshardt and M. Hesse, *Angew. Chem.* **86**, 256 (1974); *Angew, Chem., Int. Ed. Engl.* **13**, 252 (1974).
35. F. G. Fischer and H. Bohn, *Justus Liebigs Ann. Chem.* **603**, 232 (1957).
36. C. M. Gilbo and N. W. Coles, *Aust. J. Biol. Sci.* **17**, 758 (1964).
37. A. Hollerbach and G. Spiteller, *Monatsh. Chem.* **101**, 141 (1970).
38. A. A. Ryabinin and E. M. Ilina, *Dokl. Akad. Nauk SSSR* **67**, 513 (1949); *CA* **44**, 1455h (1950).
39. T. A. Smith and J. L. Garraway, *Phytochemistry* **3**, 23 (1964).
40. R. Willstätter and W. Heubner, *Ber. Dtsch. Chem. Ges.* **40**, 3869 (1907).
41. H. Ripperger, K. Schreiber, and H. Budzikiewicz, *J. Prakt. Chem.* **312**, 449 (1970).
42. A. Stoessl, *Can. J. Chem.* **45**, 1745 (1967).
43. A. Stoessl, *Tetrahedron Lett.* 2287 (1966).
44. A. Stoessl, *Tetrahedron Lett.* 2849 (1966).
45. C. R. Bird and T. A. Smith, *Phytochemistry* **20**, 2345 (1981).
46. E. Fattorusso, L. Minale, G. Sodano, K. Moody, and R. H. Thomson, *J. Chem. Soc., Chem. Commun.* 752 (1970).
47. J. A. McMillan, I. C. Paul, Y. M. Goo, K. L. Rinehart, W. C. Krueger, and L. M. Pschigoda, *Tetrahedron Lett.* 39 (1981).
48. G. Englert, K. Klinga, Raymond-Hamet, E. Schlittler, and W. Vetter, *Helv. Chim. Acta* **56**, 474 (1973).
49. S. A. Ong, T. Peterson, and J. B. Neilands, *J. Biol. Chem.* **254**, 1860 (1979).
50. G. H. Tait, *Biochem. J.* **146**, 191 (1975).
51. R. J. Bergeron, P. S. Burton, S. J. Kline, and K. A. McGovern, *J. Org. Chem.* **46**, 3712 (1981).
52. K. K. Bhargava, R. W. Grady, and A. Cerami, *J. Pharm. Sci.* **69**, 986 (1980).
53. R. J. Bergeron, S. J. Kline, N. J. Stolowich, K. A. McGovern, and P. S. Burton, *J. Org. Chem.* **46**, 4524 (1981).
54. R. J. Bergeron, K. A. McGovern, M. A. Channing, and P. S. Burton, *J. Org. Chem.* **45**, 1589 (1980).
55. Y. Nagao, K. Seno, K. Kawabata, T. Miyasaka, S. Takao, and E. Fujita, *Tetrahedron Lett.* 841 (1980).
56. E. Schlittler, U. Spitaler, and N. Weber, *Helv. Chim. Acta* **56**, 1097 (1973).
57. H.-P. Husson, C. Poupat, and P. Potier, *C. R. Hebd. Seances Acad. Sci., Ser. C* **276**, 1039 (1973).
58. J. S. McManis and B. Ganem, *J. Org. Chem.* **45**, 2041 (1980).
59. T. Peterson and J. B. Neilands, *Tetrahedron Lett.* 4085 (1979).
60. T. Peterson, K.-E. Falk, S. A. Leong, M. P. Klein, and J. B. Neilands, *J. Am. Chem. Soc.* **102**, 7715 (1980).
61. F. J. Schmitz, K. H. Hollenbeak, and R. S. Prasad, *Tetrahedron Lett.* 3387 (1979).
62. K. Chantrapromma, J. S. McManis, and B. Ganem, *Tetrahedron Lett.* 2605 (1980).
63. A. Guggisberg, M. M. Badawi, M. Hesse, and H. Schmid, *Helv. Chim. Acta* **57**, 414 (1974).
64. K. Sailer and M. Hesse, *Helv. Chim. Acta* **51**, 1817 (1968).
65. F. Schneider, K. Bernauer, A. Guggisberg, P. van den Broek, M. Hesse, and H. Schmid, *Helv.Chim. Acta* **57**, 434 (1974).
66. A. Guggisberg, P. van den Broek, M. Hesse, H. Schmid, F. Schneider, and K. Bernauer, *Helv. Chim. Acta* **59**, 3013 (1976).
67. U. Kramer, A. Guggisberg, M. Hesse, and H. Schmid, *Angew. Chem.* **89**, 899 (1977); *Angew. Chem., Int. Ed. Engl.* **16**, 861 (1977).

68. E. Stephanou, A. Guggisberg, and M. Hesse, *Helv. Chim. Acta* **62**, 1932 (1979).
69. U. Kramer, A. Guggisberg, M. Hesse, and H. Schmid. *Helv. Chim. Acta* **61**, 1342 (1978).
70. U. Kramer, A. Guggisberg, M. Hesse, and H. Schmid, *Angew. Chem.* **90**, 210 (1978); *Angew. Chem., Int. Ed. Engl.* **17**, 200 (1978).
71. Y. Nakashita and M. Hesse, *Angew. Chem.* **93**, 1077 (1981); *Angew. Chem., Int. Ed. Engl.* **20**, 1021 (1981).
72. K. Kostova, A. Lorenzi-Riatsch, Y. Nakashita, and M. Hesse, *Helv. Chim. Acta* **65**, 249 (1982).
73. H. J. Veith, M. Hesse, and H. Schmid, *Helv. Chim. Acta* **53**, 1355 (1970).
74. R. Charubala, A. Guggisberg, and M. Hesse, unpublished results.
75. W. Dietsche and C. H. Eugster, *Chimia* **14**, 353 (1960).
76. E. Glet, J. Gutschmid, and P. Glet, *Hoppe-Seyler's Z. Physiol. Chem.* **244**, 229 (1936).
77. C. H. Eugster, *Heterocycles* **4**, 51 (1976).
78. C. G. Baumann, W. Dietsche, and C. H. Eugster, *Chimia* **14**, 85 (1960).
79. E. Wälchli-Schaer and C. H. Eugster, *Helv. Chim. Acta* **61**, 928 (1978).
80. M. Ogawa, J. Nakajima, and M. Natsume, *Heterocycles* **19**, 1247 (1982).
81. C. E. Turner, M.-F. H. Hsu, J. E. Knapp, P. L. Schiff, and D. J. Slatkin, *J. Pharm. Sci.* **65**, 1084 (1976).
82. M. A. ElSohly, C. E. Turner, C. H. Phoebe, J. E. Knapp, P. L. Schiff, and D. J. Slatkin, *J. Pharm. Sci.* **67**, 124 (1978).
83. H. L. Lotter, D. J. Abraham, C. E. Turner, J. E. Knapp, P. L. Schiff, and D. J. Slatkin, *Tetrahedron Lett.* 2815 (1975).
84. C. Tamura, G. A. Sim, J. A. D. Jeffreys, P. Bladon, and G. Ferguson, *J.Chem. Soc., Chem. Commun.* **20**, 485 (1965).
85. J. A. D. Jeffreys and G. Ferguson, *J.Chem. Soc. B* 826 (1970).
86. C. Tamura and G. A. Sim, *J. Chem. Soc. B* 991 (1970).
87. P. Potier, J. Le Men, M.-M. Janot, and B. Bladon, *Tetrahedron Lett.* 36 (1960).
88. P. Bladon, M. Chaigneau, M.-M. Janot, J. Le Men, P. Potier, and A. Melera, *Tetrahedron Lett.* 321 (1961).
89. E. W. Warnhof, *Fortschr. Chem. Org. Naturst.* **28**, 162 (1970).
90. K. Wiesner, D. M. MacDonald, C. Bankiewicz, and D. E. Orr, *Can. J. Chem.* **46**, 1881 (1968).
91. K. Wiesner, Z. Valenta, D. E. Orr, V. Liede, and G. Kohan, *Can. J. Chem.* **46**, 3617 (1968).
92. H.-P. Husson, C. Poupat, B. Rodriguez, and P. Potier, *Tetrahedron* **29**, 1405 (1973).
93. S. Mizusaki, Y. Tanabe, M. Noguchi, and E. Tamaki, *Phytochemistry* **10**, 1347 (1971).
94. N. A. Pilewski, J. Tomko, A. B. Ray, R. W. Doskotch, J. L. Beal, G. H. Svoboda, and W. Kubelka, *Lloydia* **35**, 186 (1972).
95. R. W. Doskotch, A. B. Ray, W. Kubelka, E. D. Fairchild, C. D. Hufford, and J. L. Beal, *Tetrahedron* **30**, 3229 (1974).
96. R. W. Doskotch, E. H. Fairchild, and C. D. Hufford, *Tetrahedron* **30**, 3237 (1974).
97. C. Poupat, *Tetrahedron Lett.* 1669 (1976).
98. Y. Nagao, K. Seno, and E. Fujita, *Tetrahedron Lett.* 4931 (1980).
99. M. J. Humora, D. E. Seitz, and J. Quick, *Tetrahedron Lett.* 3971 (1980).
100. M. Humora and J. Quick, *J. Org. Chem.* **44**, 1166 (1979).
101. S. M. Kupchan, H. P. J. Hintz, R. M. Smith, A. Karim, M. W. Cass, W. A. Court, and M. Yatagai, *J. Chem. Soc., Chem. Commun.* 329 (1974).
102. S. M. Kupchan, H. P. J. Hintz, R. M. Smith, A. Karim, M. W. Cass, W. A. Court, and M. Yatagai, *J. Org. Chem.* **42**, 3660 (1977).
103. H. Wagner and J. Burghart, *Helv. Chim. Acta* **64**, 283 (1981).
104. H. H. Wasserman, R. P. Robinson, and H. Matsuyama, *Tetrahedron Lett.* 3493 (1980).

105. A. Guggisberg, B. Dabrowski, U. Kramer, C. Heidelberger, M. Hesse, and H. Schmid, *Helv. Chim. Acta* **61**, 1039 (1978).
106. A. Guggisberg, U. Kramer, C. Heidelberger, R. Charubala, E. Stephanou, M. Hesse, and H. Schmid, *Helv. Chim. Acta* **61**, 1050 (1978).
107. H. Yamamoto and K. Marouka, *J. Am. Chem. Soc.* **103**, 6133 (1981).
108. H. Ripperger, *Phytochemistry* **19**, 162 (1980).
109. M. Díaz and H. Ripperger, *Phytochemistry* **21**, 255 (1982).
110. H. Ripperger, private communication from July 1, 1982.
111. H. Wagner, J. Burghart, and W. E. Hull, *Tetrahedron Lett.* 3893 (1978).
112. H. Wagner and J. Burghart, *Helv. Chim. Acta* **65**, 739 (1982).
113. R. Hocquemiller, M. Leboeuf, B. C. Das, H.-P. Husson, P. Potier, and A. Cavé, *C. R. Hebd. Seances Acad. Sci., Ser. C* **278**, 525 (1974).
114. R. Hocquemiller, A. Cavé, and H.-P. Husson, *Tetrahedron* **33**, 645 (1977).
115. R. Hocquemiller, A. Cavé, and H.-P. Husson, *Tetrahedron* **33**, 653 (1977).
116. H. H. Wasserman and H. Matsuyama, *J. Am. Chem. Soc.* **103**, 461 (1981).
117. G. T. Carter and K. L. Rinehart, *J. Am. Chem. Soc.* **100**, 4302 (1978).
118. S. M. Kupchan, A. P. Davies, S. J. Barboutis, H. K. Schnoes, and A. L. Burlingame, *J. Org. Chem.* **34**, 3888 (1969).
119. S. M. Kupchan, A. P. Davies, S. J. Barboutis, H. K. Schnoes, and A. L. Burlingame, *J. Am. Chem. Soc.* **89**, 5718 (1967).
120. W. C. Evans, A. Ghani, and V. A. Woolley, *J. Chem. Soc., Perkin Trans. 1* 2017 (1972).
121. W. C. Evans and A. Somanabandhu, *Phytochemistry* **16**, 1859 (1977).
122. J. W. Blunt, M. H. G. Munro, and S. C. Yorke, *Tetrahedron Lett.* 2793 (1982).
123. J. Martin-Tanguy, F. Cabanne, E. Perdrizet, and C. Martin, *Phytochemistry* **17**, 1927 (1978).
124. S. Funayama, K. Yoshida, C. Konno, and H. Hikino, *Tetrahedron Lett.* 1355 (1980).
125. K. Chantrapromma and B. Ganem, *Tetrahedron Lett.* 23 (1981).
126. H. O. Bernhard, I. Kompiš, S. Johne, D. Gröger, M. Hesse, and H. Schmid, *Helv. Chim. Acta* **56**, 1266 (1973).
127. P. Dätwyler, H. Bosshardt, H. O. Bernhard, M. Hesse, and S. Johne, *Helv. Chim. Acta* **61**, 2646 (1978).
128. P. Dätwyler, H. Bosshardt, S. Johne, and M. Hesse, *Helv. Chim. Acta* **62**, 2712 (1979).
129. M. Tamada, K. Endo, H. Hikino, and C. Kabuto, *Tetrahedron Lett.* 873 (1979).
130. H. Bosshardt, A. Guggisberg, S. Johne, and M. Hesse, *Pharm. Acta Helv.* **53**, 355 (1978).
131. H. Bosshardt, A. Guggisberg, S. Johne, H.-J. Veith, M. Hesse, and H. Schmid, *Pharm. Acta Helv.* **51**, 371 (1976).
132. C. Konno, M. Tamada, K. Endo, and H. Hikino, *Heterocycles* **14**, 295 (1980).
133. H. Hikino, M. Ogata, and C. Konno, *Heterocycles* **17**, 155 (1982).
134. M. Pais, G. Rattle, R. Sarfati, and F.-X. Jarreau, *C. R. Hebd. Seances Acad. Sci., Ser. C* **266**, 37 (1968).
135. M. Pais, G. Rattle, R. Sarfati, and F.-X. Jarreau, *C. R. Hebd. Seances Acad. Sci., Ser. C* **267**, 82 (1968).
136. M. Pais, R. Sarfati, F.-X. Jarreau, and R. Goutarel, *Tetrahedron* **29**, 1001 (1973).
137. M. Pais, R. Sarfati, F.-X. Jarreau, and R. Goutarel, *C. R. Hebd. Seances Acad. Sci., Ser. C* **272**, 1728 (1971).
138. M. Pais, R. Sarfati, and F.-X. Jarreau, *Bull. Soc. Chim. Fr. Part 2*, 331 (1973).
139. H. H. Wasserman, G. D. Berger, and K. R. Cho, *Tetrahedron Lett.* 465 (1982).
140. K. Seifert, S. Johne, and M. Hesse, *Helv. Chim. Acta* **65**, 2540 (1982).
141. K. Wiesner, D. M. MacDonald, Z. Valenta, and R. Armstrong, *Can. J. Chem.* **30**, 761 (1952).

142. R. Brüning and H. Wagner, *Phytochemistry* **17**, 1821 (1978).
143. M. V. Kisakürek, A. J. M. Leeuwenberg, and M. Hesse, *in* "The Alkaloids: Chemical and Biological Perspectives" (S. W. Pelletier, ed.), Vol. 1, p. 211. Wiley (Interscience), New York, 1983.
144. O. J. Crocomo, L. C. Basso, and O. G. Brasil, *Phytochemistry* **9**, 1487 (1970).
145. F. Linneweh, *Hoppe-Seyler's Z. Physiol. Chem.* **205**, 126 (1932).
146. T. A. Smith and G. R. Best, *Phytochemistry* **17**, 1093 (1978).
147. A. Stoessl, R. Rohringer, and D. J. Samborski, *Tetrahedron Lett.* 2807 (1969).
148. F. Cabanne, J. Martin-Tanguy, and C. Martin, *Physiol. Veg.* **15**, 429 (1977).
149. F. Cabanne, J. Martin-Tanguy, E. Perdrizet, J.-C. Vallée, L. Grenet, J. Prévost, and C. Martin, *C. R. Hebd. Seances Acad. Sci., Ser. D* **282**, 1959 (1976).
150. J. Martin-Tanguy, C. Martin, and M. Gallet, *C. R. Hebd. Seances Acad. Sci., Ser. D* **276**, 1433 (1973).
151. J. Delétang, *Ann. Tab.: Sect. 2* **11**, 123 (1974).
152. J. G. Buta and R. R. Izac, *Phytochemistry* **11**, 1188 (1972).
153. E. I. Mbadiwe, *Phytochemistry* **12**, 2546 (1973).
154. T. A. Wheaton and I. Stewart, *Nature (London)* **206**, 620 (1965).
155. A. A. Ryabinin and E. M. Ilina, *Dokl. Akad. Nauk SSSR* **76**, 689 (1951); *CA* **45**, 8479d (1951).
156. A. Hillmann-Elies and G. Hillmann, *Z. Naturforsch., B: Anorg. Chem., Org. Chem. Biochem. Biophys. Biol.* **8B**, 526 (1953).
157. W. J. Griffin, *Australas. J. Pharm.* [N.S.] **51**, 19 (1967).
158. J. F. Coulsen and W. J. Griffin, *Planta Med.* **16**, 174 (1968).
159. S. Johne, D. Gröger, and R. Radeglia, *Phytochemistry* **14**, 2635 (1975).
160. L. Gran, *Lloydia* **36**, 209 (1973).
161. H. J. Veith, A. Guggisberg, and M. Hesse, *Helv. Chim. Acta* **54**, 653 (1971).
162. T. Potjewijd, Dissertation, Universiteit Leiden (1933).
163. R. A. Konowalowa and O. J. Magidson, *Arch. Pharm. (Weinheim, Ger.)* **266**, 449 (1928).
163a. D. L. Eng-Wilmot and D. van der Helm, *J. Am. Chem. Soc.* **102**, 7719 (1980).
164. C. Poupat, H.-P. Husson, B. Rodriguez, A. Husson, P. Potier, and M.-M. Janot, *Tetrahedron* **28**, 3087 (1972).
165. C. Poupat, B. Rodriguez, H.-P. Husson, P. Potier, and M.-M. Janot, *C. R. Hebd. Seances Acad. Sci., Ser C* **269**, 335 (1969).
166. R. W. Doskotch, E. H. Fairchild, and W. Kubelka, *Experientia* **28**, 382 (1972).
167. H. H. Wasserman, *Colloq. Chim. Heterocycl. 8th,* Abstracts, (1982).
168. R. J. Bergeron, P. S. Burton, K. A. McGovern, E. J. St. Onge, and R. R. Streiff, *J. Med. Chem.* **23**, 1130 (1980).
169. H. Wagner and J. Burghart, *Planta Med.* **32A**, 9 (1977).
170. H. Wagner, J. Burghart, and S. Bladt, *Tetrahedron Lett.* 781 (1978).
171. R. W. Doskotch, A. B. Ray, and J. L. Beal. *J. Chem. Soc., Chem. Commun.* 300 (1971).
172. M. Hesse, *in* "Biochemical Applications of Mass Spectrometry" (G. R. Waller and O. C. Dermer, eds.), 1st Suppl. Vol., p. 797. Wiley (Interscience), New York, 1980.
173. A. Guggisberg, H.-J. Veith, M. Hesse, and H. Schmid, *Helv. Chim. Acta* **59**, 3026 (1976).
174. H. G. Boit, *Chem. Ber.* **87**, 1082 (1954).
175. C. Poupat, H.-P. Husson, B. C. Das, P. Bladon, and P. Potier, *Tetrahedron* **28**, 3103 (1972).
176. Y. Nagao, S. Takao, T. Miyasaka, and E. Fujita, *J. Chem. Soc., Chem. Commun.* 286 (1981).
177. H.-P. Husson, C. Poupat, B. Rodriguez, and P. Potier, *Tetrahedron Lett.* 2697 (1971).
178. S. Huneck, *Naturwissenschaften* **49**, 233 (1962).

179. E. Hairs, *Bull. Acad. R. Belg.* 1042 (1909).
180. M.-M. Janot and J. Le Men, *Bull. Soc. Chim. Fr.* 1840 (1956).
181. O. R. Hansen, *Acta Chem. Scand.* **1,** 656 (1947).
182. E. Steinegger and T. Reichstein, *Pharm. Acta Helv.* **22,** 258 (1947).
183. M. M. Badawi, A. Guggisberg, P. van den Broek, M. Hesse, and H. Schmid, *Helv. Chim. Acta* **51,** 1813 (1968).
184. C. L. Green, C. Mayer, and C. H. Eugster, *Helv. Chim. Acta* **52,** 673 (1969).
185. C. Mayer, C. L. Green, W. Trueb, P. C. Wälchli, and C. H. Eugster, *Helv. Chim. Acta* **61,** 905 (1978).
186. P. C. Wälchli, G. Mukherjee-Müller, and C. H. Eugster, *Helv. Chim. Acta* **61,** 921 (1978).
187. T. Baytop and E. Gurkan, *Istanbul Univ. Eczacilik Fak. Mecm.* **8,** 63 (1972); *CA* **79,** 15817b (1973).
188. P. Rüedi and C. H. Eugster, *Helv. Chim. Acta* **61,** 899 (1978).
188a. P. Wälchli and C. H. Eugster, *Angew. Chem.* **85,** 172 (1973).
189. P. C. Wälchli and C. H. Eugster, *Helv. Chim. Acta* **61,** 885 (1978).
190. H. Jaggy, *Planta Med.* **33,** 285 (1978).
191. R. Tschesche and V. B. Pandey, private communication from December 10, 1980.
192. A. Guggisberg, P. Dätwyler, H. P. Ros, D. Seps, and M. Hesse, unpublished results.
193. M. Tamada, K. Endo, and H. Hikino, *Heterocycles* **12,** 783 (1979).
194. G. Misra, S. K. Nigam, and C. R. Mitra, *Phytochemistry* **10,** 3313 (1971).

—— CHAPTER 4 ——

APPLICATION OF ENAMIDE CYCLIZATIONS IN ALKALOID SYNTHESIS

ICHIYA NINOMIYA AND TAKEAKI NAITO

Kobe Women's College of Pharmacy
Kobe, Japan

I. Introduction

Alkaloids are a group of naturally occurring compounds that contain at least one nitrogen. Because of their potent and interesting pharmacological activities, alkaloids have been widely studied. The syntheses of alkaloids have provided many useful novel synthetic methodologies in the field of synthetic organic chemistry.

The most abundant and interesting groups of alkaloids are the isoquinoline and indole alkaloids, possessing at least one nitrogen-containing, six-

THE ALKALOIDS, VOL. XXII

ISBN 0-12-469522-1

membered ring. The well-known Bischler–Napieralski, Pictet–Spengler, and Pomerantz–Fritsch syntheses have been studied extensively and successfully applied to a great number of total syntheses of these groups of alkaloids.

Although a number of other effective synthetic methodologies have been developed for the synthesis of isoquinoline alkaloids (*1,2*), there still remain quite a number of alkaloids which have not been synthesized, even by using newer methods. Improvement of existing methodologies and the introduction of new ones should provide more efficient and convenient syntheses of alkaloids.

Stimulated by the recent introduction of photochemical means to synthetic organic chemistry, the reaction and application of enamines and their derivatives have been studied (*3*). Although enamines have been regarded as one of the most useful synthetic weapons developed in modern organic chemistry (*3*), these particular groups are usually too unstable and decompose rather readily under photochemical conditions. Therefore, *N*-acylenamines, which are readily prepared from imines by simple acylation, were the compounds of choice to study for photochemical reaction.

The first observation was N→C acyl migration on irradiation of *N*-acylenamines (*4–10*). However, when *N*-benzoylenamines were irradiated, a new photocyclization reaction was discovered (*9,10*), and since has been developed for use as a synthetic tool with general applicability to the synthesis of various alkaloids. As a result, it is now firmly established as enamide photocyclization.

Started in the late 1960s, studies on enamide photocyclization and its application to the synthesis of alkaloids have been successfully accumulated and have achieved a number of total syntheses of various types of isoquinoline and indole alkaloids and related heterocyclic compounds (*11*).

The compound chosen for study in this chapter is Enamide, a conventional name for the group of compounds having the structures of *N*-α,β-unsaturated acylenamines and *N*-α,β-unsaturated acylanilides. These enamides are readily prepared by acylation of the corresponding imines or anilines with α,β-unsaturated acid chloride and are found to be considerably stable compounds, although losing reactivity to some extent when compared with the parent enamines.

Introduction of a double bond in the acyl portion of *N*-acylenamines or *N*-acylanilides forms a six π-electron conjugated system and therefore, an electrocyclic cyclization under photochemical conditions is expected, as in the case of photocyclization of 1,3,5-hexatrienes and stilbenes.

This expectation was visualized through a number of examples of enamide photocyclization and the results obtained clearly suggest that this enamide photocyclization is a general cyclization reaction with wide applicability, particularly to the synthesis of alkaloids (*11*).

Taking particular consideration of the application to alkaloid synthesis, enamides are divided into four groups depending on the nature of the double bonds involved (Scheme 1).

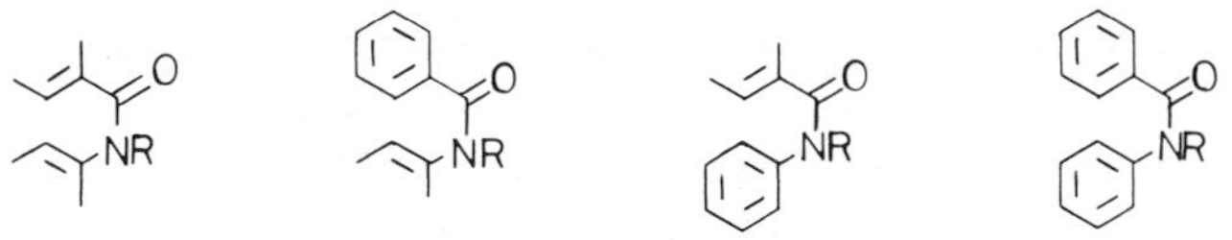

-Acylenamine   *N*-Benzoylenamine   *N*-Acylanilide   *N*-Benzanilide

SCHEME 1

Of these four types of enamides, the most prolific and useful compounds are undoubtedly the *N*-benzoylenamines, judging from their synthetic application.

After 10 years of research on enamide photocyclization, it has now been established that any type of enamide can undergo smooth photocyclization to afford the corresponding lactam in good yields. The characteristic points of their photocyclizations are summarized as follows:

1. Enamide photocyclization is basically of a nonoxidative nature, yielding the lactam isomeric to the starting enamide.
2. In the presence of an oxidizing agent, the enamide also undergoes smooth photocyclization to yield the dehydrolactam with the loss of two hydrogens.
3. In the presence of metal hydride, the enamide can undergo reductive photocyclization to give the dihydrolactam in good yield.
4. Asymmetric photocyclization to a chiral lactam can be achieved by the use of a chiral metal hydride complex under reductive conditions, showing the enantio-differentiated nature of this type of cyclization.
5. Asymmetric photocyclization starting from a chiral enamide also proceeds smoothly under nonoxidative conditions to give a diastereo-differentiated product.
6. The orientation of photocyclization can be controlled by using a substituent, particularly at the ortho position where the cyclization is directed.
7. Enamides can even undergo cyclization under thermal and acylating conditions, depending on the structure of enamide.

According to the reaction conditions, nonoxidative, oxidative, and reductive and the variety of the structures of enamides, several types of lactams shown in Scheme 2 can be synthesized by the use of this cyclization.

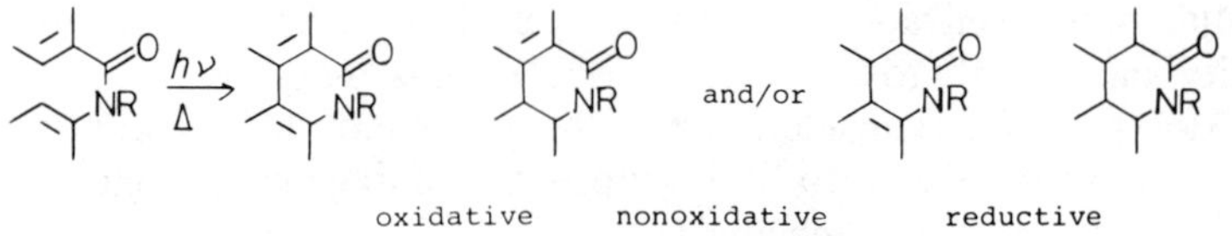

oxidative   nonoxidative   reductive

SCHEME 2

Clearly these structures show the superiority of enamide photocyclization as a useful synthetic method for the synthesis of alkaloids.

Several reviews on enamide photocyclization (11–13) and photochemical synthesis of isoquinoline alkaloids (14) have been published.

II. Photocyclization of Enamides

Enamide photocyclization is briefly summarized in Scheme 3. On irradiation of any one of four types of enamides, the photocyclized products are always formed as six-membered lactams with different oxidation degrees depending on the reaction condition applied. The didehydrolactam E is obtained by irradiation in the presence of an oxidizing agent, whereas the monodehydrolactam **D**, whose structure is isomeric to the starting enamide, is obtained under nonoxidative conditions (11). Finally, saturated lactams are obtained by irradiation in the presence of a hydride agent; that is, under reductive conditions (15).

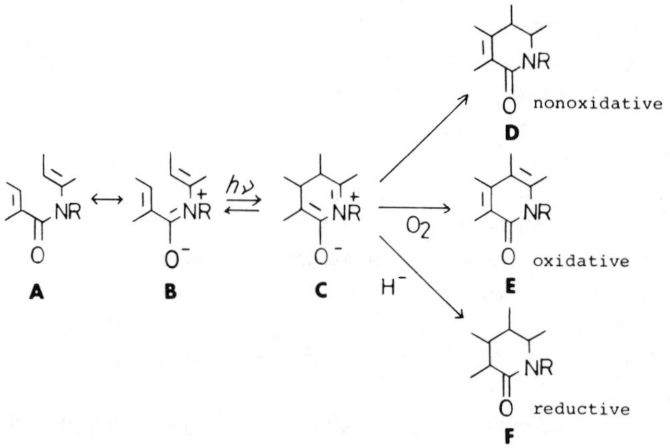

SCHEME 3

Furthermore, the chiral metal hydride complex, which is readily prepared from a metal hydride and chiral amine or amino alcohol, has been successfully applied to reductive photocyclization, thus allowing visualization of asymmetric photocyclization (16).

All these photocyclizations are well explained in terms of an electrocyclic mechanism of nitrogen-containing, six π-electron conjugated system, according to the Woodward–Hoffmann rule, by postulating the intermediacy of a common trans cyclic structure from which respective types of products are formed depending on the reaction conditions: either a nonoxidative, oxidative, or reductive condition.

A. Nonoxidative Photocyclization of Enamides

Most enamides, irrespective of their structures, are known to undergo photocyclization under nonoxidative conditions giving rise to lactams, whose structures are isomeric to the starting enamides, in good yields. One exception is the photocyclization of benzanilides that cyclize only under oxidative conditions to afford dehydrolactams (11). However, the stereochemical and regiochemical aspects of these photocyclizations are different, depending on the types of enamides, mainly either on N-α,β-unsaturated acylenamine type or N-α,β-unsaturated acylanilide type.

1. Photocyclization of Enamides of N-α,β-Unsaturated
   Acylenamine Type

This type of enamide is prepared from the corresponding imines by acylation with an α,β-unsaturated acid chloride in the presence of triethylamine. The enamides thus obtained are quite stable crystalline compounds in most cases and are dissolved in an appropriate solvent such as methanol, benzene, or ether, depending on the enamide, and are irradiated using a low- or high-pressure mercury lamp at room temperature, or preferably at a low temperature (below 10°C). The reaction is traced by checking TLC and irradiation is stopped at the point when the spot corresponding to the starting enamide disappears. Evaporation of the solvent leaves a residual oil or crystals that can be readily purified by simple chromatography or recrystallization. TLC and GLC of the reaction mixture clearly show the homogeneity of the photocyclized lactam, which has a trans ring juncture readily assignable from its NMR spectrum; thus, the stereospecific nature of photocyclization of this type of enamide is established (11).

Since the enamide **A** affords the dehydrolactam **E** on irradiation in the presence of iodine, and no conversion from **D** to **E** is observed under nonoxidative conditions, the nonoxidative nature of the original enamide photocyclization is clearly established (Scheme 4).

The above stereochemical and nonoxidative nature of photocyclization of this type of enamide is enough to propose the mechanism of this photocyclization as that of the electrocyclic nature of a six π-electron

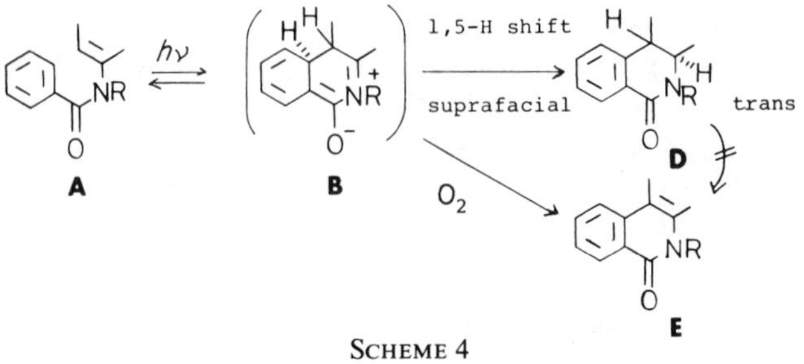

SCHEME 4

conjugated system as follows. According to the Woodward – Hoffmann rule of conrotatory cyclization of a six π-electron conjugated system under photochemical conditions, an enamide cyclizes to form a trans cyclic intermediate **B** from which a hydrogen migrates suprafacially in a 1,5 manner under thermal condition to give a trans lactam **D** exclusively.

Based on the establishment of photocyclization of enamides of N-α,β-unsaturated acylenamine type, its application to alkaloid synthesis has been very fruitful, particularly in the synthesis of skeletal structures of natural alkaloids. Which alkaloids are synthesized is a matter of choosing the appropriate starting ketones and acids for the preparation of enamides.

The phenanthridine skeleton is synthesized by photocyclization of the enamides prepared from cyclohexanonimines and benzoyl chlorides (*17,18*). The benzo[*c*]phenanthridine skeletons are formed from the enamides prepared from 2-tetralonimines and benzoyl chlorides (*19,20*). More conveniently, the skeletons of protoberberine alkaloids are readily synthesized from the enamides prepared by simple acylation of 1-methyl-3,4-dihydroisoquinolines with benzoyl chlorides (*21 – 24*). This berbine synthesis is one of the most typical examples of the application of enamide photocyclization to alkaloid synthesis and can be further extended to the facile synthesis of the skeletons of the yohimbine group of indole alkaloids (*25,26*).

a. The Substituent Effects in the Photocyclization of *N*-Benzoylenamines (Control in the Regiochemistry of Photocyclization). For the application of enamide photocyclization to total synthesis of alkaloids, the regiocontrol, that is, the control over the direction of cyclization, had to be established. After studies of the substituent effects, the use of an electron-donating group such as a methoxyl group is found to be most effective for this purpose (*27,28*). A regiospecific photocyclization to the root of the *o*-methoxyl group introduced in the enamide, followed by 1 → 5 migration of the methoxyl group and its elimination, is the method of choice. Therefore, in order to prepare a specifically substituted ring system, an enamide having an additional *o*-methoxyl group in the benzene ring is prepared and applied to its photocyclization.

Similarly, introduction of halogen or some other groups is also found to be effective for this regiocontrol in some cases of cyclizations (*29–32*).

These regiospecific photocyclizations to the root of the *o*-methoxyl group can be explained in terms of a nucleophilic attack from the enamine portion to the somewhat electron-deficient position at the root of the methoxyl group, as in the case of nucleophilic aromatic photosubstitution (*33*), and followed by a 1,5-sigmatropic shift of the substituent and its elimination to afford the didehydrolactam (**1**) as the final product (Scheme 5).

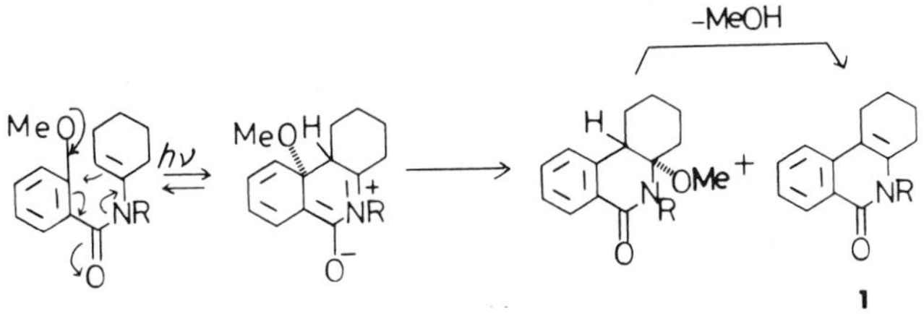

SCHEME 5

As a result of studies on the enamide photocyclization under nonoxidative conditions, this type of photocyclization of *N-α,β*-unsaturated acylenamines becomes one of the most useful and prolific methodologies for the synthesis of alkaloids as shown in the following section.

Among other substituent effects, as summarized in Table I, introduction of an amino group to an ortho position brought about an opposite effect as shown (*27,28*). The presence of a primary amino group next to the carbonyl group of the enamide would form a strong hydrogen bond, thus fixing the conformation of the enamide suitable for the cyclization at the position opposite the substituent (Scheme 6).

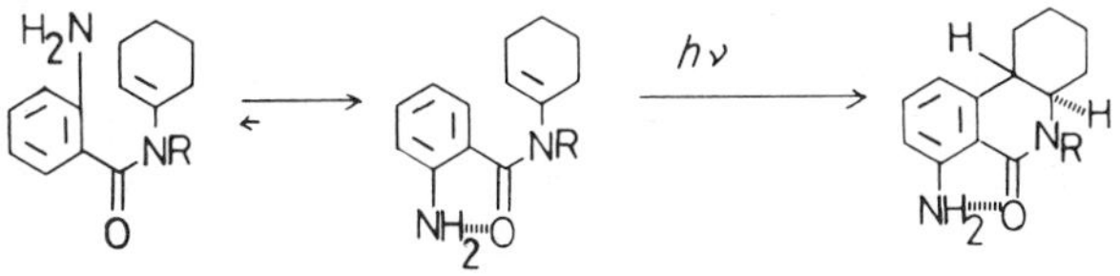

SCHEME 6

In general, introduction of an electron-withdrawing group into the benzene ring of enamides seems to retard the cyclization, giving only poor yields of products, whereas the presence of an electron-donating group, such as methoxyl and amino groups, gives good yields of the cyclized products in addition to the above-mentioned regiospecific effect (*27,28*).

TABLE I

EFFECT OF SUBSTITUENTS ON PHOTOCYCLIZATION PRODUCTS

$$\text{(scheme)} \xrightarrow[\text{R}=CH_2Ph]{h\nu} \text{Product I} + \text{Product II}$$

Substituents	Product I (%)	Product II (%)
$R^1 = R^3 = R^4 = H, R^2 = OMe$	$R^1 = R^3 = H, R^2 = OMe$ (42)	
$R^1 = R^2 = R^4 = H, R^3 = OMe$	$R^1 = R^2 = H, R^3 = OMe$ (29)	
$R^1 = R^2 = R^4 = H, R^3 = COOMe$	$R^1 = R^2 = H, R^3 = COOMe$ (10)	$R^1 = R^2 = R^4 = H, R^3 = COOMe$ (15)
$R^1 = R^2 = R^4 = H, R^3 = NO_2$		$R^1 = R^2 = R^4 = H, R^3 = NO_2$ (14)
$R^2 = R^3 = R^4 = H, R^1 = OMe$	$R^1 = R^2 = R^3 = H, 4a\text{-}OMe$ (30)	$R^1 = R^2 = R^3 = R^4 = H$ (18)
$R^3 = R^4 = H, R^1 = R^4 = OMe$		$R^2 = R^3 = R^4 = H, R^1 = OMe$ (10)
$R^3 = R^4 = H, R^1 = R^2 = OMe$		$R^1 = R^2 = R^3 = H, R^4 = OMe$ (15)
$R^3 = R^4 = H, R^1 = OCH_2O = R^2$		$R^1 = R^2 = R^3 = H, R^4 = OH$ (14)
$R^2 = R^3 = R^4 = H, R^1 = NH_2$	$R^2 = R^3 = H, R^1 = NH_2$ (71)	
$R^2 = R^3 = R^4 = H, R^1 = COOMe$		$R^2 = R^3 = R^4 = H, R^1 = COOMe$ (2.5)
		$R^1 = R^2 = R^3 = R^4 = H$ (1.5)

2. Photocyclization of Enamides of N-(α,β-Unsaturated Acyl)anilide Type

Contrary to the photocyclization of N-benzoylenamines, the photocyclization of N-acylanilides proceeds quite differently but very smoothly to give a mixture of *cis*- and *trans*-3,4-dihydro-2-quinolone-type compounds. A strong solvent effect was observed that affected the ratio of stereoisomeric products as well as the different substituent effects, the accelerating effect by electron-withdrawing groups and the retarding effect by electron-donating groups (*34,35*).

Although this type of photocyclization gives good yields of the cyclization products, its synthetic application remains relatively unexploited because of a lack of attractive natural products as targets. However, some quinoline alkaloids with additional oxygen-containing heterocycles are known, thus offering a challenge to chemists.

a. The Stereochemistry of Photocyclization of N-Acylanilides. Although the first examples of photocyclization of N-acylanilides, which gave a mixture of the cyclized lactams with undetermined stereochemistry, were reported independently by Chapman *et al.* (*36,37*) and Ogata *et al.*, (*38*), Ninomiya *et al.* established the stereochemistry of photocyclization of this type of enamide in addition to the solvent and substituent effects by using the N-cyclohexenoylanilides (*34,35*). These N-acylanilides (**2**) underwent smooth photocyclization to afford a mixture of trans and cis lactams (**3** and **4**), and their ratios depended on the solvent employed. The trans lactam (**3**) was obtained predominantly when an aprotic solvent such as ether and benzene was employed, whereas the cis lactam (**4**) became a major product when irradiated in a protic solvent such as methanol. These ratios did not change under thermal and photochemical conditions. The above solvent effect was also observed in the photocyclization of various substituted N-cyclohexenoylanilides (**2**) as summarized in Scheme 7 (*34,35*).

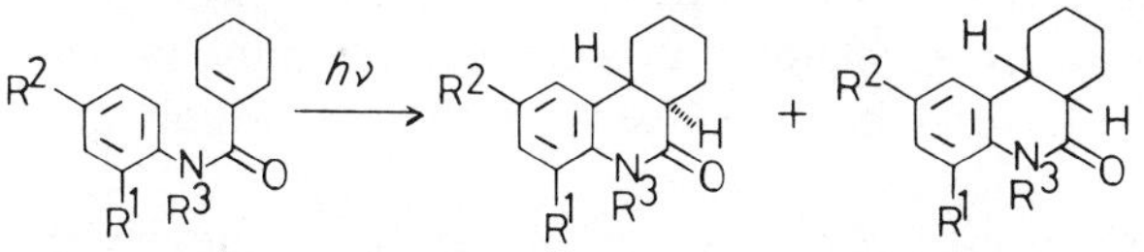

Bz = CH₂Ph	Yield (%)	Time (hr)	Solvent effect (trans/cis)		
	(cis + trans)		Et₂O	benzene	MeOH
R^1=H, R^2=OMe, R^3=Bz	60-65	40	9.2	1.2	0.2
R^1=H, R^2=COOMe, R^3=Bz	50-60	9	34.5	8.7	0.5
R^2=H, R^1=OMe, R^3=Bz	52-65	35	20.8	1.9	0.3
R^2=R^3=H, R^1=COOMe	60	10	2.0	0.5	—

SCHEME 7

b. The Control of Regiochemistry of Photocyclization of *N*-Acylanilides.
The orientation of photocyclization of *N*-acylanilides is also controlled by
the use of the ortho substituent, particularly by the introduction of an
electron-withdrawing group such as ester and ketone as summarized in
Scheme 8 *(27,35)*.

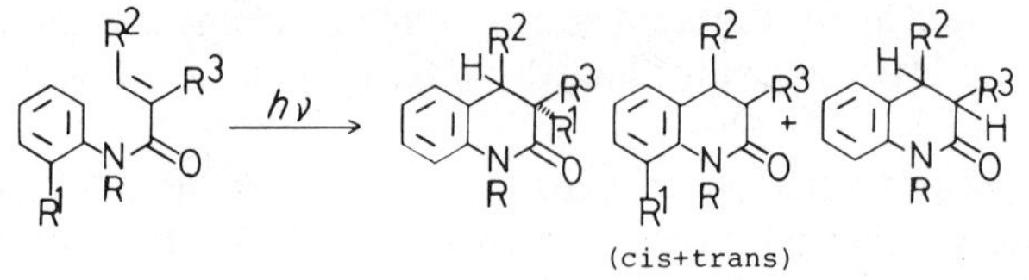

R^2=H,R^1=COOMe,R^3=Me,R=Bz   (40%)

R^2=H,R^1=Ac,R^3=Me,R=Bz   (25%)

R^1=COOMe,R^2=$(CH_2)_4$=R^3,R=Bz (71%)

R^1=Ac,R^2=$(CH_2)_4$=R^3,R=Bz   (45%)

R^1=CN,R^2=$(CH_2)_4$=R^3,R=Bz   (40%)

R^2=R=H,R^1=COOMe,R^3=Me   (61%)

R=H,R^1=COOMe,R^2=$(CH_2)_4$=R^3 (60%)

R^2=R=H,R^1=COOH,R^3=Me   (33%)

R^2=H,R^1=COOH,R^3=Me,R=Bz   (41%)

R^1=COOH,R^2=$(CH_2)_4$=R^3,R=Bz   (70%)

R^2=H,R^1=CONH$_2$,R^3=Me,R=Bz   (50%)

R^1=CONH$_2$,R^2=$(CH_2)_4$=R^3,R=Bz   (53%)

SCHEME 8

Apparently, the above results were quite contrary to those observed in the
photocyclization of *N*-benzoylenamines described in Section II,A,1,a. *N*-
Acylanilides carrying an electron-withdrawing group in the ortho position
underwent smooth photocyclization exclusively to the root of the substituent that was then migrated in a 1,5 manner suprafacially to afford the
lactam (**5**), having the substituent in the 3 position of the quinolone
structure.

This regiospecific photocyclization can be tentatively explained as in
Scheme 9. On the other hand, failure of migration by a substituent in the
photocyclization of *N*-noracylanilide (**6**) can be ascribed to a strong hydro-

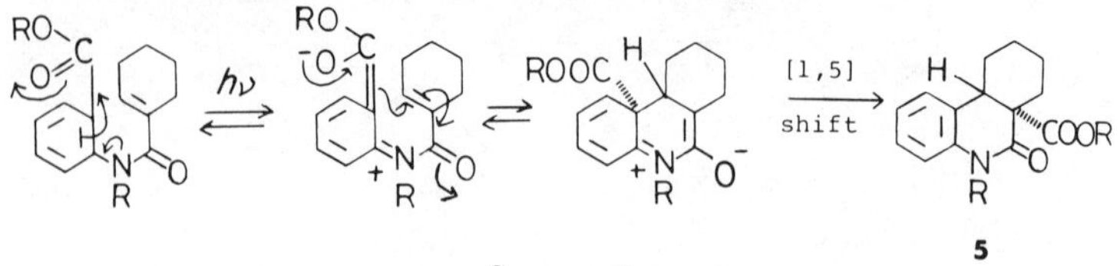

SCHEME 9

gen bond formed between a hydrogen on nitrogen and the carbonyl group in the substituent, thus fixing a conformation of the enamide in a form favorable to cyclize to the opposite site of the substituent (*35*) (Scheme 10).

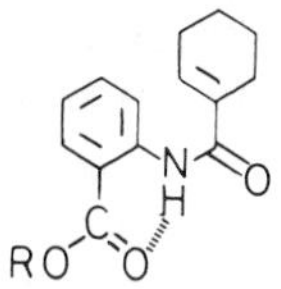

6

SCHEME 10

Interestingly, from a synthetic point of view, the photocyclization of some *N*-alkylacylanilides having an *o*-carboxyl group occurs only to the root of the substituent, which is then eliminated to yield the decarboxylated lactam. A similar elimination of the ortho substitutent was observed in the photocyclization of the enamides having an *o*-carbamoyl group, although this is less useful synthetically (*35*).

The reaction mechanism of these *N*-acylanilide photocyclizations was investigated by using deuterated solvents (*34*). Quite contrary to the result of nonincorporation of deuterium from the deuterated solvent in the photocyclization of *N*-benzoylenamines, over 90% deuterium incorporation was observed in the photocyclized cis lactam (**E**) formed as the major product in the photocyclization of *N*-acylanilides in deuterium methoxide. There was only 12% incorporation in the trans lactam (**D**), formed predominantly from photocyclization in ether with small amount of D_2O (*34*). This result clearly suggests the major contribution of a step-wise mechanism in *N*-acylanilide photocyclization and the small contribution of a concerted mechanism, which is in constrast to the major role it plays in the photocyclization of *N*-benzoylenamines (Scheme 11). Another view of the photocyclization of benzanilides involving a heterocyclic aromatic ring is described by Kanaoka (*39*).

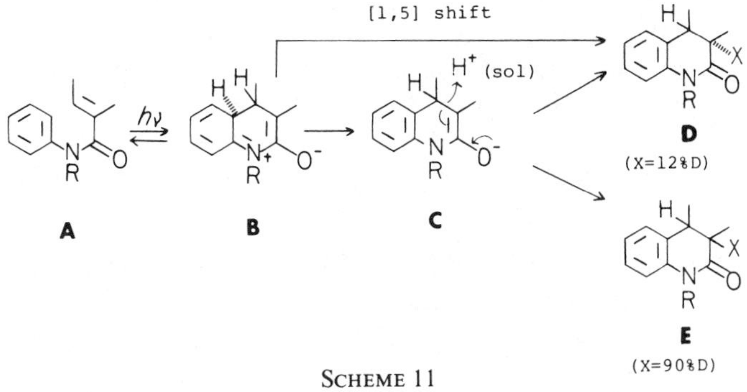

SCHEME 11

B. Oxidative Photocyclization of Enamides

Enamide photocyclization was first observed in the photocyclization of benzanilides, which afforded the phenanthridone derivatives on irradiation in the presence of an oxidizing agent (*39–42*). The role of the oxidizing agent in this photocyclization is to abstract two hydrogens from the cyclic intermediate, thus giving rise to a stable cyclic dehydrolactam as the product (*39–42*). It was also reported that under irradiation of benzanilides in the absence of the oxidizing agent, the formation of a cyclic intermediate was detected only spectrographically (*43*). However, only the starting anilide was recovered by interrupting irradiation (Scheme 12).

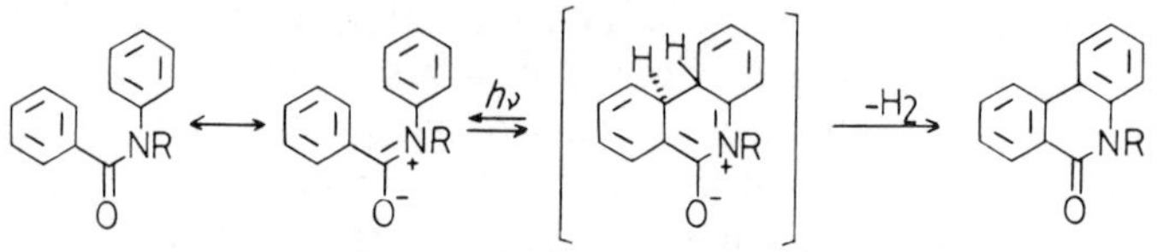

Scheme 12

In the photocyclization of enamides of the *N*-benzoylenamine type and the *N*-acylanilide type, which are both capable of undergoing nonoxidative photocyclization, the presence of an oxidizing agent causes the abstraction of two hydrogens, therefore forming the dehydrolactam having a double bond, usually at the ring juncture.

However, as data on the photocyclization of benzanilide-type enamides including a heteroaromatic ring have accumulated (*39*), the formation of the cyclized lactam by irradiation of this type of enamide seems to depend on the aromaticity of the ring involved. Although oxidative photocyclization can be applied to almost all types of enamides, the role of the oxidizing agent has not been precisely determined.

Oxidative Photocyclization of *N*-Benzoylenamines and *N*-Benzanilides

From the synthetic point of view, oxidative photocyclization of these two types of enamides has a definite advantage in the construction of a polycyclic ring system with a double bond at the ring juncture in one step from the starting compounds. The photocyclized product can be readily converted to the cis-fused ring system by a facile catalytic reduction or to the fully aromatized lactam on dehydrogenolysis (*19,20*) as in the case of *N*-benzoylenamines.

Oxidative Photocyclization of *N*-α,β-Unsaturated Acylenamines. Under irradiation in the presence of an oxidizing agent such as oxygen or iodine, enamides of the *N*-benzoylenamine type (*11,18*) and *N*-α,β-unsaturated acylanilide type (*11,34*) underwent smooth photocyclization accompanied by simultaneous dehydrogenolysis to afford the dehydrolactams (**1** and **7**) that contain a double bond mainly at the ring juncture (Scheme 13).

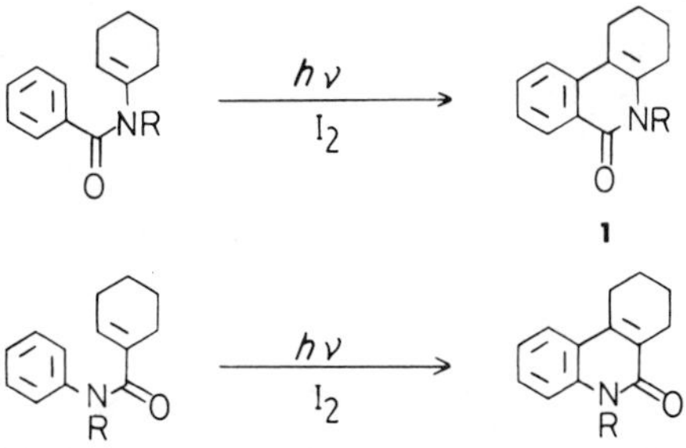

SCHEME 13   **7**

However, a dehydrogenolysis step seems to be dependent largely on either the susceptibility of the cyclized product or the reaction condition, that is, the presence of oxygen or the temperature.

In the photocyclization of 2-aroyl-1-methylene-3,4-dihydroisoquino-lines, a complete expulsion of oxygen from the solvent is required in order to avoid the formation of the dehydrolactam (*31,32,44*). Under unrestricted conditions using commercially available solvent that is not completely degassed, the dehydrolactams (**9**) are predominantly formed in addition to a small amount of the nonoxidative product (*21–24*) (Scheme 14).

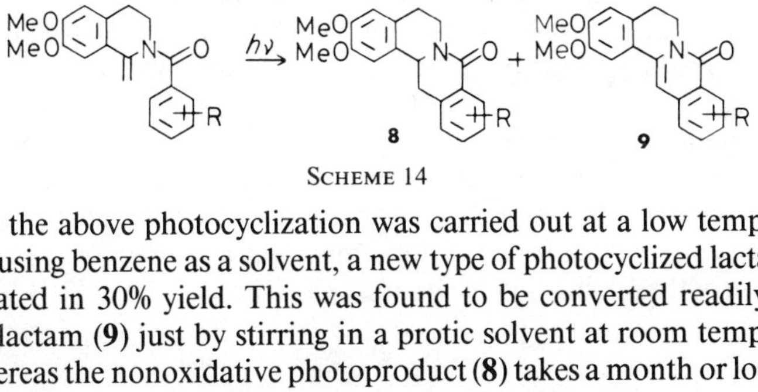

SCHEME 14

When the above photocyclization was carried out at a low temperature ($\approx 6°C$) using benzene as a solvent, a new type of photocyclized lactam (**10**) was isolated in 30% yield. This was found to be converted readily to the dehydrolactam (**9**) just by stirring in a protic solvent at room temperature (*45*), whereas the nonoxidative photoproduct (**8**) takes a month or longer for its conversion to the dehydrolactam (*24,44*) (Scheme 15).

Therefore, the formation of a dehydrolactam by irradiation of enamides under nonoxidative or unrestrictive conditions can be assumed to proceed via the route involving a 1,5-hydrogen shift of the H_b proton from the cyclic intermediate **A** to afford a thermally very unstable lactam **10** with a dihy-drobenzene structure. This compound (**10**) would then undergo facile dehydrogenation even at room temperature to afford the dehydrolactam (**9**) as the final product, although the actual role of the oxidizing agent remains to be clarified.

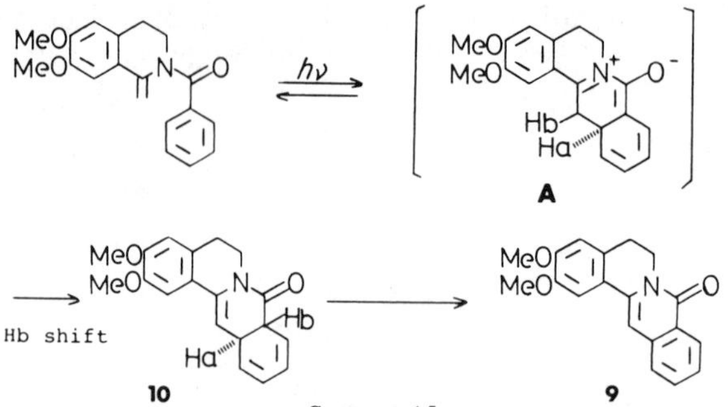

SCHEME 15

Therefore, the formation of dehydrolactam under nonoxidative photocyclization seems to depend on the migratory aptitude of either one of two hydrogens, H_a, or H_b, in the cyclic intermediate.

In addition to the works on enamide photocyclization described in this chapter, there are some interesting reports on the mechanistic study (*46,47*) and also on the application of nonoxidative cyclization to the synthesis of pharmacologically active heterocyclic compounds (*39,48–52*).

C. REDUCTIVE PHOTOCYCLIZATION OF ENAMIDES

Various types of enamides are now shown to undergo ready photocyclization according to a mechanism of a six π-electron conrotatory cyclization via a trans cyclic intermediate **B**, from which the photocyclized product is formed depending on the reaction condition, either oxidative or nonoxidative.

The structure of this cyclic intermediate **B**, which contains an immonium structure, suggests the possibility of undergoing a facile reduction by hydride, if present during the course of photocyclization. This expectation was visualized as expected on the various enamides and therefore opened up a new phase of the application of enamide photocyclization (*15*) (Scheme 16).

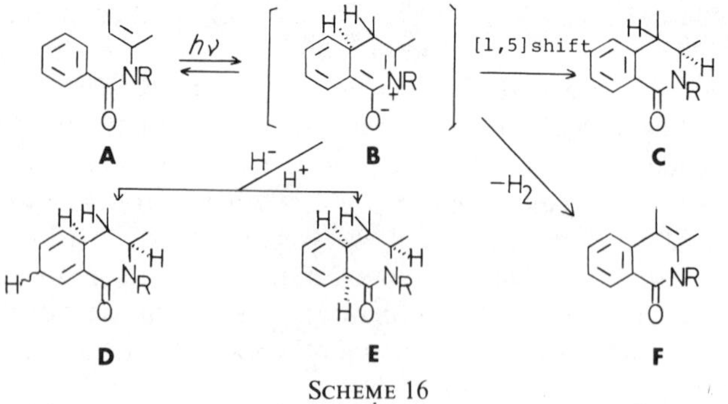

SCHEME 16

Enamide **11** was irradiated in benzene–methanol (10 : 1) in the presence of a large excess of sodium borohydride at low temperature (*15*). A mixture of two hydrogenated lactams, **12** and **13**, was obtained. Therefore, the reaction was proved to proceed as expected by the manner of hydride attack on the immonium moiety, followed by protonation from methanol at two positions on the benzene ring. When acetonitrile was used as solvent, only the unconjugated lactam **12** was obtained in an excellent yield (*15*) (Scheme 17).

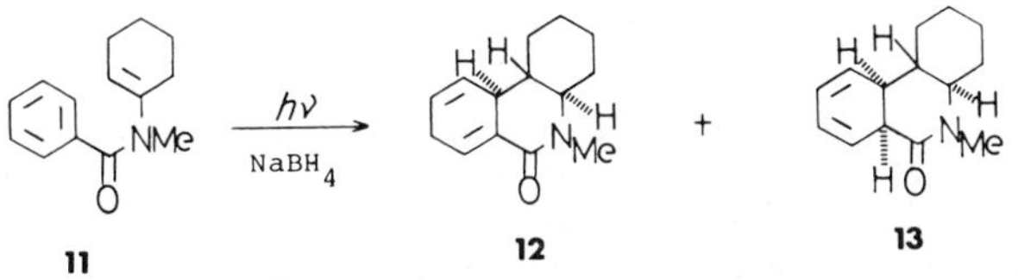

Solvent	12	13
Et$_2$O-MeOH	43%	11%
MeCN-MeOH	78%	—

SCHEME 17

The reaction course of the reductive photocyclization of enamides in Scheme 17 was firmly established by the experiments using deuterated reagents (*15*). When enamide **14** was irradiated in the presence of sodium borodeuteride in acetonitrile–methanol (10 : 1) solution, the isolated product **15** in 93% yield was deuterated at the 3 position, whereas irradiation in the presence of sodium borohydride in acetonitrile–deuterium methoxide solution afforded lactam **16**, deuterated in the benzene ring as shown in Scheme 18 (*53*).

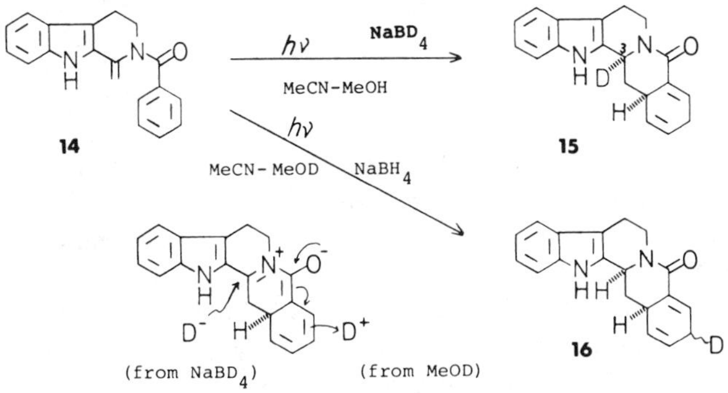

SCHEME 18

So far, reductive photocyclization has been successfully applied to the synthesis of yohimbine-type compounds (*54*), benzo[*f*]quinolines, and ergot alkaloids including lysergic acid (*55–57*) (Scheme 19).

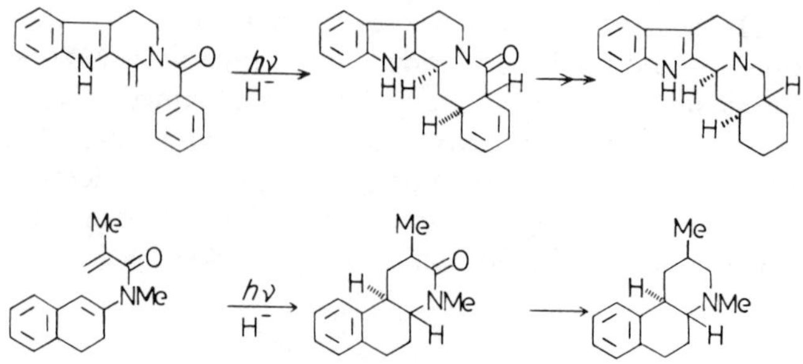

SCHEME 19

D. ASYMMETRIC PHOTOCYCLIZATION OF ENAMIDES

Discovery of reductive photocyclization of enamides has certainly increased its importance as a general and useful synthetic tool.

Irradiation of enamides in the presence of a hydride (e.g., sodium borohydride) in a solution containing a protic solvent such as methanol, brought about reductive photocyclization (*15,54–57*). However, it is assumed that irradiation in the presence of aprotic solvent affords the photocyclized product (**17**) identical to that formed by irradiation under nonoxidative conditions according to the route suggested in Scheme 20.

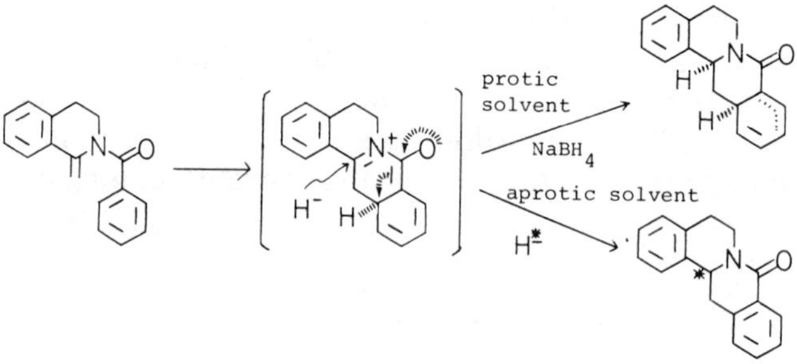

SCHEME 20   **17**

The use of a chiral hydride complex has been central to the asymmetric reduction of ketones such as acetophenone (*58*). A number of excellent chiral metal hydride complexes have been introduced by many researchers, including Noyori (*59,60*), Meyers (*61*), Mukaiyama (*62,63*), Terashima (*64,65*), and others (*58*). It is apparent that there is a close similarity in structure between acetophenone and the proposed intermediate in enamide photocyclization, therefore suggesting the possibility of undergoing photocyclization in an asymmetric manner.

When irradiation of enamide **18** was carried out in the presence of a chiral metal hydride complex prepared from lithium aluminum hydride (LAH) and a chiral amino alcohol such as quinine, quinidine, N-methylephedrine, or chirald, the photocyclized lactam **17** showed optical activity to some extent, thus proving the enantio-differentiated nature of this reductive photocyclization (*16*) (Scheme 21).

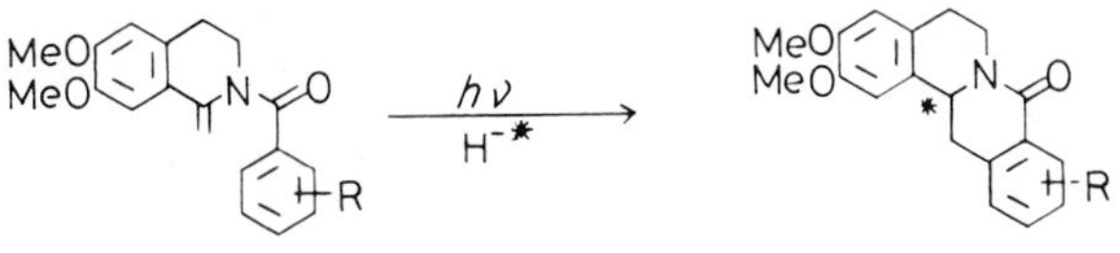

18

SCHEME 21

17

Furthermore, an asymmetric synthesis of natural xylopinine (**20**) by the route involving reductive photocyclization was successfully accomplished (*16*). Independently, Kametani *et al.* reported another asymmetric total synthesis of natural xylopinine via the route involving a diastereo-differentiated photocyclization of the chiral enamide (**19**) under nonoxidative conditions (*66*). Therefore, it is now established that enamide photocyclization is a reaction capable of not only undergoing a wide variety of cyclizations under various conditions but also having a high quality (Scheme 22).

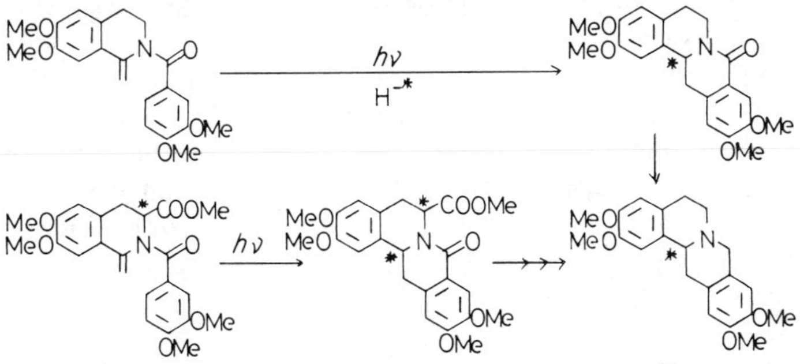

19

20

SCHEME 22

III. Thermal Cyclization of Enamides

As knowledge of the reactivity of enamides accumulated during the course of a 10-year study on enamides, attention was focused on the cyclizability of enamides also under thermal conditions. It was first thought that electron deficiency in the aromatic ring of enamides caused by the introduction of an electron-withdrawing factor would bring about nucleophilic attack from the enamine portion of the molecule, therefore giving rise to the cyclized product, even under thermal conditions as described in Scheme 23.

As expected, it has been shown that some enamides having an electron deficiency in their aromatic moiety can undergo cyclization under thermal

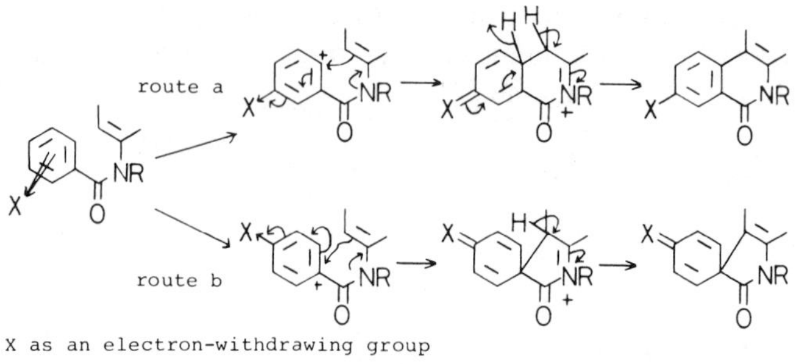

SCHEME 23

conditions. Introduction of a nitro group and the formation of a pyridinium structure in the aromatic ring of enamides were enough to bring about thermal cyclization.

A. CYCLIZATION OF ENAMIDES HAVING A NITRO GROUP ON BENZENE

As an electron-withdrawing factor, the nitro group shows some effect on the cyclization of enamides (53,67). When a nitro group was introduced into the benzene ring of enamide **21**, irrespective of its position, only N→C acyl migration was observed. However, thermal cyclization was observed when an additional methoxyl group was introduced in the ortho position (53).Thermal cyclization becomes the normal reaction when two nitro groups are present in the benzene ring, regardless of their positions (53,67). In all these cases, the cyclized products are dehydrogenated lactams as shown in Scheme 24.

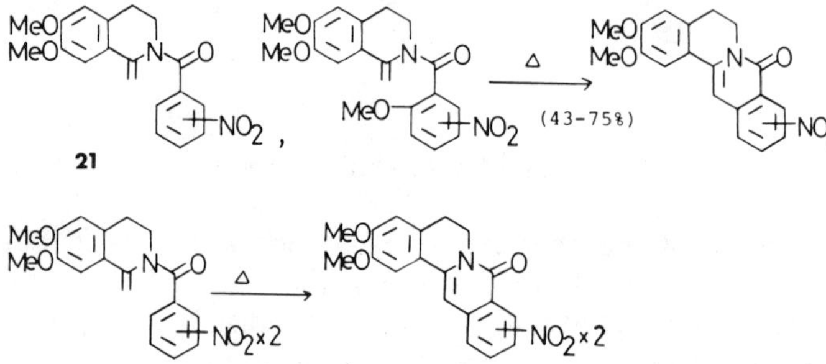

SCHEME 24

B. CYCLIZATION OF ENAMIDES HAVING A PYRIDINE MOIETY

Electron deficiency occurs readily during acylation of the enamides of the N-pyridinecarbonylenamine type (67–70). Acylation of this type of enamide yields the already cyclized compounds.

1. Acylation of Imines with Isonicotinoyl Chloride (Route b)

Acylation of imine **22** with one mole of isonicotinoyl chloride gave enamide **23** in good yield. However, the product obtained from acylation of imine **22** with an excessive amount of the acid chloride was identical to that obtained from acylation of enamide **23**, also with an excessive amount of the acid chloride, and the product was the already cyclized spirodihydropyridine **24** (Scheme 25) (*68,69*).

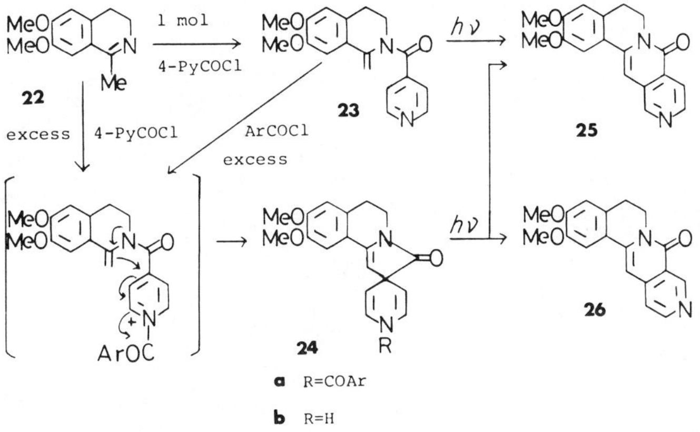

4-PyCOCl = Isonicotinoyl chloride

SCHEME 25

The spirodihydropyridine **24** thus obtained was converted to the corresponding azaberbines (**25** and **26**) by alkaline hydrolysis, followed by photolysis (*68*). Thus, this spirodihydropyridine formation became a useful synthetic route for the alkaloids such as nauclefine (*68*). Similarly, the formation of spirodihydropyridines was observed from acylation of harmalane and the open-chain 3-aminocrotonate (*70*).

2. Acylation of Imines with Nicotinoyl Chloride (Route a)

Similarly, acylation of imine **22** with nicotinoyl chloride afforded enamide **27** when one mole of the acid chloride was used, whereas acylation with an excessive amount of the acid chloride yielded the already cyclized dihydroazaberbine (**28**) in good yield, which was also obtained from the enamide by acylation (*67*). Hydrolysis of **28** afforded the corresponding azaberbine (**26**) (Scheme 26) (*67*).

By applying this facile thermal cyclization of enamides, nauclefine (**29**) was synthesized (Scheme 27) (*69,71,72*).

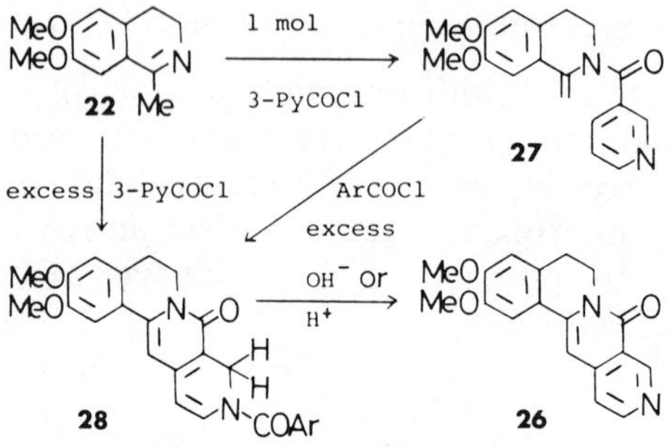

3-PyCOCl = Nicotinoyl chloride

SCHEME 26

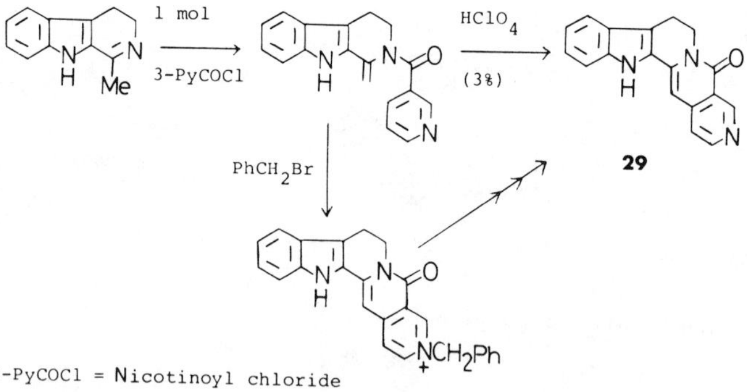

3-PyCOCl = Nicotinoyl chloride

SCHEME 27

C. COMPARISON OF THE CYCLIZABILITY OF ENAMIDES

In the course of study on thermal cyclization of enamides of N-nicotinoy-lenamine type (27), the cyclizability of enamides under various conditions was investigated (73). The results are summarized in Table II.

Apparently, all the enamides studied, irrespective of their substituent on the pyridine ring, underwent cyclization under both photochemical and thermal conditions to afford the cyclized lactams, without showing a partic-ular regioselectivity to the orientation of cyclization. However, in the case of enamide 27a, unsubstituted on the pyridine ring, the cyclization was not observed under thermal conditions but under both photochemical and acylating conditions. Introduction of an additional electron-withdrawing group in the pyridine ring of enamides 27b and c facilitated the cyclization under any conditions mentioned to afford the corresponding azaberbines 26 and 30.

TABLE II

THERMAL CYCLIZATION OF ENAMIDES

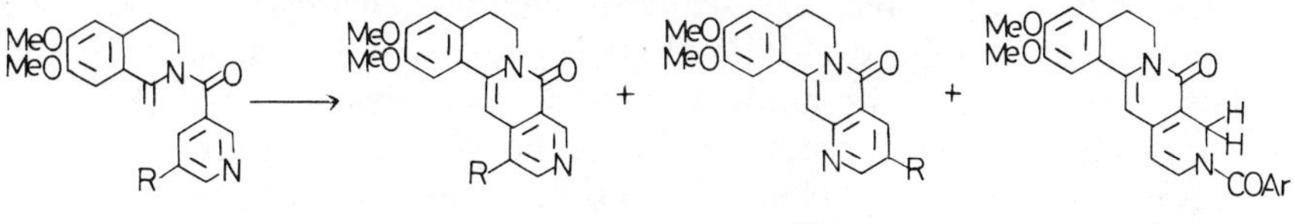

Conditions	26	30	28
Photochemical	20[a]	10	—
	43[b]	22	—
	39[c]	20	—
Thermal	—[a]	—	—
	55[b]	7	—
	54[c]	10	—
Under Acylation	10[a]	—	20
	4[b]	34	—
	3[c]	24	—

[a] R = H.
[b] R = COOEt.
[c] R = Ac.

Based on the above result, a newly isolated alkaloid alamarine (**33**, structure, see p. 240) was synthesized by the route involving enamide cyclization under both thermal and photochemical conditions, giving predominantly the desired lactam (**32**) (Scheme 28) (*73*).

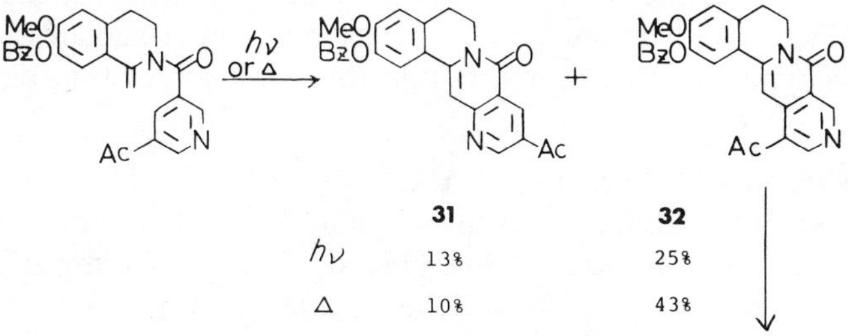

	31	**32**
hν	13%	25%
Δ	10%	43%

Bz=CH₂Ph

33 Alamarine

SCHEME 28

Recently, Mujumdar and Martin (*74*) and Atta-ur-Rahman and Ghazala (*75*) independently reported thermal cyclization of the enamides of *N*-nicotinoylenamine type.

IV. Alkaloid Syntheses: Application of Enamide Cyclization

One of the major areas of research in enamide chemistry has been the application of enamide cyclization to the synthesis of natural alkaloids and related heterocyclic compounds. Ever since Ninomiya and co-workers discovered nonoxidative photocyclization of the N-benzoylenamine-type enamides in 1969 (9), several types of alkaloids have been synthesized, starting from enamides of rather simple structures, by applying enamide cyclization under both photochemical and thermal conditions.

An advantage of this methodology is that a strained or complicated molecule of the natural alkaloids can be built from simple compounds, such as commercially available carboxylic acids or amines, from which enamides are prepared. In addition, on cyclization, the skeletons of complex polycyclic and multifunctionalized natural products or their precursors, which are easily converted to the target alkaloids, can be formed in one step.

By taking advantage of these features, enamide cyclization has shown its great versatility in the synthesis of a wide variety of isoquinoline and indole alkaloids.

In this chapter, we first summarize more or less basic works on the synthesis of the common and basic skeletons of various types of natural alkaloids, aiming at their application to total synthesis, and also those works that seem applicable to the above syntheses, and then collected works on total syntheses of alkaloids via the route involving enamide cyclization.

A. Amaryllidaceae Alkaloids

The use of enamide photocyclization in the synthesis of Amaryllidaceae alkaloids has remained a basic study and so far limited only to the synthesis of the skeletons of lycorine and crinine, as well as intermediates in the total synthesis of haemanthidine and nortazettine, and some of the degradation products of these alkaloids.

The Skeleton of Lycorine-Type Alkaloids. In the course of photochemical studies on 1-aroylindoles, Carruthers and Evans (76) synthesized pyrrolo[3,2,1-d,e]phenanthridin-7-one (35), the basic skeleton of lycorine, in 10% yield by irradiation of 1-o-iodobenzoylindole (34). However, photocyclization of the corresponding indole analogs proceeded quite differently as shown in Scheme 29.

Photocyclization of the 2-bromobenzoylindolines also proceeded smoothly to give anhydrolycorine, a degradation product of lycorine, or the apoerysopine skeleton (77). Contrary to the aforementioned results of Carruthers, Hara et al. (77) synthesized anhydrolycorine by photocycliza-

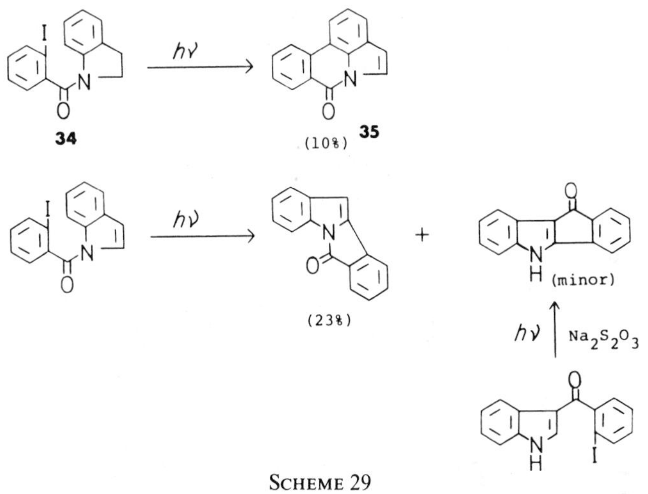

SCHEME 29

tion of the 2-bromobenzoylindoline **36**, followed by reduction with LAH. Since irradiation of the ortho-unsubstituted enamide **38** gave only the photo-Fries product **39**, this cyclization is considered to proceed by a carbon–carbon bond homolysis, followed by free radical cyclization. This synthesis meant a formal synthesis of γ-lycorane (**37**) (*77*) (Scheme 30).

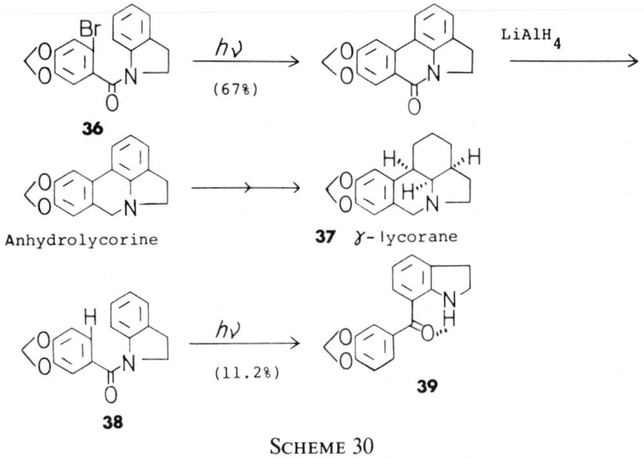

SCHEME 30

γ-Lycorane. In addition to the above synthesis (*77*), Iida *et al.* (*78*) reported a facile synthesis of γ-lycorane (**37**) along with (±)-α-anhydrodihydrocaranine (**42**). Irradiation of the enamide (**40**), readily prepared in three steps from 3-chloro-4-methoxyphenethylamine, gave the dehydrogenated lactam **41** in 69–70% yield, which was then reduced to afford γ-lycorane and (±)-α-anhydrodihydrocaranine (**42**). In this cyclization, the presence of a carbonyl group next to an olefinic double bond of the enamide (**40**) showed a marked accelerating effect on the photocyclization (Schemes 31 and 32).

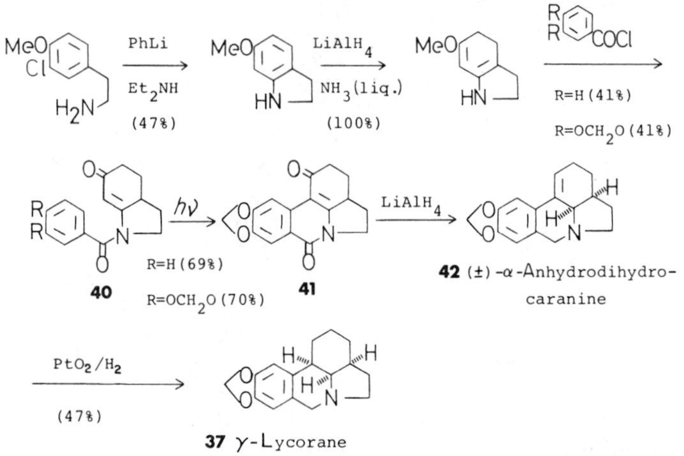

40 R=H (69%)
R=OCH$_2$O (70%) **41**

42 (±)-α-Anhydrodihydro-
caranine

37 γ-Lycorane

SCHEME 31

Anhydrolycorine

Dimethylapoerysopine

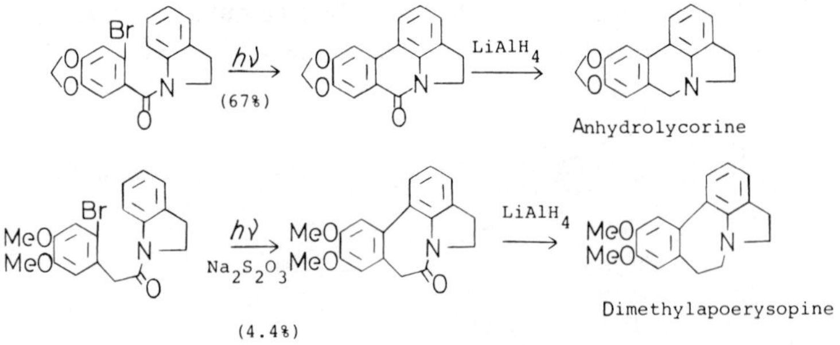

SCHEME 32

Narciprimine and Arolycoricidine. According to the route described in the synthesis of anhydrolycorine (*77*), arolycoricidine (**44**) and narciprimine (**46**) were synthesized by the routes of oxidative photocyclization of enamide **43** and free radical cyclization of *o*-bromoenamide **45** under irradiation (*42*) (Scheme 33).

Crinane. A stereoselective synthesis of (±)-crinane was reported as the first application of nonoxidative photocyclization of an enamide to alkaloid synthesis (*79,80*). Acylation of the benzylimine of 2-allylcyclohexanone with piperonyloyl chloride gave a mixture of two isomeric enamides, **47** and **48**, which were converted to the homogeneous enamide **48** by simple heating or irradiation with a high-pressure mercury lamp. Irradiation of the

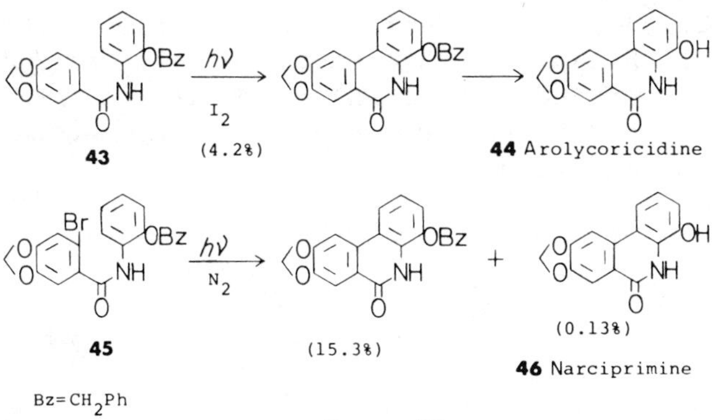

SCHEME 33

enamide **48** with a low-pressure mercury lamp afforded the trans lactam **49** in 20% yield in addition to a small amount of the regioisomer **50**. Ozonation of the allyl group in **49**, followed by LAH reduction gave the amino alcohol **51** that was then debenzylated and treated with thionyl chloride to afford (±)-crinane (**52**) (*79,80*) (Scheme 34).

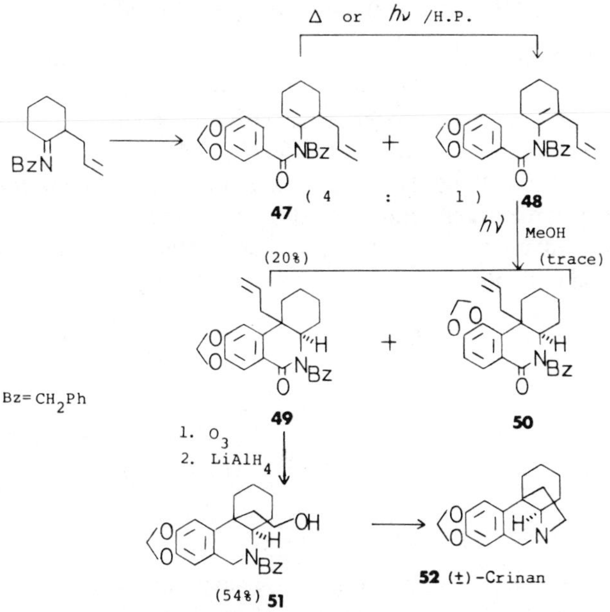

SCHEME 34

Ninomiya *et al.* have also synthesized a key intermediate (**55**) in the Hendrickson's synthesis (*81*) of haemanthidine and nortazettine from the enamide (**53**) (*12*). Irradiation of the enamide (**53**) afforded the lactam **54** in

40% yield, and the subsequent hydrogenation gave the desired trans lactam in 80% yield. This was further converted to the key intermediate **55** (*81*). Again, photocyclization of the enamide **53** proceeded very smoothly because of the presence of an electron-withdrawing group in the *β*- position of the enamide (Scheme 35).

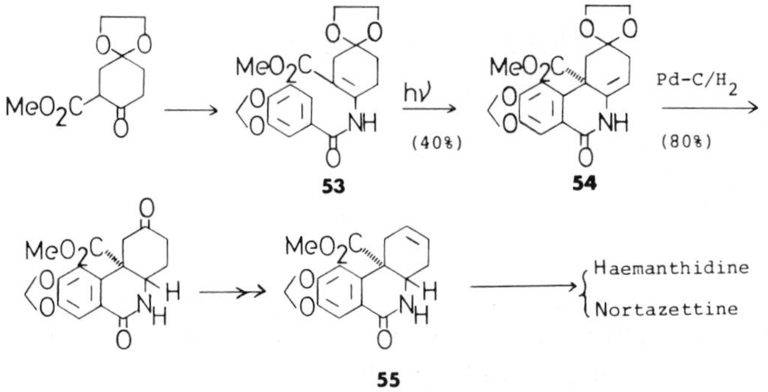

53   **54**

55

SCHEME 35

B. BENZO[*c*]PHENANTHRIDINE ALKALOIDS

Benzo[*c*]phenanthridine alkaloids, which are divided into two groups represented by nitidine and chelidonine, possess various levels of oxidation in rings B and C, where nitrogen exists either as lactam, tertiary amine, or quaternary salt (*1,2*). Recently, because of the discovery of marked potencies in some of the alkaloids against leukemia (*82,83*), the need for a large supply has become urgent and presents a challenge to synthetic chemists to establish a convenient and large-scale synthetic route to this group of alkaloids (Scheme 36).

Nitidine   Chelidonine

SCHEME 36

1. Synthesis of the Basic Skeleton

The enamide **56** is readily prepared by acylation of the 1-tetraloneimine with benzoyl chloride in the presence of triethylamine, resulting in good yield. Photocyclization of the enamide **56** proceeds straightforward and is of an nonoxidative and stereospecific nature, yielding a homogeneous trans lactam (**57**) in over 50% yield (*19,20*). Lithium aluminum hydride reduction of the trans lactam (**57**) gave the corresponding trans amine **58**. When the lactam (57) was treated with selenium at an elevated temperature in order to identify the structure of the photocyclized lactam (57), the major product obtained was the cis lactam (61), which is isomeric with the photoproduct (57), along with a small amount of the known benzo[c]phenanthridone (62) (84).

Reduction of the cis lactam 61 with LAH afforded the cis amine 60, which was also obtained from either oxidative photocyclization of the enamide 56 or nonoxidative photocyclization of the bromoenamide 56, followed by successive reductions. Oxidative photocyclization of the enamide 56 in the presence of iodine afforded the corresponding dehydrolactam 59 in good yield, which is a useful intermediate for further conversion to various aromatized benzo[c]phenanthridines, the basic structure of many aromatic alkaloids (19,20).

Regiochemistry of enamide photocyclization in this synthesis can also be controlled by the use of an ortho substituent that acts as an eliminating group on cyclization to the root of the substituent. Thus, the o-methoxyl and bromo groups are used for this regioselective cyclization as exemplified by a number of total syntheses of alkaloids (19,20) (Scheme 37).

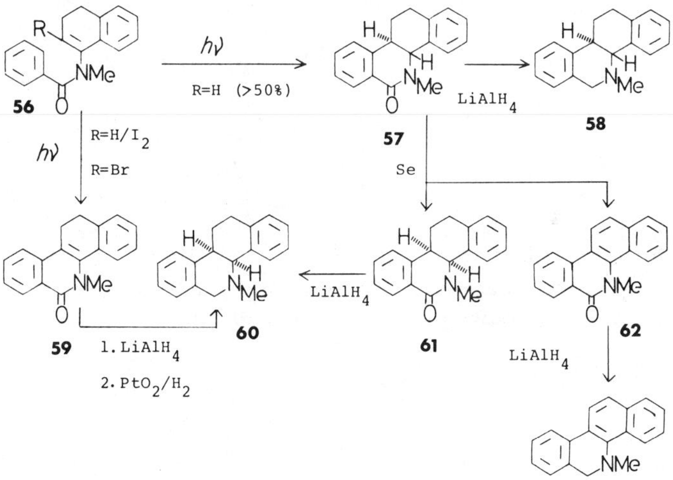

SCHEME 37

The substituent effects that have either an accelerating or retarding effect on photocyclization have been studied by Ninomiya et al. (28,85,86) (Scheme 38).

Prolonged irradiation (20 hr) on the enamides 63a and b, having an electron-withdrawing group in the para position, brought about cyclization and isomerization to afford the cis lactam 65 as the sole product in 20% yield, whereas 7-hr irradiation afforded the trans lactam 64. Furthermore,

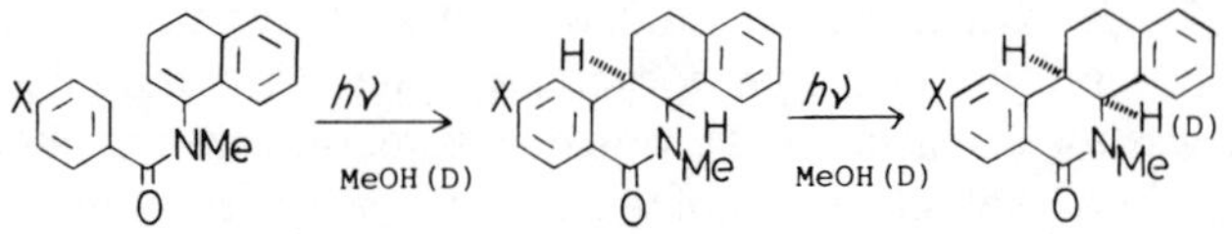

63

a X=COOMe

b X=CN

c X=OMe

64

65

SCHEME 38

the trans lactam (64) was isomerized to the cis lactam 65 quantitatively, when irradiated in methanol (28,87). The p-methoxy-substituted enamide 63c gave only the trans lactam 64, even after prolonged irradiation (28,87).

The presence of a substituent (R = Me) on the double bond of the enamide 66 lowered the yield, probably due to steric crowdedness, whereas an electron-withdrawing group such as ester or amide in the same position raised the yield considerably (85,86) (Scheme 39).

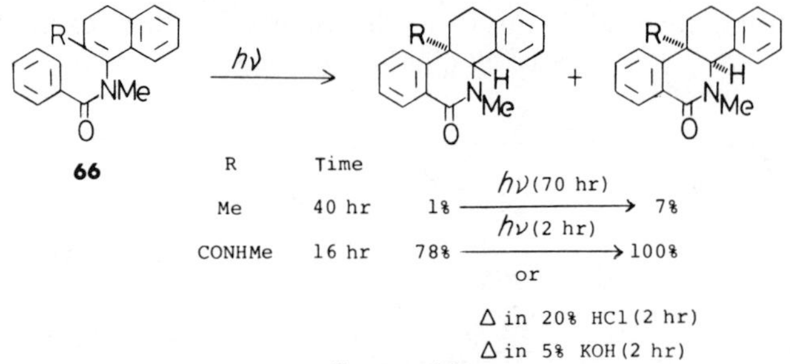

66

R	Time		
Me	40 hr	1% $h\nu$(70 hr)	7%
CONHMe	16 hr	78% $h\nu$(2 hr)	100%

or

Δ in 20% HCl (2 hr)

Δ in 5% KOH (2 hr)

SCHEME 39

Kessar et al. (88) described the first photocyclization of (2-bromobenzoyl)-1-naphthylamine, which was successfully applied to the synthesis of some aromatic alkaloids of this group (Scheme 40).

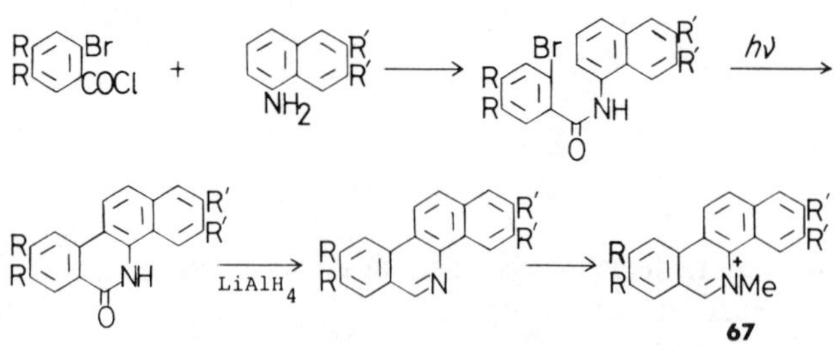

67

	R	R'	Yield(%) of 67
a	OCH$_2$O	H	48
b	OMe	OCH$_2$O	70
c	OCH$_2$O	OCH$_2$O	70

SCHEME 40

2. Total Synthesis of Fully Aromatized Alkaloids

Synthesis of the fully aromatized alkaloids can be classified by the patterns of substituents, mostly the methoxyl and methylenedioxy groups, namely, the 2,3,7,8-tetrasubstituted and 2,3,8,9-tetrasubstituted alkaloids.

Chelerythrine. Ninomiya *et al.* (*89*) in 1977 applied enamide photocyclization to the enamide **68** on their way to total synthesis of homochelidonine. The cyclization did not show any regioselectivity, thus giving two lactams (**70**) and (**69**), of which the former (**70**) was converted to oxychelerythrine (**71**) for the assignment of the structure of the photoproduct in this synthesis (Scheme 41).

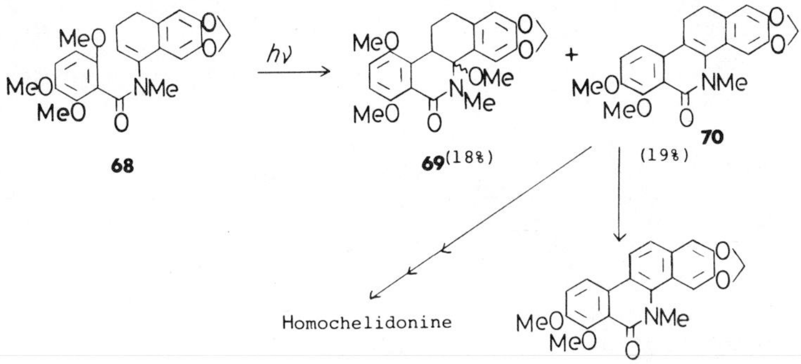

SCHEME 41

Sanguinarine. As in the case of oxychelerythrine, a similar approach was used to synthesize oxysanguinarine (**72**) (*11*) (Scheme 42).

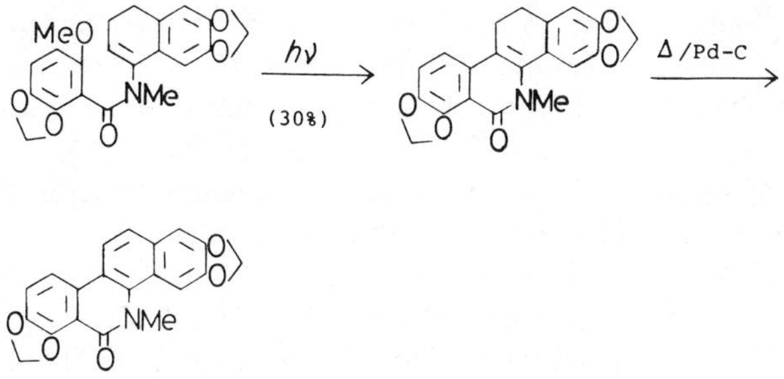

72 Oxysanguinarine

SCHEME 42

Chelirubine. Although all of the 2,3,7,8,*x*-pentasubstituted alkaloids have been synthesized only recently, their synthetic routes had been hampered by the ambiguity of the structures of these alkaloids, particularly with respect to the positions of the substituents on the rings (*90*). The structure of chelirubine was finally established by the total synthesis of the proposed structure by Ishii *et al.* collaborating with Ninomiya's group (*91,92*).

The enamide **73** was prepared from 2,3-dimethoxy-5,6-methylenedioxybenzoic acid and 6,7-methylenedioxy-1-tetralone, and was then irradiated to afford the lactam **74** in 38% yield. On acid treatment **74** was converted to the dehydrolactam **75** as a result of the elimination of the migrated methoxyl group. Dehydrogenation with 30% Pd/C, reduction with vitride, followed by treatment with DDQ, completed total synthesis of chelirubine (**76**) (*91,92*) (Scheme 43).

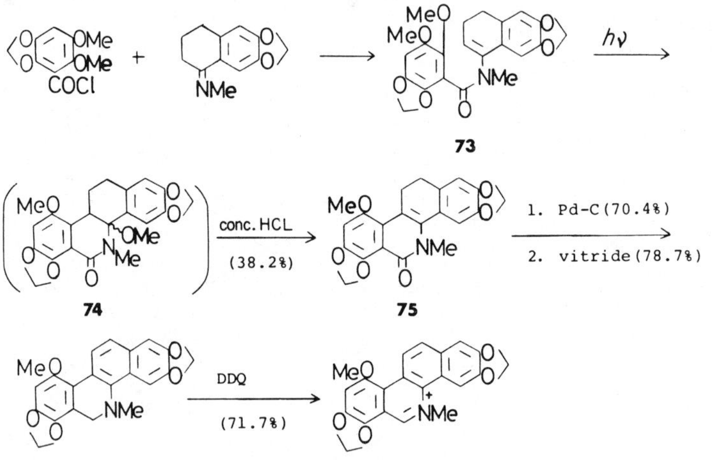

76 Chelirubine
SCHEME 43

Avicine and Nitidine. Among many 2,3,8,9-tetrasubstituted alkaloids, nitidine (**77**) and fagaronine (**79**) have been shown to possess very potent antileukemic activity, therefore becoming targets for their synthesis (*82,83*).

Nitidine (**77**) and avicine (**78**) were synthesized by photocyclization of two types of enamides **80** and **85**, and **90** of which the former two were used for irradiation by Ninomiya *et al.* (*93*), and the latter by Kessar *et al.* (*88*).

Photocyclization of the enamide **80** and the *o*-methoxy-substituted enamide **85** under nonoxidative conditions afforded the trans lactams **81** (52%) and **86** (53%) and the dehydrolactams **82** (41%) and **87** (50%), respectively. Since dehydrogenation of the trans lactams **81** and **86** with 30% Pd/C gave lactams **83** and **88**, although in low yields, the dehydrolactams (**82**) and (**87**) were used for ready dehydrogenation with the same reagent, followed by reduction to the desired alkaloids dihydroavicine (**84**) and dihydronitidine (**89**), natural alkaloids that had been converted already to nitidine and avicine, respectively (*1,2*) (Schemes 44 and 45).

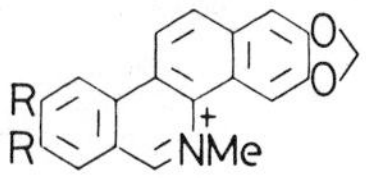

77 R=OMe   Nitidine

78 R=OCH$_2$O Avicine

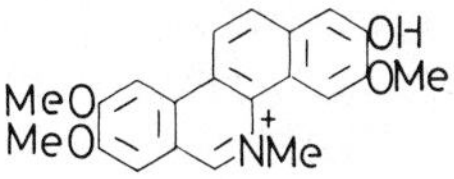

79 Fagaronine

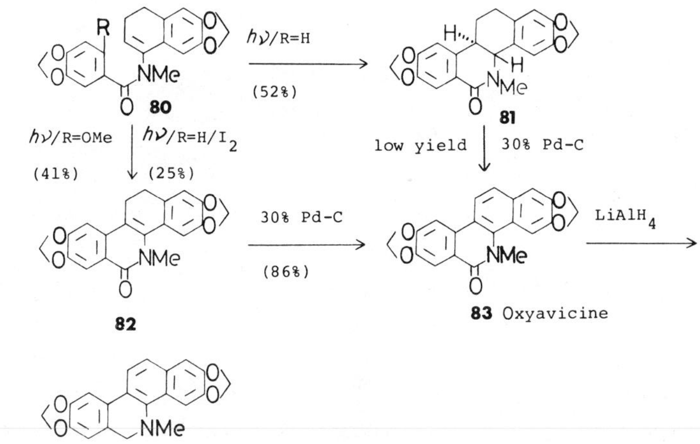

80

hν/R=OMe   hν/R=H/I$_2$

(41%)   ↓ (25%)

81

low yield   30% Pd-C

82

30% Pd-C

(86%)

83 Oxyavicine

LiAlH$_4$

84 Dihydroavicine

SCHEME 44

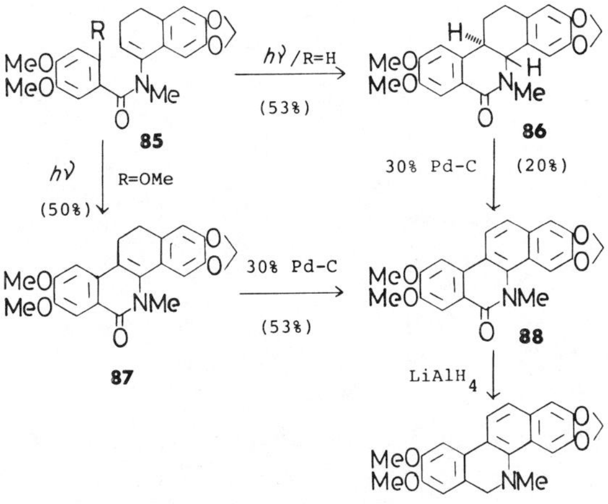

85

hν   R=OMe

(50%) ↓

86

30% Pd-C   (20%)

87

30% Pd-C

(53%)

88

LiAlH$_4$

89 Dihydronitidine

SCHEME 45

A more direct approach to nitidine and avicine was reported by Kessar *et al.* (*88*) who showed that *o*-bromonaphthalide (**90**) was directly converted to the lactam **91** in 70% yield. Subsequently, they also demonstrated that when the enamide was reduced to the amine, photocyclization and dehydrogenation occurred to give the fully aromatized compound **92** (*88*) (Scheme 46).

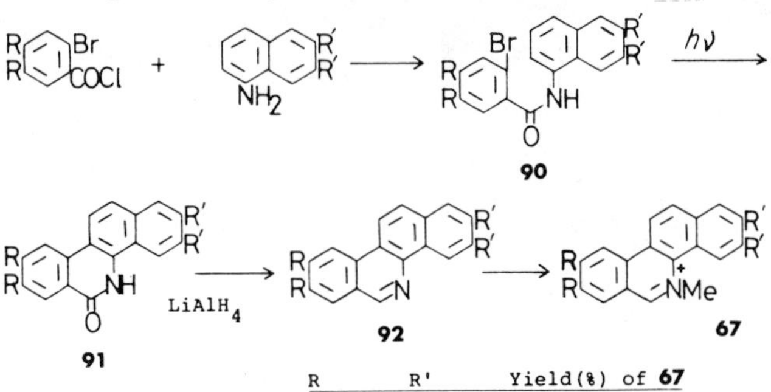

	R	R'	Yield(%) of **67**
a	OCH$_2$O	H	48
b	OMe	OCH$_2$O	70
c	OCH$_2$O	OCH$_2$O	70

SCHEME 46

Begley and Grimshaw (*94*) reported a modified synthesis of nitidine (**77**) by shortening the steps in Kessar's route (*88*) by using the *N*-methylbenzamide (**93**) under photochemical conditions. However, they failed to obtain benzo[*c*]phenanthridone by electrochemical means (*94*) (Scheme 47).

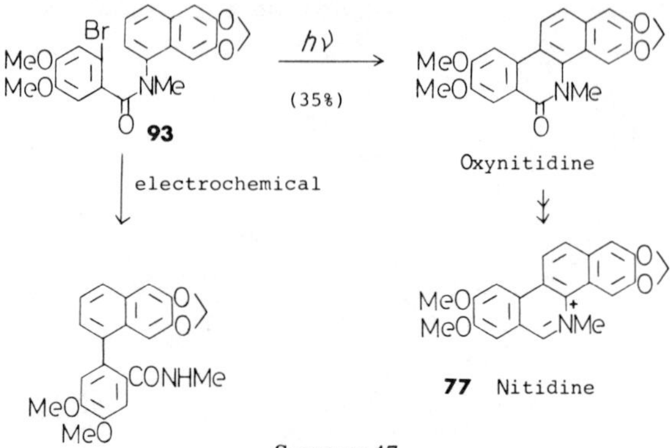

SCHEME 47

Fagaronine. Fagaronine (**79**) is an alkaloid possessing a potent antileukemic activity now under clinical evaluation (*82*). However, the presence of a hydroxyl group at the 2 position of the skeleton made its synthesis very troublesome. Using the methodology developed by Kessar *et al.* (*88*), Ninomiya *et al.* (*95*) synthesized demethylfagaronine (**96**) via the lactam **95**, which was readily prepared from the *o*-bromoenamide **94** on irradiation (Scheme 48).

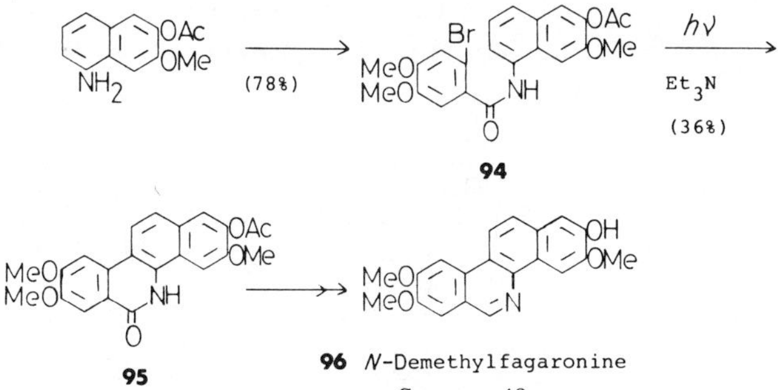

SCHEME 48

3. Total Synthesis of 4b,5,6,10b,11,12-Hexahydro Alkaloids

Hexahydrobenzo[*c*]phenanthridine alkaloids consist of two groups: the corynolines and the chelidonines (*1,2*). They differ only on the presence or absence of a 10b-methyl group at the B/C ring junction, and otherwise have the common structural feature of a B/C-*cis*-4b,5,6,10b,11,12-hexahydro-benzo[*c*]phenanthridine skeleton with at least one hydroxyl group at the 11 position. Therefore, the stereochemistry of the substituent in ring C and the mode of ring juncture are the major problems in their total synthesis.

Homochelidonine. Based on the stereoselective synthetic route developed by using model compounds (*96*), Ninomiya *et al.* (*89*) applied enamide photocyclization to the enamide **86**, which was prepared from 6,7-methy-lenedioxy-1-tetralone and 2,3,6-trimethoxybenzoic acid, and obtained a mixture of two photocyclized lactams **70** and **69** in yields of 19 and 18%, respectively, as a result of no regioselectivity due to the presence of two *o*-methoxyl groups, one of which was introduced to control the cyclization to the root of the substituent. The structure of the desired lactam **70** was confirmed by the conversion to oxychelerythrine (*71*). Introduction of the oxygen function into ring C was performed by oxidation with lead tetraace-tate to give the 12-acetoxy derivative **97** in good yield, which was then converted by hydrolysis to the phenolic compound **98**. A further oxidation of this 12-hydroxy derivative (**98**) with chromic trioxide afforded the 11,12-*o*-quinone **99**, which was first reduced with LAH, followed by catalytic hydrogenation of the remaining 4b,10b double bond to yield the B/C-cis-11,12-trans-glycol **100**, though in only 12% yield. The conversion of this glycol (**100**) to homochelidonine (**103**) was accomplished by solvolytic cleavage of the 12-hydroxyl group via an 11,12-epoxide. Acetylation fol-lowed by mesylation afforded the diester **101**, which on hydrolysis with base was converted to the 11-hydroxy-12-methoxy derivative **102**, presumably via the intermediary formation and cleavage of an 11,12β-epoxide. Hydrogenolysis of the 12-methoxyl group of **102** in the presence of palladium on carbon completed the first total synthesis of homochelidonine (**103**) (*89*) (Scheme 49).

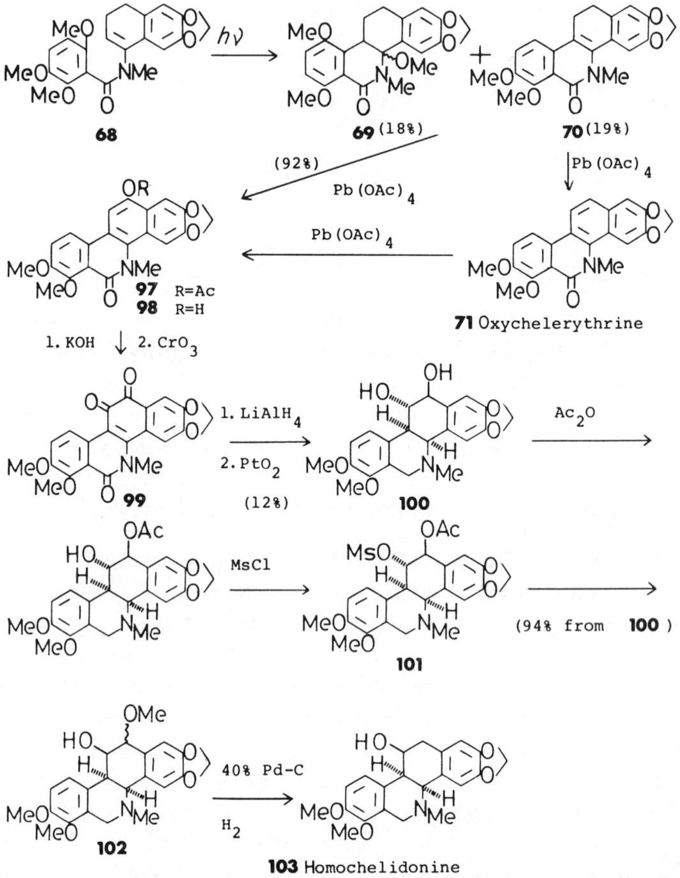

SCHEME 49

Corynoline, 12-Hydroxycorynoline, and 11-Epicorynoline. By taking the reported interconversions (*97–99*) of the corynoline group of alkaloids into consideration, Ninomiya *et al.* (*85,86,100–102*) accomplished the first total synthesis of most of the corynoline group **114**, **116**, and **117** by applying enamide photocyclization. They first prepared two important key intermediates, the lactam **109** and the amine **110**, which were proved to be the potential starting compounds for stereoselective introduction of the functional groups into the 11 and 12 positions.

The enamide 104 was readily prepared from two starting blocks, 6-methoxy-2,3-methylenedioxybenzoic acid and 6,7-methylenedioxy-2-methyl-1-tetralone. Irradiation of the enamide 104 in methanol solution afforded a mixture of two lactams, 105 and 106, in yields of 20 and 10% respectively, as a result of poor regioselectivity in the cyclization. One o-methoxyl group migrated in a 1,5 manner to the 4b position after cyclization to the root of this methoxyl group and was readily eliminated under hydrogenolytic conditions with palladium on carbon. The products obtained were a mixture of the B/C-trans and -cis lactams 108 and 107 in yields of 60 and 21%, respectively, of which the former (108) can be used for the synthesis of isocorynoline (1,2). The desired B/C-cis lactam 107 alone was obtained by hydrogenolysis with sodium borohydride in the presence of boron trifluoride etherate in 70% yield. Dehydrogenolysis with DDQ converted the lactam 107 to the 11,12-dehydrolactam 109 in 28% yield. This was further reduced with LAH to the corresponding dehydroamine 110; thus the first target compounds for total synthesis were prepared (Scheme 50).

Glycol formation from the dehydrolactam 109 by oxidation with performic acid, followed by hydrolysis afforded a mixture of the 11,12-cis 111 and 11,12-trans (112) glycols, which was reduced with LAH to give the corresponding dihydroxyamines 113a and 113b, respectively. Elimination of the 12-hydroxyl group of 113a and 113b was accomplished by hydrogenolysis with palladium on carbon to afford 11-epicorynoline (114) (101,102).

On the other hand, the dehydroamine 110 was converted on treatment with performic acid to the 11,12-trans glycol 115, identical to natural 12-hydroxycorynoline (115) (100,101). Hydrogenolytic cleavage of the 12-hydroxyl group of 12-hydroxycorynoline (115) with palladium on carbon afforded corynoline (117) along with the isomeric cis glycol 116 (100,101).

C. PROTOBERBERINE AND RELATED ALKALOIDS

There are two types of enamides readily available for the simple synthesis of protoberberine alkaloids via the route involving photocyclization. One is the 2-acyl-1-benzylideneisoquinoline-type enamide that has a stilbene structure in the molecule and is known to undergo a six π-electron conrotatory electrocyclic cyclization to a dihydrophenanthrene and subsequent dehydrogenation (103), and thus provides a useful synthetic route to aporphine alkaloids. The second is the 2-aroyl-1-methyleneisoquinoline-type enamide that gives the 8-oxoberbine in good yields on irradiation.

Photocyclization of the former type of enamides for the synthesis of aporphine alkaloids will be summarized in the following section.

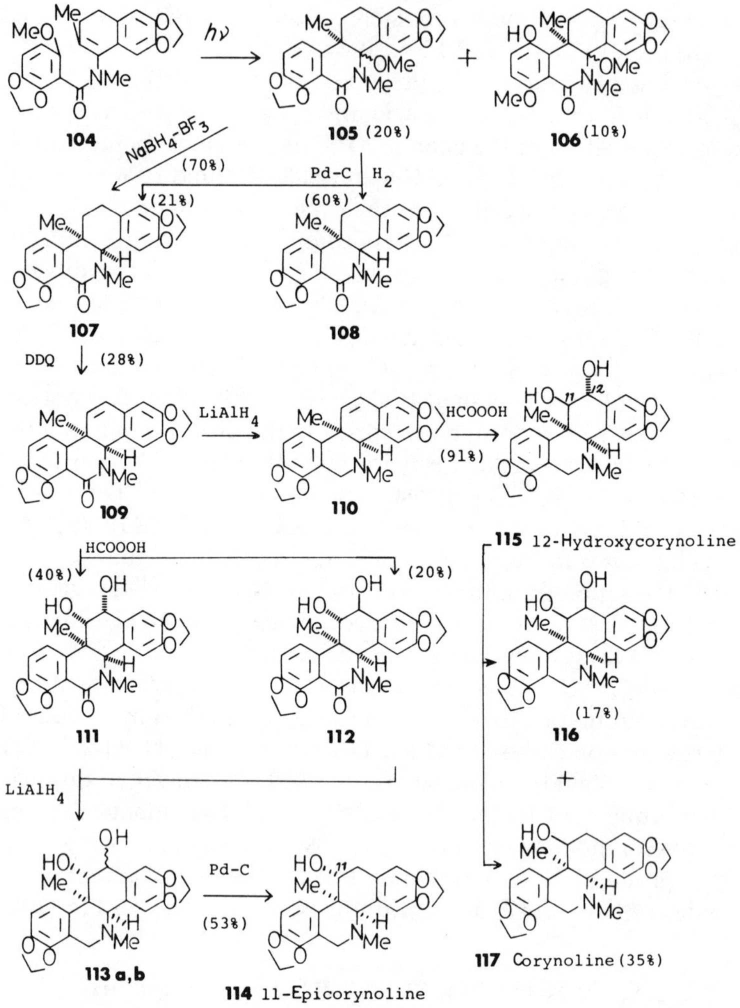

104

105 (20%)   +   106 (10%)

NaBH₄·BF₃ (70%)   (21%)   Pd–C H₂ (60%)

107   108

DDQ ↓ (28%)

109   LiAlH₄   110   HCOOOH (91%)   115 12-Hydroxycorynoline

HCOOOH (40%)   (20%)

111   112   116 (17%)

LiAlH₄   +

113 a,b   Pd–C (53%)   114 11-Epicorynoline   117 Corynoline (35%)

SCHEME 50

1. Synthesis of the Basic Skeleton

Acylation of the well-known Bischler–Napieralski product, 1-benzyl-3,4-dihydroisoquinoline, with ethyl chloroformate gave a mixture of two carbamates, E isomer 118 and Z isomer 119 in the ratio of 1 : 2, which were readily assigned from their UV and NMR spectra and underwent cyclization on irradiation under either nonoxidative or oxidative conditions to afford the 8-oxoberbine 120 or aporphine skeleton 121 (104,105). Irradiation of either the E or Z enamide, (118 or 119) or their mixture quickly reaches photoequilibrium between the two geometrical isomers and also a second photochemical equilibrium between the E isomer and the photocyclized aporphine **121**.

When the enamide **118** was irradiated in the presence of an oxidizing agent such as iodine, both the expected aporphine **121** in 65% yield and the 8-oxoberbine **120** in low yield were obtained; the latter became the major product when enamide **118** was irradiated under nonoxidative conditions (Scheme 51).

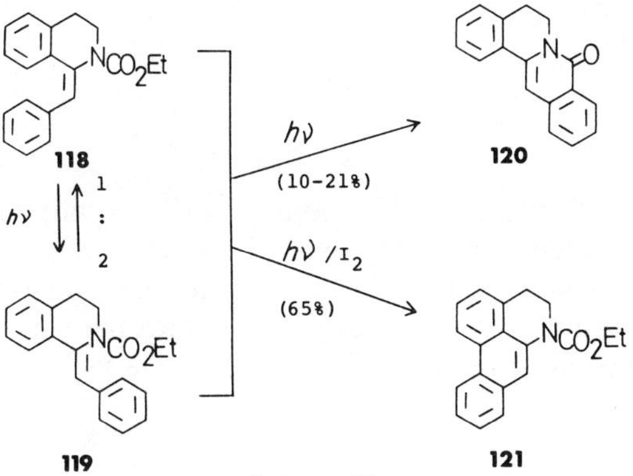

SCHEME 51

This nonoxidative photocyclization reaction was extended to the N-acetyl derivative **122**, which was irradiated through a Vycor filter in the presence of iodine and hydriodic acid to give the product **123** in 42% yield (*106*). Lenz and Yang (*106*) have synthesized a protoberberine alkaloid, β-coralydine, by applying this type of photocyclization (Schemes 52 and 53).

A similar photocyclization was reported on the corresponding formate **124**, which exists in the Z configuration predominantly or sometimes exclusively (*107*). Irradiation in the presence of hydriodic acid afforded the protoberberine iodide **126** in good yields, which was further converted to ($\pm$)-xylopinine by sodium borohydride reduction.

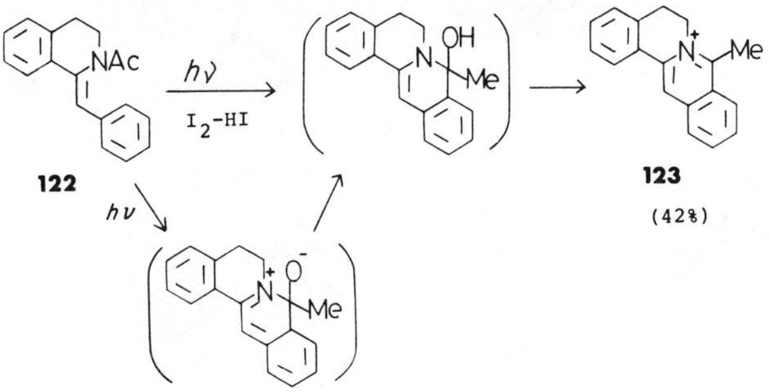

SCHEME 52

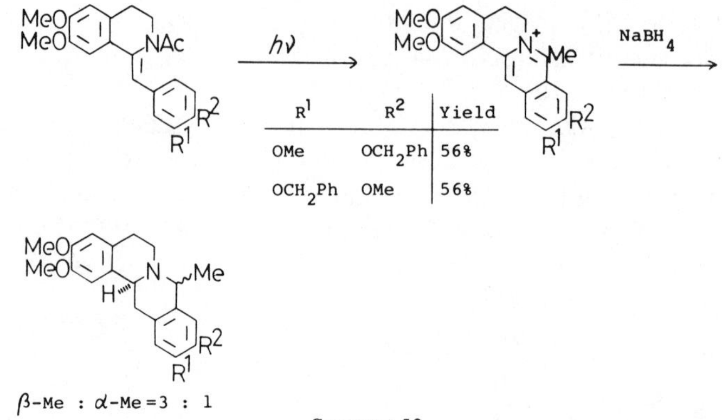

R^1	R^2	Yield
OMe	OCH$_2$Ph	56%
OCH$_2$Ph	OMe	56%

β-Me : α-Me = 3 : 1

SCHEME 53

The enamide **127** that is substituted in an ortho position of the benzylidene group was prepared to investigate regioselective cyclization under nonoxidative conditions. However, irradiation brought about an exclusive cyclization to the unsubstituted position to afford the 2,3,9,12-tetramethoxy-substituted protoberberine iodide **128** (Scheme 54).

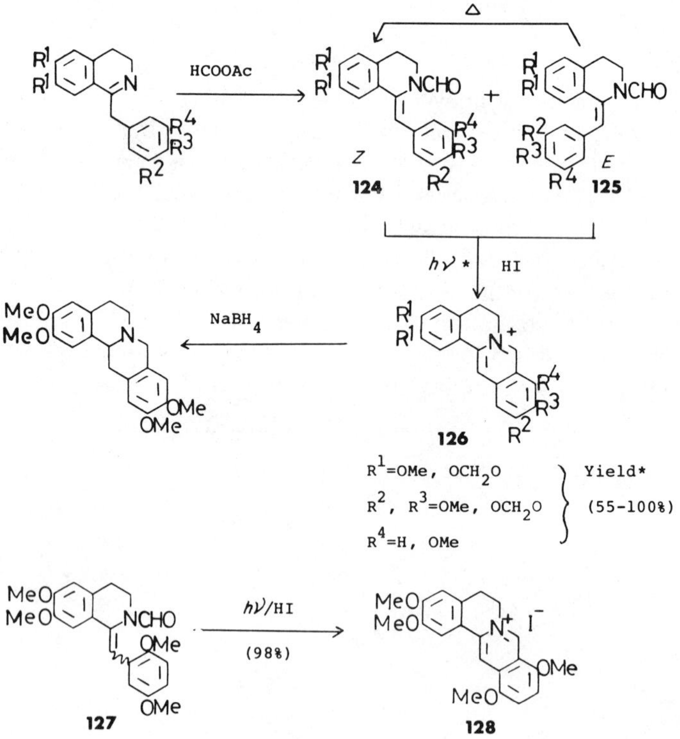

R^1=OMe, OCH$_2$O   Yield*
R^2, R^3=OMe, OCH$_2$O   (55-100%)
R^4=H, OMe

SCHEME 54

Cava and Havlicek (*108*) found that irradiation of the *N*-benzoate **129** in the presence of iodine and cupric acetate yielded the 13-phenyloxoprotoberberine **130** exclusively.

Interestingly, photocyclization of the diphenylenamide **131** afforded a mixture of the 13-phenyloxoprotoberberine **130** and the aporphine skeleton **132** in yields of 9 and 60%, respectively, under oxidative conditions (iodine), whereas the former (**130**) was the exclusive product under nonoxidative conditions (*108*) (Scheme 55).

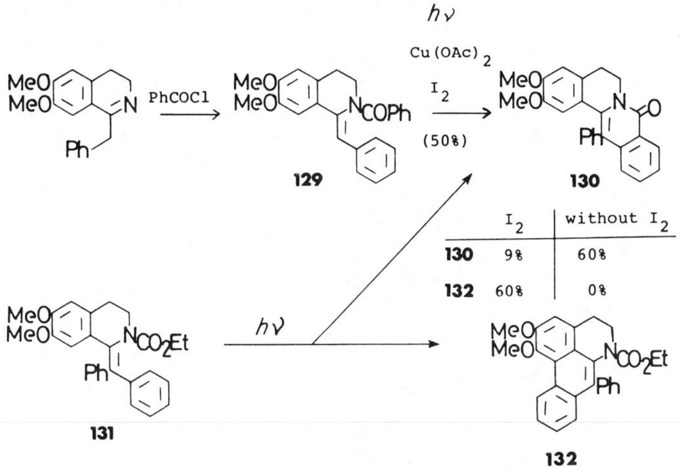

	I_2	without I_2
130	9%	60%
132	60%	0%

SCHEME 55

Based on the photocyclization of 2-acyl-1-benzylideneisoquinolines, Ninomiya *et al.* (*21–24*) and Lenz (*31,32,44,109*) independently reported the synthesis of the oxoberbines from the 2-aroyl-1-methyleneisoquinoline **133** under nonoxidative conditions.

Enamides **133a–k** were readily prepared from their corresponding benzoic acids and the available 1-methyl-3,4-dihydroisoquinolines. Irradiation afforded the 8-oxoberbines **134** and **135** in excellent yields. A wide variety of the substituents are capable of being accommodated both on the isoquinoline nucleus and on the benzene ring of the benzoic acids. When irradiation was conducted in a thoroughly degassed solution, the 8-oxoberbines (**134**) were formed in excellent yields (*31,32,44*). However, when irradiation was run simply by passing an inert gas (nitrogen) through the solution, a considerable amount of the dehydrooxoprotoberberines (**135**) were formed (*21–24*) (see Table III).

Ninomiya *et al.* (*45*) reinvestigated photocyclization of the enamide **18** at low temperature and found that irradiation in benzene at 6°C afforded two types of lactams: the H_a-shifted lactam **8** and the H_b-shifted lactam **10**; the latter is readily converted to the corresponding dehydrolactam **9** simply by exposure to air. These results are comparable to the previous reports

TABLE III

Preparation of 8-Oxoberbines from Enamides

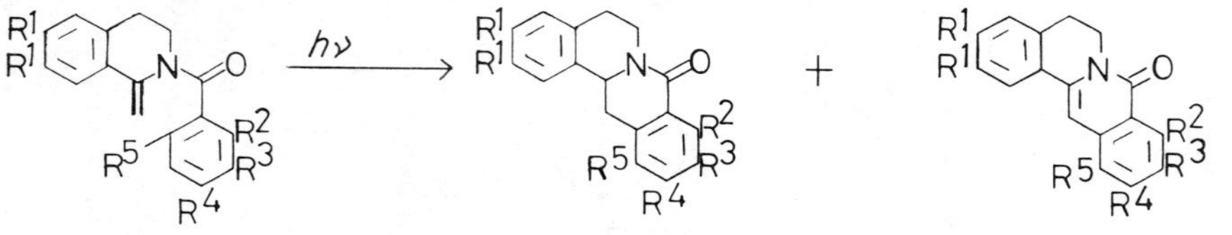

$$133 \xrightarrow{h\nu} 134 + 135$$

133   **134**(%)   **135**(%)

Substituents (133)

a $R^1 = R^2 = R^3 = R^4 = R^5 = H$

b $R^2 = R^3 = R^4 = R^5 = H, R^1 = OMe$

c $R^2 = R^5 = H, R^1 = R^3 = R^4 = OMe$

d $R^1 = R^3 = R^4 = R^5 = H, R^2 = OMe$

e $R^3 = R^4 = R^5 = H, R^1 = R^2 = OMe$

f $R^3 = R^4 = H, R^1 = R^2 = R^5 = OMe$

g $R^1 = R^4 = H, R^2 = R^3 = R^5 = OMe$

h $R^4 = H, R^1 = R^2 = R^3 = R^5 = OMe$

i $R^4 = H, R^1 = R^5 = OMe, R^2 = OCH_2O = R^3$

j $R^2 = R^3 = R^5 = H, R^1 = OMe, R^4 = NO_2$

k $R^3 = R^4 = R^5 = H, R^1 = OMe, R^2 = COOMe$

134(%)

$R^1 = R^2 = R^3 = R^4 = R^5 = H$ (40)

$R^2 = R^3 = R^4 = R^5 = H, R^1 = OMe$ (18)

$R^2 = R^5 = H, R^1 = R^3 = R^4 = OMe$ (5)

135(%)

$R^1 = R^2 = R^3 = R^4 = R^5 = H$ (20)

$R^2 = R^3 = R^4 = R^5 = H, R^1 = OMe$ (24)

$R^2 = R^5 = H, R^1 = R^3 = R^4 = OMe$ (40)

$R^2 = R^3 = R^4 = R^5 = H$ (5)

$R^1 = R^2 = R^3 = R^4 = R^5 = H$ (50)

$R^2 = R^3 = R^4 = R^5 = H, R^1 = OMe$ (46)

$R^3 = R^4 = R^5 = H, R^1 = R^2 = OMe$ (56)

$R^1 = R^4 = R^5 = H, R^2 = R^3 = OMe$ (25)

$R^1 = R^4 = H, R^2 = R^3 = R^5 = OMe$ (10)

$R^4 = R^5 = H, R^1 = R^2 = R^3 = OMe$ (20)

$R^3 = R^4 = H, R^1 = R^2 = R^5 = OMe$ (8)

$\left\{ \begin{array}{l} R^4 = R^5 = H, R^1 = OMe, \\ \quad R^2 = OCH_2O = R^3 \ (44) \\ R^3 = R^4 = H, R^1 = R^2 = OMe, R^5 = OH \\ \quad (21) \\ R^2 = R^3 = R^5 = H, R^1 = OMe, R^4 = NO_2 \\ \quad (8) \\ R^3 = R^4 = R^5 = H, R^1 = OMe, \\ \quad R^2 = COOMe \ (5) \end{array} \right.$

$R^2 = R^3 = R^4 = R^5 = H, R^1 = OMe$, 13a-COOMe (8)

$R^2 = R^3 = R^4 = R^5 = H, R^1 = OMe$
$R^2 = COOMe$

(*21–24*) on the predominant formation of the dehydrolactam, even under nonoxidative irradiation, in this type of photocyclization.

Also, Ninomiya *et al.* (*45*) overcame the disadvantage of preferential formation of dehydrolactams by adding 1.5 mol of sodium borohydride in the reaction mixture during the course of irradiation and succeeded in the convenient synthesis of (±)-xylopinine (**20**) in the overall yield of 52% starting from the Bischler–Napieralski product (Scheme 56).

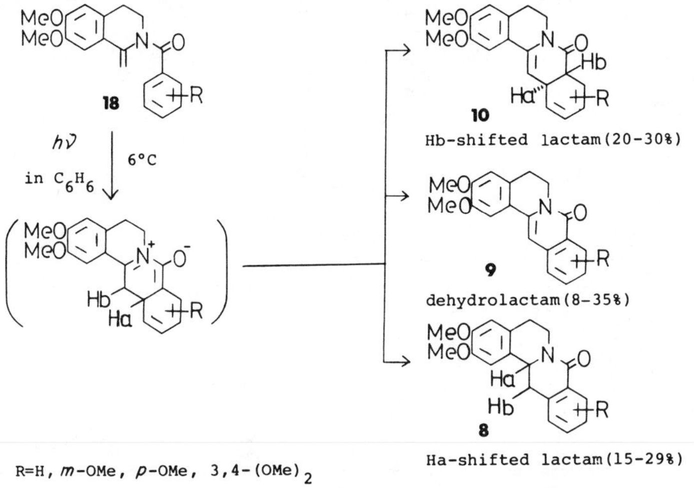

R=H, *m*-OMe, *p*-OMe, 3,4-(OMe)$_2$

SCHEME 56

Irradiation of the *o*-methoxyenamide **133e** yielded the dehydrolactam **135** exclusively, as a result of a regioselective photocyclization to the root of the methoxyl group, followed by elimination of the substituent (*24*).

Ninomiya *et al.* (*24,25*) have presented some evidence indicating a competition between 1,5 migration and direct elimination of a methoxyl group of the enamides. Irradiation of an assortment of the ortho-substituted enamides (**133**) led to the discovery that a variety of both electron-withdrawing and -donating groups in the enamides can function as the leaving groups, and that their photocyclization occurs exclusively to the carbon atom bearing these substituents (*32*).

The stereochemistry and photocyclization of the corresponding 1-ethylideneenamide **136** has also been studied (*24,109*). Acylation of the 1-ethylisoquinoline with benzoyl chloride afforded a mixture of two enamides of geometrical isomers, the unstable *E* (**136**) and stable *Z* isomers (**137**), which were easily assigned from the nuclear Overhauser effect data in their NMR spectra (*24*).

Irradiation of the stable *Z* isomer (**137**) established a photoequilibrium with the unstable *E* isomer. However, the mixture of two geometrical

TABLE IV
ISOMERIZATION OF 137

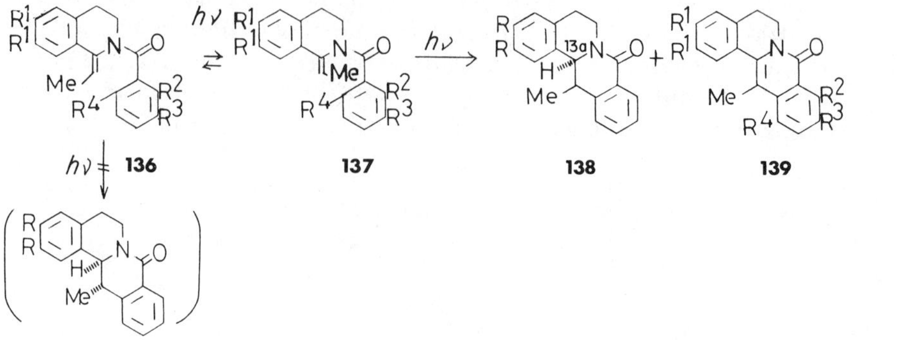

Substituents	138(%)	139(%)
$R^1 = R^2 = R^3 = R^4 = H$	R = H (30)	$R^1 = R^2 = R^3 = R^4 = H$ (4)
$R^1 = R^3 = R^4 = H, R^2 = OMe$	R = H, 13a-OMe	
$R^2 = R^3 = R^4 = H, R^1 = OMe$	R = OMe (37)	$R^2 = R^3 = R^4 = H, R^1 = OMe$ (9)
$R^3 = R^4 = H, R^1 = R^2 = OMe$		$R^2 = R^3 = R^4 = H, R^1 = OMe$ (35)
$R^3 = H, R^1 = R^2 = R^4 = OMe$		$R^3 = R^4 = H, R^1 = R^2 = OMe$ (34)
$R^1 = R^4 = OMe, R^2 = OCH_2O = R^3$		$\begin{cases} R^4 = R^1 = OMe, R^2 = OCH_2O = R^3 (41) \\ R^3 = H, R^1 = R^2 = OMe, R^4 = OH (29) \end{cases}$

isomers was readily converted homogeneously to the stable Z isomer during the purification process by chromatography, whereas the unstable E isomer could only be isolated by careful recrystallization of the mixture (24). Photocyclization of either one of the E or Z enamides and also a mixture of these isomers afforded the identical products 138 and 139, whereas a possible stereoisomer of 140 was not isolated. Therefore, this indicates that a complete isomerization occurred during the photocyclization process (24). On standing in the open air or in solution, the 13-methyllactams 138 are slowly converted to the corresponding dehydrolactams (139) (24) (see Table IV).

2. Synthesis of Berbine Derivatives

As a further extension of photocyclization of the enamides prepared from 1-methylisoquinolines, Naito and Ninomiya (67) and Lenz (48) synthesized aza analogs of berbine, azaberbines, by irradiation of pyridine analogs of the parent enamides (27) (Scheme 57).

Reductive photocyclization of the enamide 18 in ether–ethanol solution in the presence of an excessive amount of sodium borohydride afforded two

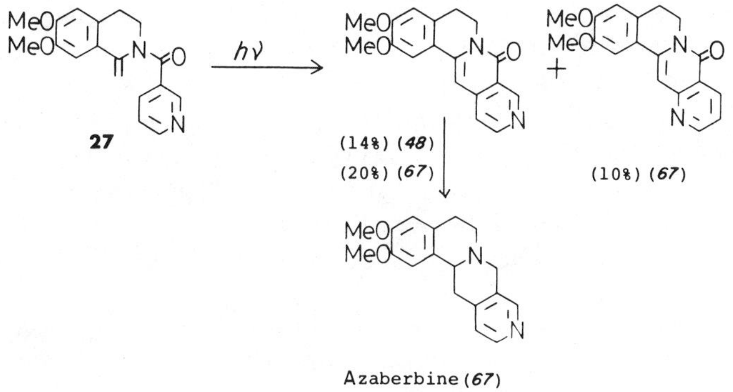

MeO
MeO
27
hν
(14%) (48)
(20%) (67)
(10%) (67)
Azaberbine (67)

SCHEME 57

hydrogenated lactams, **141** and **142**; the lactam with the unconjugated diene structure (**142**) was obtained as a sole product when acetonitrile was used as solvent (*15*) (Scheme 58).

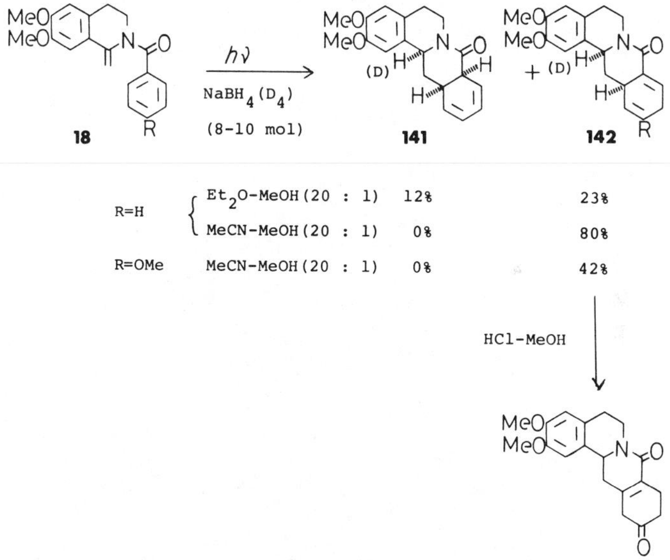

18
hν
NaBH4 (D4)
(8-10 mol)
141
142
R=H
Et2O-MeOH (20 : 1) 12% 23%
MeCN-MeOH (20 : 1) 0% 80%
R=OMe
MeCN-MeOH (20 : 1) 0% 42%
HCl-MeOH

SCHEME 58

The enamides of *N*-aroylenamine type (**143–146**), which contain an electron-deficient aromatic ring caused by the introduction of some electron-withdrawing factor, undergo a facile cyclization not only photochemically but also thermally to afford the dehydrolactams (**147–150**) in good yields (*53,67*). The dinitro-substituted enamide (**143**) was thermally cyclized when refluxed in benzene to afford the corresponding dehydrolactam (**147**) in 95% yield. Compound **147** was also obtained from the 1-methylisoquinoline **22** by acylation with 3,5-dinitrobenzoyl chloride in benzene under refluxing temperature in 70% yield. Although the mononitro-substituted *N*-benzoylenamines (**144**) gave no cyclized product even at elevated temperature, the introduction of an additional methoxyl group to the ortho position of enamides **144** and **145** furnished smooth thermal cyclization to give the corresponding nitroberbine derivatives **148** and **149** in yields of 43 and 75%, respectively (*53*), however with the sacrifice of the methoxyl group as in the case of a regioselective photocyclization of the *o*-methoxy-substituted enamides (*23,32*).

These thermal cyclizations offer a promising route for the synthesis of berbine compounds with electron-withdrawing groups in ring D (Scheme 59).

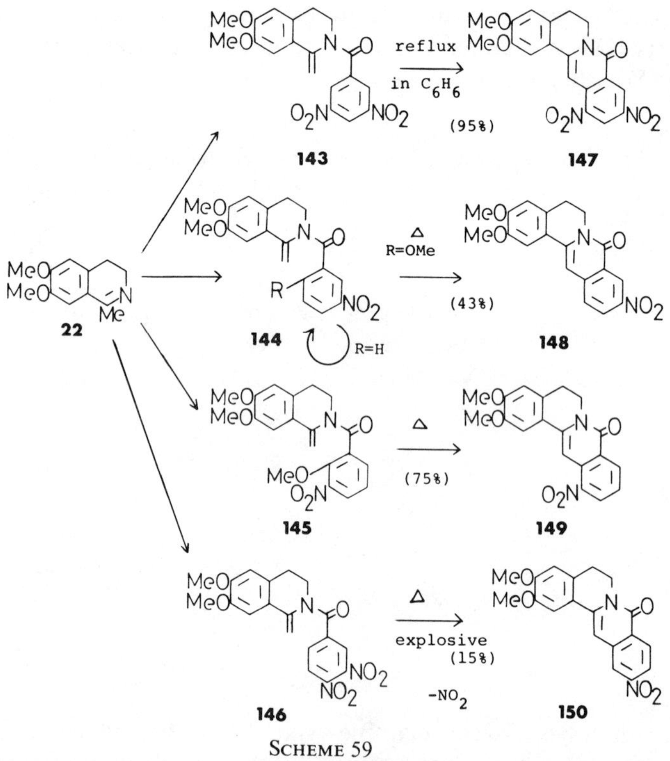

SCHEME 59

Based on the facile thermal cyclization of the nitro-substituted enamides, Naito and Ninomiya (*67*) have further developed thermal cyclization for the synthesis of azaberbines under acylating conditions.

Acylation of either 1-methylisoquinolines with an excessive amount of nicotinoyl chloride or the enamide with the same acid chloride yielded the cyclized azaberbine **28** in good yield, which was readily deacylated with hydrochloric acid or 5% KOH/CH$_3$OH to afford the corresponding 8-oxoberbine **26** (Scheme 60).

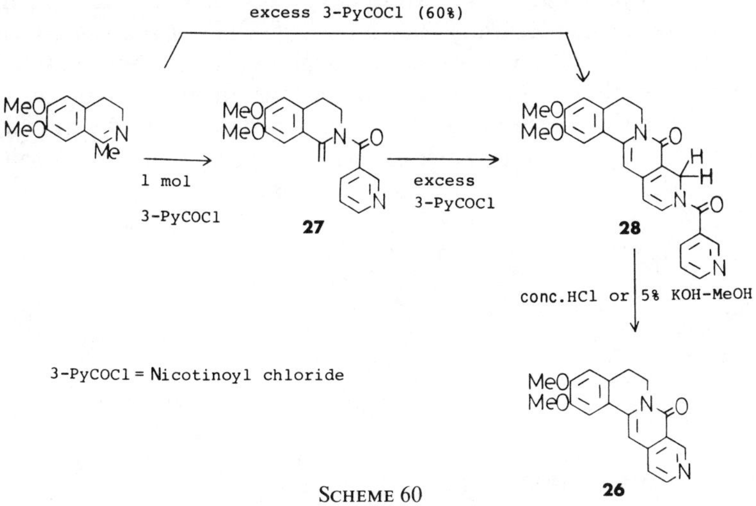

SCHEME 60

The cyclization of 1-methylene-2-nicotinoylisoquinolines **27** under three different conditions, photochemical, thermal, and acylating conditions, are compared (*73*). Starting from the above enamide (**27**), Ninomiya *et al.* (*73*) synthesized 10-azaberbine as the major product by simple heating at 195–200°C, whereas 12-azaberbine was the major product under acylating condition as shown in Table II. Further, total synthesis of alamarine (**33**), the only alkaloid having an azaberbine structure, was synthesized via the route involving enamide cyclization under both thermal and photochemical conditions (*73*) as shown in Scheme 71.

On the other hand, the cyclization of the 2-isonicotinoylenamines under acylating conditions afforded the spirodihydropyridine derivatives in good yields; these were converted to two types of azaberbines on deacylation and subsequent photorearrangement as shown in the Scheme 25.

3. Total Synthesis of Alkaloids

Total synthesis of protoberberine alkaloids via the route involving enamide photocyclization consists of nonoxidative photocyclization of the 2-aroyl-1-methylene-3,4-dihydroisoquinolines and the subsequent metal-hydride reduction of the photocyclized lactams. This simple combination of reactions for alkaloid synthesis provides one of the most convenient synthetic routes to this group of popular alkaloids.

β-Coralydine. Lenz and Yang (*106*) discovered that photocyclization of the *N*-acetyl-1-benzylideneenamide **151** gives 8-methylprotoberberine iodide **152** in excellent yield and successfully applied this procedure to the total synthesis of coralyne, known to have a potent antileukemic activity, especially against L-1210 and P-388 in mice (*110*). Coralyne was reduced to the corresponding tetrahydroberberine alkaloid, β-coralydine (**153**).

Photocyclization of this type of enamide is assumed also to proceed by a six π-electron ring closure to the cyclic intermediate (**A**) under irradiation, followed by elimination of water to form the cyclized quaternary salt **152** (Scheme 61).

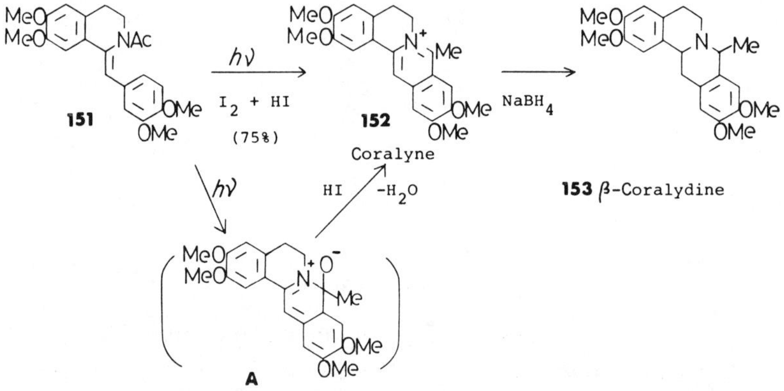

SCHEME 61

(±)-Corytenchirine. Kametani *et al.* (*111*) applied Lenz and Yang's method (*106*) to the synthesis of (±)-corytenchirine (**154**), an 8-methylprotoberberine alkaloid, thereby confirming the stereochemistry of the alkaloid (Scheme 62).

Xylopinine. Among many popular protoberberine alkaloids, xylopinine has been of interest to synthetic chemists because of its pharmacological activity (*112*). Three groups (*21,23,29–32,44,107*) have independently described total synthesis of this alkaloid; Ninomiya *et al.* (*16*) and Kametani *et al.* (*66*) succeeded in the asymmetric synthesis of this alkaloid by enamide photocyclization.

Nonoxidative photocyclization of the enamide **133** in thoroughly degassed *tert*-butanol solution gave the lactam **155** in 93.5% yield as the sole product (*31,44*). Ninomiya *et al.* (*21,23*) carried out the same cyclization of **133** by passing an inert gas during the course of irradiation and obtained three different types of products **156**, **155**, and **157** in yields of 40, 5, and 5%, respectively.

Regioselective photocyclization of the *o*-methoxy-substituted enamide **133** (*32*) or the *o*-bromoenamide **133** (*29,30*) gave only the dehydrolactam **156** in 85 or 80% yield, respectively. Reduction of the lactam **155** with LAH

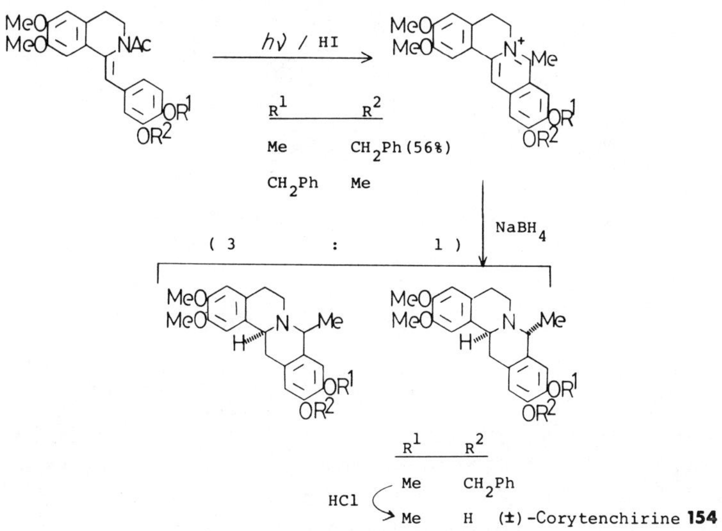

R^1	R^2
Me	CH$_2$Ph (56%)
CH$_2$Ph	Me

(3 : 1)

R^1	R^2
Me	CH$_2$Ph
Me	H (±)-Corytenchirine **154**

SCHEME 62

or reduction of the dehydrolactam **156** with LAH and sodium borohydride successively afforded xylopinine. Kametani *et al.* (*29,30*) reported a convenient procedure for the conversion of the dehydrolactam **156** to (±)-xylopinine by using phosphorous oxychloride and subsequent reduction with sodium borohydride (Scheme 63).

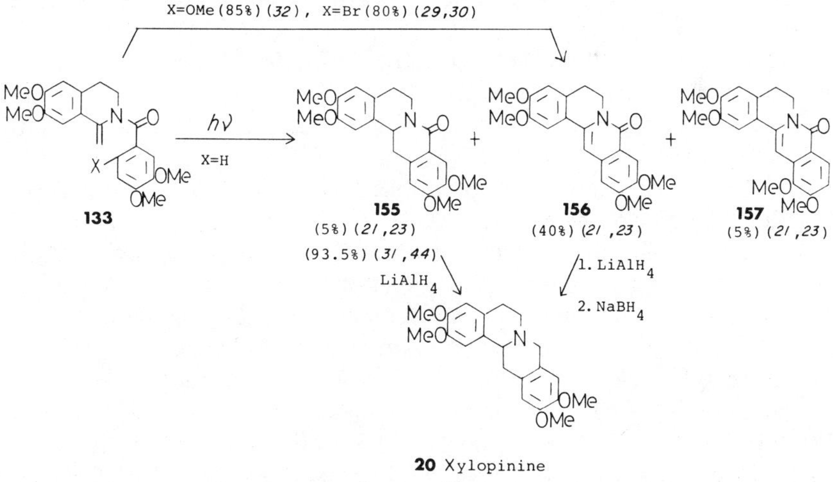

20 Xylopinine

SCHEME 63

Lenz (*107*) also demonstrated the synthetic utility of photocyclization of the formate **124** followed by sodium borohydride reduction to achieve a simple synthesis of (±)-xylopinine (Scheme 64).

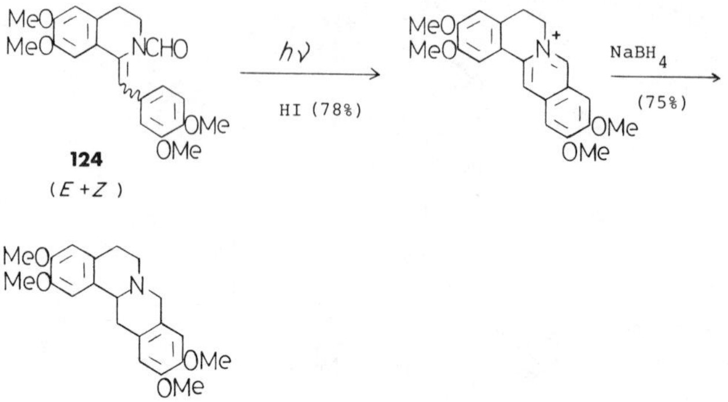

236   ICHIYA NINOMIYA AND TAKEAKI NAITO

20 Xylopinine

SCHEME 64

On the assumption that asymmetric reduction of an intermediate by a chiral metal hydride complex could occur during the course of photocyclization of the enamide **133**, as exemplified by reductive photocyclization, Ninomiya *et al.* (*16*) have undertaken and completed the photochemical asymmetric synthesis of (−)-xylopinine.

After a metal hydride complex was prepared from LAH and quinine (1 : 1), irradiation of a mixture of the resulting solution containing the above chiral hydride agent and the enamide (**133**) led to the formation of two optically active lactams **158** [6%, $[\alpha]_D$ −63° (c = 0.48, CHCl$_3$)] and **155** [13%, $[\alpha]_D$ −102° (c = 0.44, CHCl$_3$)] with 37% optical purity. Reduction of the lactam **155** with LAH furnished (−)-xylopinine (**20**) in 48% chemical yield.

Interestingly, a similar irradiation of the enamide **133** in a dilute solution of the above mixture afforded the optically active lactam (**155**) [10%,$[\alpha]_D$ −71° (c = 0.63, CHCl$_3$)] and the optically active amine (**20**) [38%,$[\alpha]_D$ −16; (C = 1.75, CHCl$_3$)]. The amine was identical to (−)-xylopinine, thus providing a one-step synthesis of this alkaloid from the corresponding enamide (Scheme 65).

Kametani *et al.* (*66*) also succeeded in the asymmetric synthesis of xylopinine by applying 1,3-asymmetric induction to enamide photocyclization. Acylation of L-3,4-dimethoxyphenylalanine, followed by the Bischler–Napieralski cyclization of the resulting amide, gave the chiral 3,4-dihyroisoquinoline **159**, which was then acylated to afford the chiral enamide **19**.

Irradiation of the chiral enamide (**19**) under nonoxidative conditions gave the chiral lactam **160** in 73.3% yield; (**160**) was then converted to the amino ester **161** by the method developed by Kametani's group (*29,30*). Cleavage of the ester group was performed by Yamada's procedure (*113*), with slight modification, to give the amine **20** [66.1%,$[\alpha]_D$ −281° (c = 0.19, CHCl$_3$)], which was identical to natural xylopinine, providing the first successful

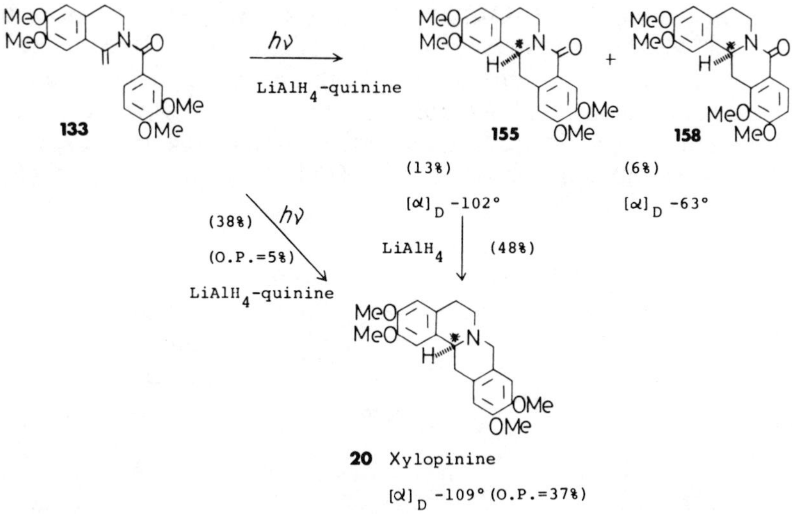

133

$h\nu$

LiAlH$_4$-quinine

155 (13%) [α]$_D$ −102°

+

158 (6%) [α]$_D$ −63°

(38%) $h\nu$ (O.P.=5%)

LiAlH$_4$-quinine

LiAlH$_4$ (48%)

20 Xylopinine

[α]$_D$ −109° (O.P.=37%)

SCHEME 65

1,3-asymmetric synthesis of xylopinine with an optical purity of 94.7% (Scheme 66).

Tetrahydropalmatine and Sinactine. Irradiation of the enamide **133** afforded a mixture of two lactams **162** and **163** in 20 and 8% yields, respectively, as a result of cyclization to the roots of each one of two o-methoxyl

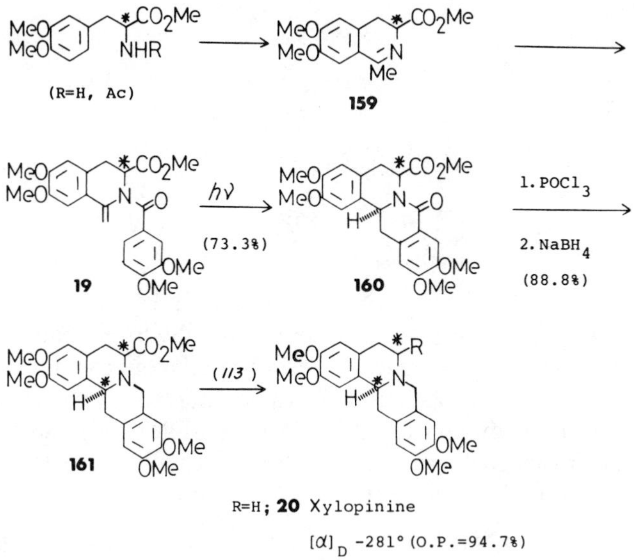

(R=H, Ac)

159

19 $h\nu$ (73.3%) **160** 1. POCl$_3$ 2. NaBH$_4$ (88.8%)

161 (113) R=H; **20** Xylopinine

[α]$_D$ −281° (O.P.=94.7%)

SCHEME 66

groups in the enamide (133) (23). Analogously, photocyclization of the enamide (133) afforded two types of photocyclized lactams, (165) and (166) in yields of 44 and 21%, respectively (22,23). The major products from the above two cyclizations 162 and 165, were reduced successively with LAH and sodium borohydride to give the tertiary amines 164 and 167, which were identical to tetrahydropalmatine (164) (23) and sinactine (167) (22,23), respectively (Scheme 67).

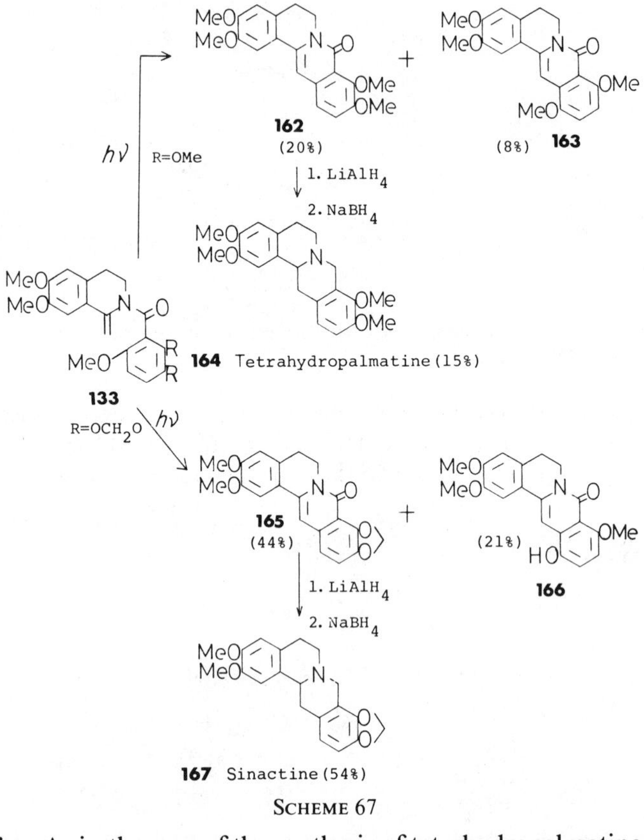

SCHEME 67

Cavidine. As in the case of the synthesis of tetrahydropalmatine (23) and sinactine (22,23), the 13-methylprotoberberine alkaloid cavidine (170) was synthesized via the dehydrolactam 169, which was readily prepared from the enamide 136 on irradiation (22,24). Since cavidine (170) had been converted (114) to thalictrifoline, this synthesis formally completed the first total synthesis of thalictrifoline (171) (Scheme 68).

(±)-Shefferine, (±)-Nandinine, (±)-Corydaline, and (±)-Thalictricavine. The 9,10-disubstituted protoberberine alkaloids were synthesized by regio-

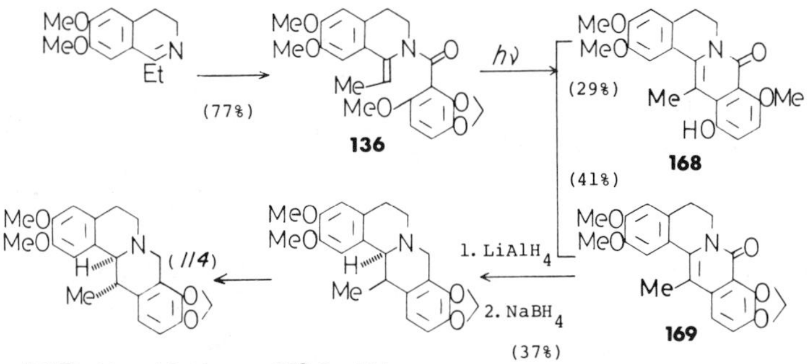

171 Thalictrifoline   **170** Cavidine
SCHEME 68

selective photocyclization of the *o*-bromoenamide **172** in high yields
(*29,30*). In all cases, the major products were the 9-hydroxylactams **173**,
which were converted to the respective alkaloids: (±)-shefferine, (±)-nan-
dinine, (±)-corydaline, and (±)-thalictricavine (Scheme 69).

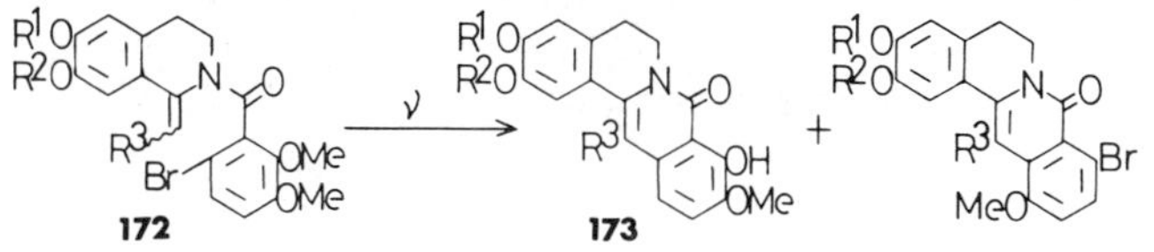

R^1	R^2	R^3			
Me	Me	H	65%		20%
			1. $\longrightarrow$ (±)-Shefferine		
-CH$_2$-		H	60%		30%
			1. $\longrightarrow$ (±)-Nandinine		
Me	Me	Me	60%		—
			2. $\longrightarrow$ (±)-Corydaline		
-CH$_2$-		Me	45%		30%
			2. $\longrightarrow$ (±)-Thalictricavine		

1. LiAlH$_4$,NaBH$_4$
2. LiAlH$_4$, NaBH$_4$, CH$_2$N$_2$
SCHEME 69

2-*O*-Methylcoreximine. The *p*-acetoxy-substituted enamide **174** was irradiated in a thoroughly degassed *tert*-butanol solution to give the 11-acetoxy-
lactam **175** in 45% yield; this was then converted to (±)-2-*o*-methylcorexi-
mine (**176**) on reduction with LAH (*44*), and thus demonstrated the
possibility of introducing a phenolic hydroxyl group in the alkaloids by
masking it as an acetate in the photocyclization (Scheme 70).

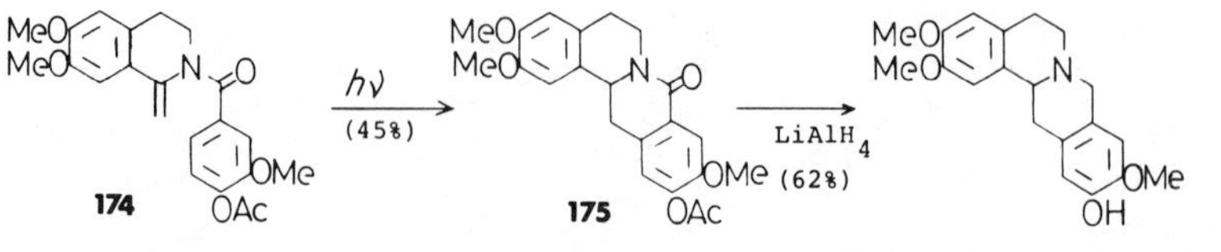

176 2-*O*-Methylcoreximine

SCHEME 70

Alamarine. Pakrashi *et al.* (*115*) isolated several new types of alkaloids alamarine, alangimarine, alangimaridine, and isoalamarine from the Indian medicinal herb, *Alangium lamarckii* Thw. (Alangiaceae). Ninomiya *et al.* (*73*) reported the first total synthesis of alamarine (**33**) via the route involving enamide cyclization under both thermal and photochemical conditions.

The enamide **177** was heated at 180–200°C for 20 min to give a mixture of two lactams **31** and **32** in 43 and 10% yields, respectively. On the other hand, irradiation of the enamide (**177**) with a low-pressure mercury lamp for 8 hr gave the lactams **31** and **32** in 25 and 13% yields, respectively. The major product (**31**) from both cyclizations was then converted to alamarine (**33**) and its acetate. Since alamarine had been converted to alangimarine (*115*), this synthesis formally completed total synthesis of alangimarine (Scheme 71).

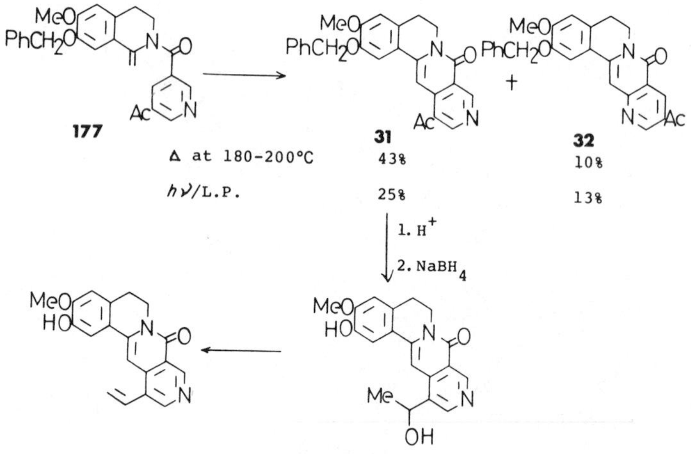

Alangimarine   **33** Alamarine

SCHEME 71

Ipecac Alkaloids. As an extension of the synthetic work of protoberberine alkaloids, Ninomiya *et al.* (*116*) reported that the 2-acryloyl-1-methylenetetrahydroisoquinolines (**178**) underwent ready photocyclization to yield the benzo[*a*]quinolizines (**182**).

Acylation of isoquinoline **22** with substituted acryloyl chlorides gave the enamide **178** which was so unstable that it was irradiated without isolation. Due to instability of the photocyclized lactam **179**, it was identified as the corresponding amine **182**. The methoxy-substituted lactams **179b** and **c** were then converted to the unsaturated amines **181** and **182** via the dehydrolactams **180b** and **c**. Conversion of the unsaturated amines **181b** and **c** to the corresponding ketones **183b** and **c** was achieved by hydroboration and the subsequent oxidation, and Pfitzner–Moffat oxidation. The ketone **183c** thus obtained is well-known as a key intermediate for the synthesis of various ipecac alkaloids (*1,2*) (Scheme 72).

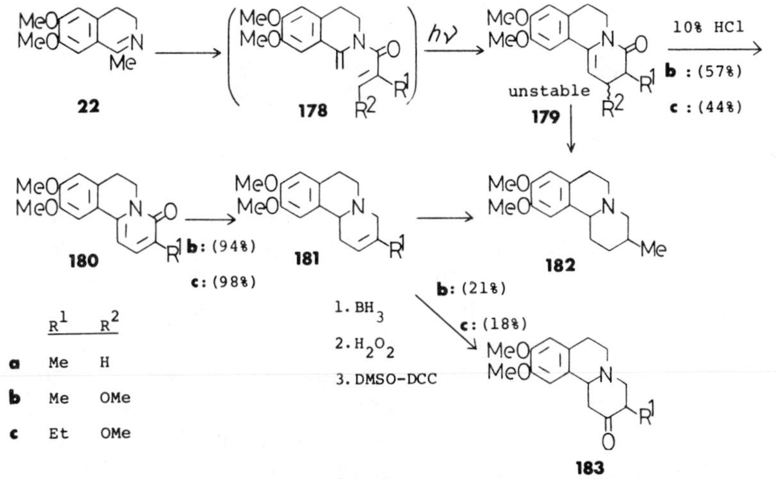

SCHEME 72

D. YOHIMBINE AND RELATED ALKALOIDS

Successful berbine synthesis summarized in Section IV,C prompted Ninomiya's group (*25,26,117,118*) to extend enamide photocyclization to harmalane, therefore giving rise to a novel and facile synthesis of polycyclic heterocycles such as the yohimbine group of compounds. Before reductive photocyclization was introduced, the use of nonoxidative photocyclization with indole alkaloids was limited to simple systems and those possessing a large degree of aromaticity.

1. Synthesis of the Basic Skeleton

Tryptamine was acetylated followed by cyclization to give harmalane in good yield; harmalane was then acylated with benzoyl chloride to afford the enamide **14** that underwent smooth photocyclization under nonoxidative conditions to give the dehydrolactam **184** in 37% yield, even when passing an inert gass. The dehydrolactam **184**, a basic skeleton of yohimbine, was converted to demethoxycarbonyldihydrogambirtannine (**185**) in 52% yield (*25,26*) (Scheme 73).

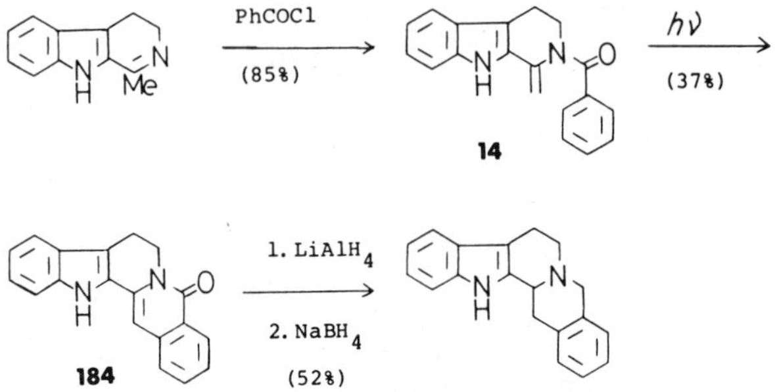

14

184   (52%)

185 Demethoxycarbonyldihydrogambirtannine

SCHEME 73

Yohimbans. Yohimbans, the basic structures of the alkaloid yohimbine, were synthesized by Ninomiya *et al.* (*54,117*) by applying enamide photocyclization under two different conditions, nonoxidative and reductive. Harmalane was acylated with 1-cyclohexene-1-carbonyl chloride to yield the enamide **186** which was so unstable that it was irradiated without purification. The resulting reaction mixture showed two spots on TLC, suggesting the formation of two stereoisomers with respect to the ring juncture (D/E). Successive reductions of the lactam **187** with LAH and sodium borohydride furnished yohimban (**189**) and alloyohimban (**191**) in the ratio of 3:2 (*117*). On the other hand, catalytic hydrogenation of the resulting lactam **187** over platinum dioxide, followed by reduction with LAH, gave yohimban (**189**), epiyohimban (**190**), and alloyohimban (**191**) in the ratio of 2:1:2 with 27–35% yields from harmalane as the starting compound (*117*) (Scheme 74).

Recently, a simple synthesis of yohimban (**189**) and alloyohimban (**191**) and alloyohimbone (**196**) was achieved by reductive photocyclization of the enamides **14** and **192** in the presence of sodium borohydride (*54*). Reductive photocyclization of the enamide **14** proceeded smoothly to give a mixture of two hydrogenated lactams (**193** and **194**), the ratios depending on the hydride and the solvent employed. The unconjugated lactam **194** was exclusively obtained in excellent yield when acetonitrile was used as solvent. Either the conjugated lactam **193** or unconjugated lactam **194** was catalytically hydrogenated over platinum dioxide to give the saturated lactams **195**, each of which was identified with authentic samples (*117*) and converted to yohimban (**189**) and alloyohimban (**191**), respectively. Thus, a simple four-step synthesis of yohimban and alloyohimban was achieved in overall yields of 38 and 34% respectively (*54*).

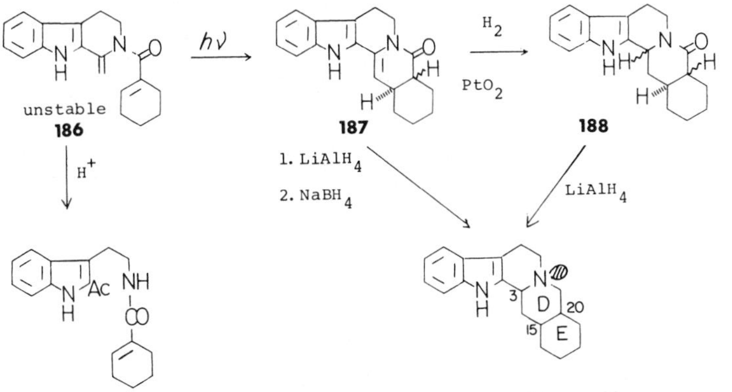

unstable
186

$h\nu$

187

H_2

PtO_2

188

1. LiAlH$_4$

2. NaBH$_4$

LiAlH$_4$

	3	15	20
189 Yohimban	---H	---H	—H
190 Epiyohimban	—H	---H	—H
191 Alloyohimban	---H	---H	---H

SCHEME 74

Alternatively, yohimban and alloyohimban were also synthesized via the route of reduction with lithium aluminum hydride of the conjugated lactam (**193**), followed by catalytic hydrogenation.

By using the *p*-methoxy-substituted enamide (**192**), Ninomiya *et al.* (*54*) succeeded in a simple synthesis of alloyohimbone (**196**) via the unconjugated lactam **194**, which has an enol ether structure. Lithium aluminum hydride reduction of the lactam **194**, followed by hydrolysis with hydrochloric acid and subsequent catalytic hydrogenation over platinum dioxide, yielded alloyohimbone (**196**) stereoselectively in an overall yield of 59% from harmalane (*54*); this was the most convenient synthesis of alloyohimbone (**196**) so far reported (Scheme 75).

2. Total Synthesis of Alkaloids

Yohimbine. Ninomiya *et al.* (*118*) reported a formal total synthesis of yohimbine by applying enamide photocyclization. Harmalane was acylated with 4-methoxy-3-methoxycarbonylbenzoyl chloride to yield the enamide **198**, which was found to be so unstable that it was irradiated without purification. Of two photocyclized products thus obtained, **199** in 32% yield and **200** in 5% yield, the latter product was identical to a known key intermediate (*119*) of yohimbine when their spectral data were compared (Scheme 76).

Azayohimbine-Type Alkaloids. Recently, some new types of alkaloids nauclefine (**29**), nauclétine (**210**), angustidine (**213**), and angustoline (**215**), were isolated (*71,120–122*) from the *Nauclea* and *Strychnos* plants, which

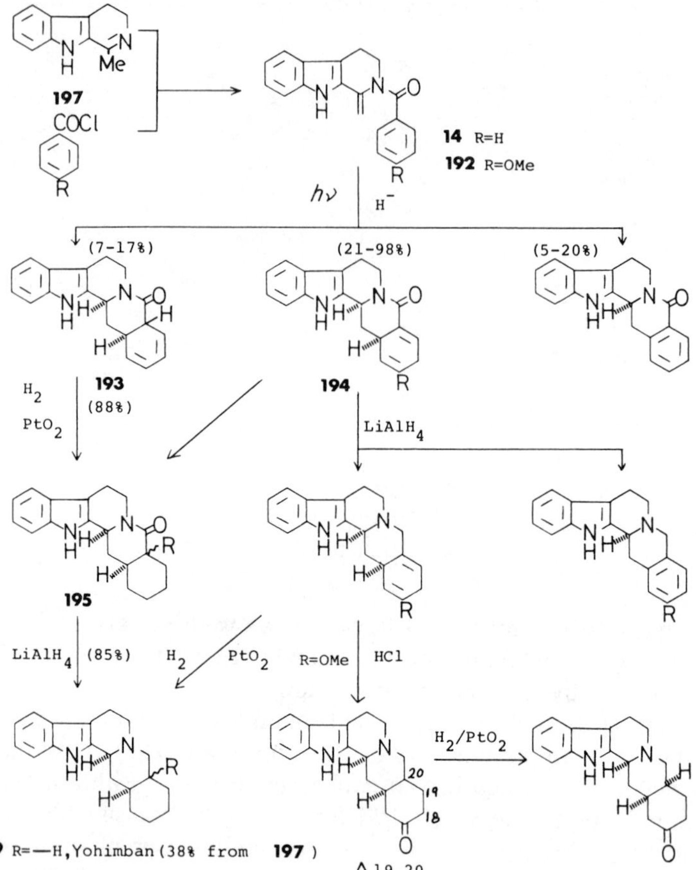

189 R=—H,Yohimban(38% from **197**)
191 R=ᴵᴵᴵᴵᴵH,Alloyohimban(34% from
    197)

Δ19,20
Δ18,19

196 Alloyohimbone
    (59% from **197**)

SCHEME 75

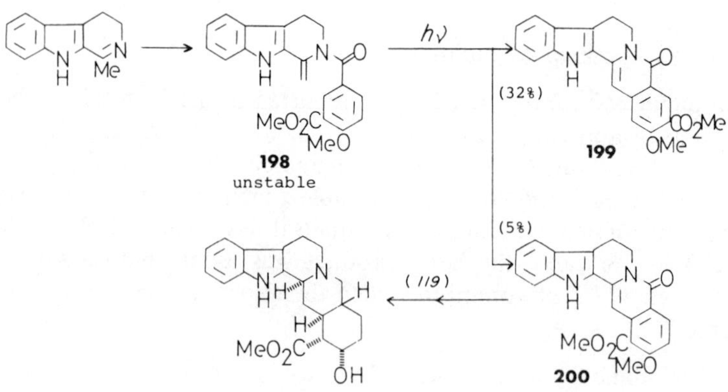

198
unstable

199

200

Yohimbine

SCHEME 76

have been known for their antileukemic activity (*120.*) These alkaloids have a common azayohimbine structure, the aza analogs of yohimbine, which are well suited as targets for synthesis by enamide photocyclization.

Nauclefine. In 1975 Sainsbury and Webb (*122*) and Hotellier *et al.* (*71*) reported the isolation and synthesis of nauclefine. Hotellier *et al.* (*71*) obtained nauclefine in 6% by treating harmalane with an excessive amount of nicotinoyl chloride. Alternatively, Sainsbury and Webb (*122*) carried out photocyclization of the enamide **201**, prepared from harmalane and the above acid chloride, to obtain a mixture of two lactams, **29** and **202**, in the ratio of 100:9, with 48% yield. The major product thus obtained was identical to parvine, whose name was first proposed by Sainsbury (*122*) but later changed to nauclefine (*123*) (Schemes 77 and 78).

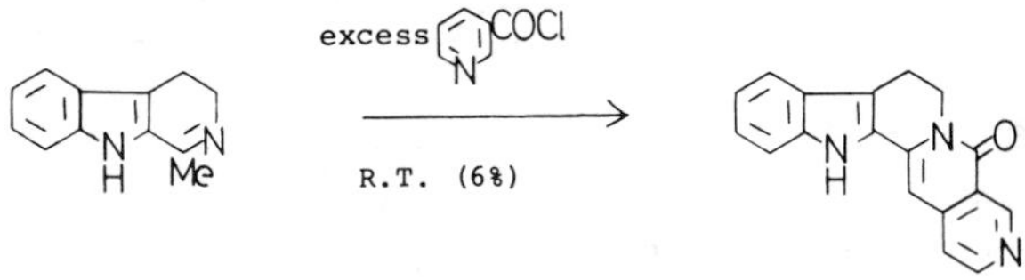

SCHEME 77

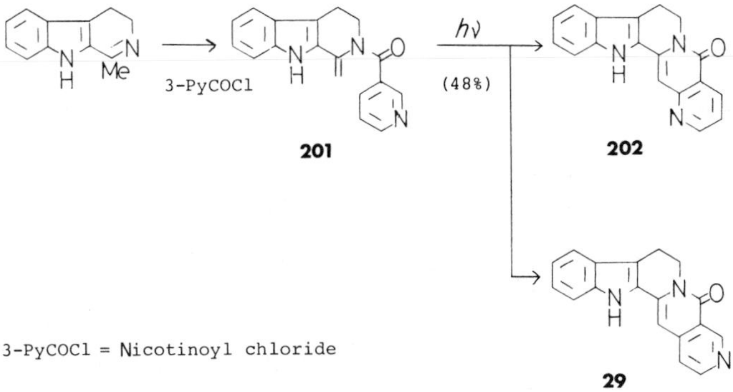

3-PyCOCl = Nicotinoyl chloride

$$\left(\begin{array}{c} \textbf{202 : 29} \\ 9 : 100 \end{array} \right)$$

SCHEME 78

In continuation of the nauclefine synthesis, Sainsbury and Uttley (*123*) unsuccessfully investigated the regioselectivity of the photocyclization of the enamide **201** and the bromoenamide **203** though a regioselective formation of the nauclefine-type product (Scheme 79).

However, it was demonstrated that either acylation or alkylation of the pyridine nitrogen of the enamide **201** brought about a regioselective thermal cyclization to give the cyclized lactam **204** or the quaternary salt **205**. The

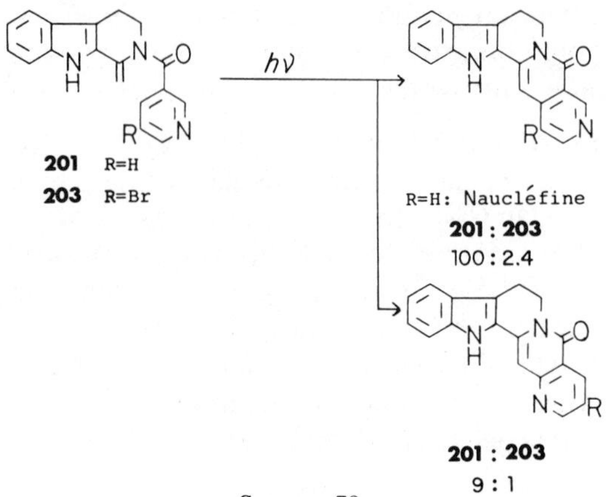

201 R=H
203 R=Br

R=H: Naucléfine
201 : 203
100 : 2.4

201 : 203
9 : 1

SCHEME 79

latter was converted to the desired naucléfine (**29**) in 65% yield from the enamide (**201**) on catalytic hydrogenation followed by dehydrogenation and debenzylation (*69,72*) (Scheme 80).

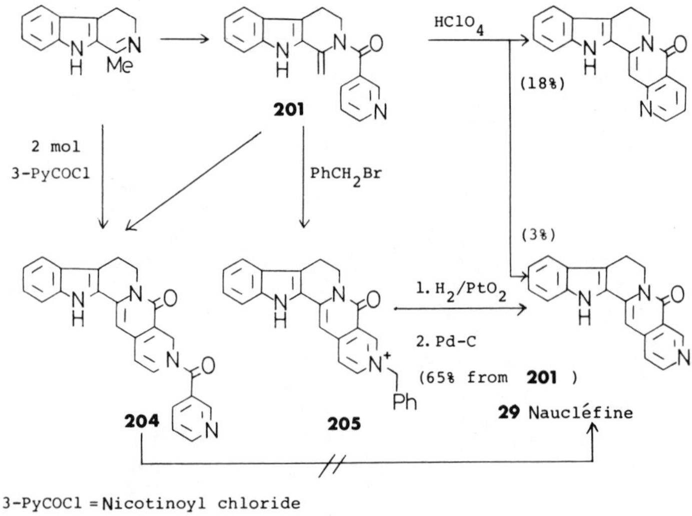

201

HClO₄

(18%)

2 mol
3-PyCOCl

PhCH₂Br

(3%)

204

205

1. H₂/PtO₂
2. Pd-C

(65% from **201**)

29 Naucléfine

3-PyCOCl = Nicotinoyl chloride

SCHEME 80

As in the azaberbine synthesis described in Section IV,C, Ninomiya *et al.* (*68*) prepared the spirodihydropyridine **206** by acylation of harmalane with an excessive amount of isonicotinoyl chloride and then synthesized naucléfine (**29**) by photorearrangement of the corresponding *N*-nor derivative **207** (Scheme 81).

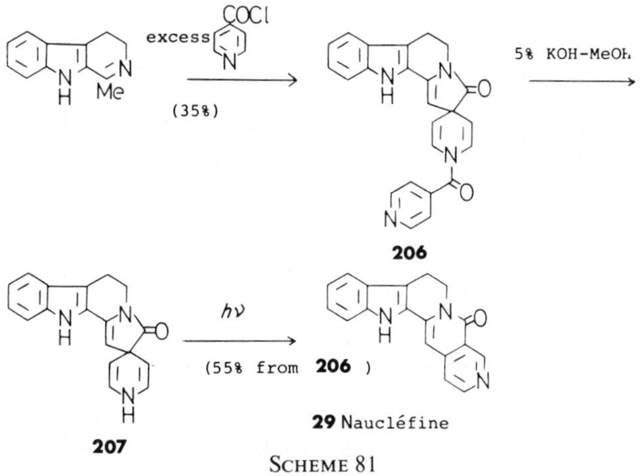

206

29 Nauclefine

207

SCHEME 81

Atta-ur-Rahman and Ghazala (75) synthesized 11-methoxynauclefine (**208**), an unnatural analog of the alkaloid nauclefine, by applying Sainsbury's procedure (69) to harmaline (Scheme 82).

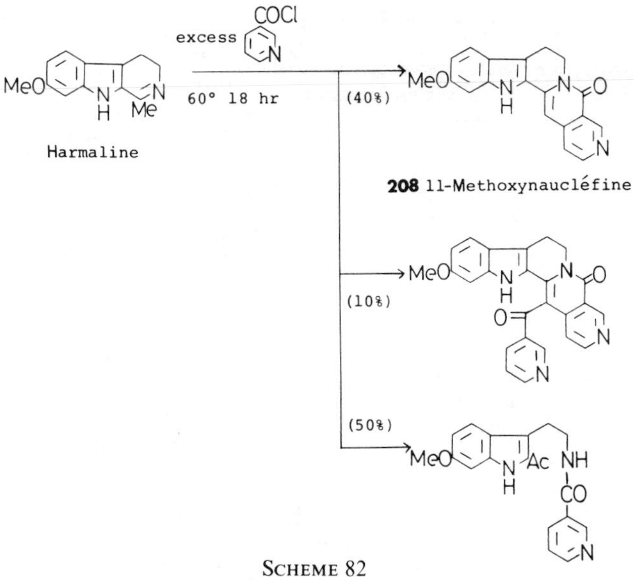

Harmaline

208 11-Methoxynauclefine

SCHEME 82

Angustidine, Naucletine, and Angustoline. In 1973 Cheung et al. (120) isolated new Corynanthe alkaloids angustoline (**215**), angustine (**214**), and angustidine (**213**) from *Strychnos angustiflora* Benth., a medicinal plant indiginous to South China. Two new additional alkaloids having similar

TABLE V

PHOTOCYCLIZATION OF HARMALANE DERIVATIVES

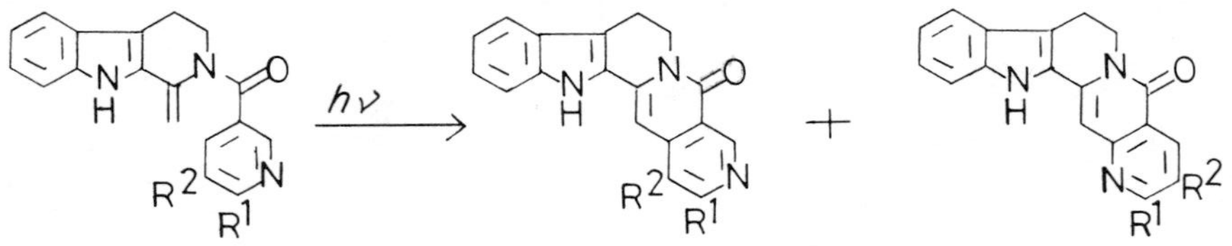

Substituents	Product I (%)	Product II (%)
20 R¹ = R² = H	**29** R¹ = R² = H (nauclefine) (48)	R¹ = R² = H (12)
R¹ = Me, R² = H	**213** R¹ = Me, R² = H (angustidine) (21)	R¹ = Me, R² = H (13)
209 R¹ = H, R² = Ac	**210** R¹ = H, R² = Ac (nauclétine) (30)	**216** R¹ = H, R² = Ac (8)
211 R¹ = H, R² = CH(Cl)Me	**212** R¹ = H, R² = CH(OMe)Me (5)	R¹ = H, R² = Et (6)
		R¹ = H, R² = CH(OMe)Me (1)
	214 R¹ = H, R² = CH = CH₂ (angustine)	
	215 R¹ =, R² = CH(OH)Me (angustoline)	

skeletal structure, naucléfine and naclétine (**210**), were also isolated from *Nauclea latifolia* Sm. by Hotellier *et al.* (*71*) in 1975.

By applying enamide photocyclization, Ninomiya *et al.* (*25,26,124*) completed the synthesis of angustidine (**213**), angustoline (**215**), and naclétine (**210**) in a few steps. Irradiation of the enamide **209** derived from harmalane and 5-acetylnicotinoyl chloride gave a mixture of two lactams, **210** and **216**, in 30 and 8% yields, respectively; the former (**210**) has the same structure as naclétine (**210**) and also was converted to angustoline (**215**) by sodium borohydride reduction. They also investigated photocyclization of the l-chloroethyl enamide **211** and obtained angustoline methyl ether (**212**) (*26,124*) (see Table V).

From a pharmacological point of view, Shafiee (*125*) investigated an effective synthesis of thia analogs of angustidine, angustoline, naucléfine, and naclétine (Scheme 83).

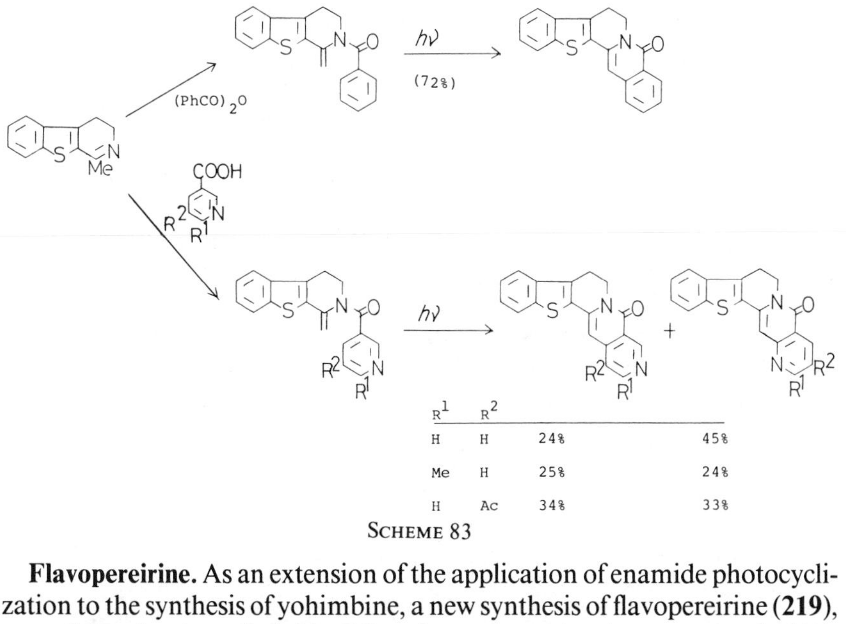

R^1	R^2		
H	H	24%	45%
Me	H	25%	24%
H	Ac	34%	33%

SCHEME 83

Flavopereirine. As an extension of the application of enamide photocyclization to the synthesis of yohimbine, a new synthesis of flavopereirine (**219**), one of the simplest alkaloid of *Strychnos* plants, has been reported (*117*). Harmalane, acylated with ethacryloyl chloride and the corresponding β-methoxy-substituted acid chloride, gave the enamides **217a** and **b**, which were so unstable that they were irradiated without further purification to afford the unstable photocyclized lactams **218a** and **b**. The lactam **218b** was treated with 10% HCl to remove a methanol moiety and then reduced with LAH, followed by dehydrogenation with palladium on carbon by heating at 280–300°C to give flavopereirine (**219**) as its perchlorate (*117*) (Scheme 84).

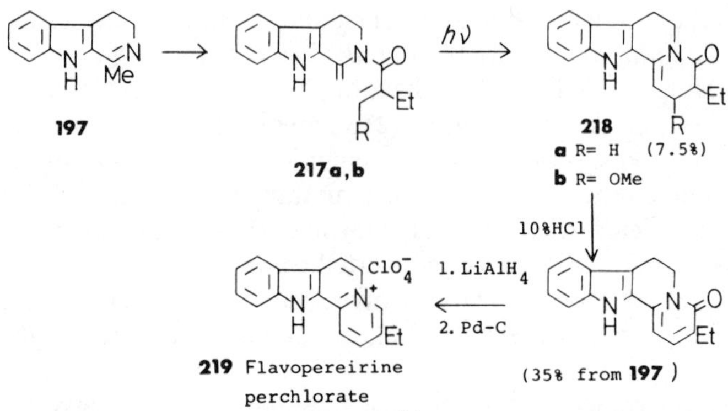

197   **217a,b**   **218**
                           a R= H   (7.5%)
                           b R= OMe

219 Flavopereirine   (35% from **197**)
perchlorate
SCHEME 84

Eburnane Alkaloids. A short and highly efficient synthesis of a key intermediate "Wenkert Enamine" (**223**) for the synthesis of Eburnane alkaloids such as vincamine has been reported (*126,127*).

Utilizing the ability of the imine–enamine equilibrium to act as an ambident nucleophile, Danieli *et al.* (*126,127*) treated the carboline **220** with acrylic acid in refluxing xylene to give the presumed intermediate **221**, which is actually an enamide, and this underwent intramolecular cyclization to give the lactam **222** in 91% yield. The suggested intermediacy of **221** was supported by N^b-acryloylation of **220** either with acrylic acid–diphenylphosphoryl azide or with acryloyl chloride in the presence of 4-dimethylaminopyridine. The lactam **222** was then converted to the target intermediate **223** by LAH reduction, or more efficiently, by sequential thionation followed by desulfurization with Raney nickel. These routes are flexible for the preparation of either the ring D-functionalized eburnanes or corynantheine-related alkaloids (Scheme 85).

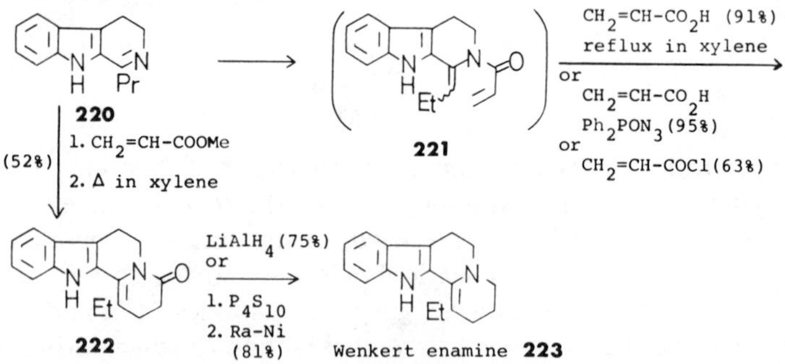

220   **221**   Wenkert enamine **223**
222
SCHEME 85

E. ERGOT ALKALOIDS

Nonoxidative photocyclization of enamides of benzanilide and acryloylenamine types has offered useful routes for the synthesis of ergot alkaloids, particularly 6,8-dimethylergolines (*53,55–57,128,129*). In addition, reductive photocyclization, a recent discovery by Ninomiya's group (*15,55–57*), is found to be the method of choice for the efficient synthesis of the ergoline group of alkaloids as shown by a novel total synthesis of (±)-lysergic acid (*57*).

1. Synthesis of the Basic Skeleton

N-Methylacrylnaphthalide, which was prepared from naphthylamine and methacryloyl chloride, was readily photocyclized by irradiation using a low-pressure mercury lamp to afford the corresponding benzo[*f*]quinolone **225** (*130*), having a skeletal structure similar to 6,8-dimethylergoline but lacking only a pyrrole ring.

Similarly, irradiation of the tricyclic enamide **226** for over 96 hr afforded the lactam **227**, in 42% yield, which was reduced with LAH and then acetylated to give the compound **228**, with an ergoline skeleton in good yield. This method provided a simple route for constructing the clavine skeleton (*130*) (Scheme 86).

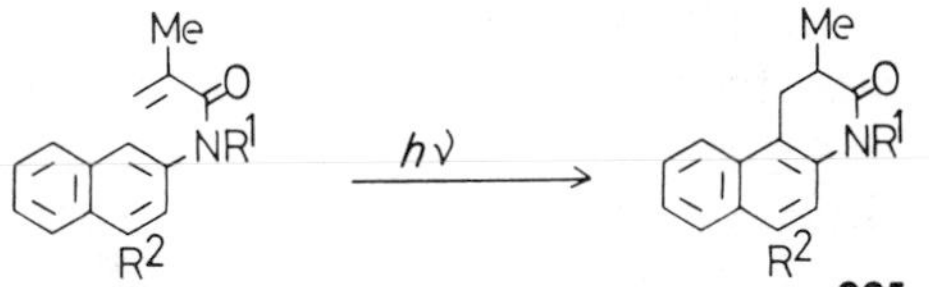

	R^1	R^2	Yield(%)
a	H	H	59
b	Me	H	50
c	H	COOMe	42.5
d	Me	COOMe	48

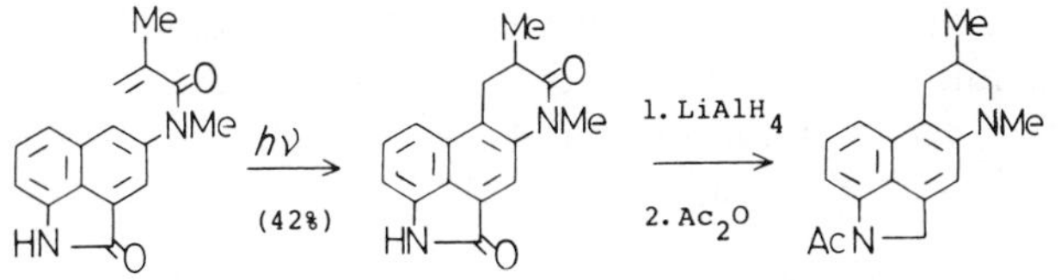

228 Clavine structure

SCHEME 86

Depyrrole analogs of clavines were readily synthesized by applying reductive photocyclization of the enamides of acryloyl and furoylenamine types, **229** and **233** (*55,56*). Irradiation of the enamide **229**, prepared from 2-tetralonimine and methacryloyl chloride in the presence of triethylamine, under nonoxidative conditions, afforded the lactams **230** and **225** in yields of 45 and 10%, respectively. Their reductive photocyclization in the presence of sodium borohydride at a low temperature of 4–5°C led to the formation of the hydrogenated lactam **231**, which was then reduced with LAH to afford the corresponding amine **232**. These amines have structures corresponding to the depyrrole analogs of three clavines: pyroclavine, festuclavine, and epicostaclavine with respect to their skeletal structure and stereochemistry, determined by comparisons of their spectral data (*55*).

Similarly, photocyclization of the β-methoxy-substituted enamide **229**, followed by two-step hydride reduction with LAH and sodium borohydride, furnished the depyrrole analog of agroclavine (*55*) (Scheme 87).

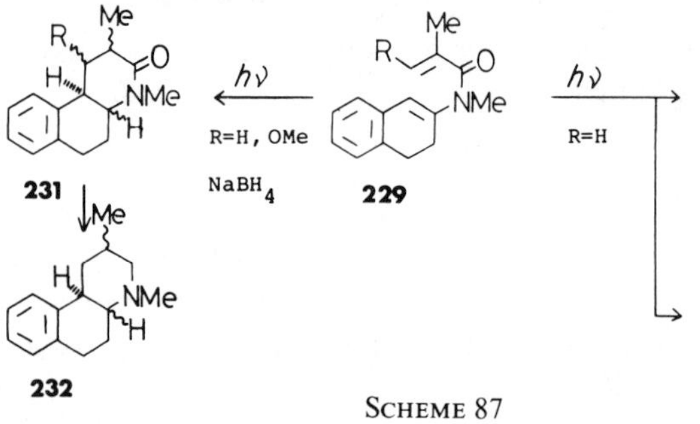

SCHEME 87

Synthetic utility of reductive photocyclization of enamides has been demonstrated by the synthesis of the depyrrole analog of isofumigaclavine A (*56*). The enamide of *N*-(3-furoyl)enamine type **233** was irradiated in the presence of sodium borohydride at 4–5°C to give a stereoisomeric mixture of two hydrogenated lactams, **234a** and **b**, with respect to the B/C ring junction. Their ratios depended on the solvent used. Osmylation of the trans amine, followed by usual workup with hydrogen sulfide, afforded a mixture of two cis glycols **235a** and **b**. Without separation, cleavage of these glycols with sodium metaperiodate afforded the unstable hydroxyaldehyde **236**, which was then subjected to Wolff–Kishner reduction to yield the C_2-methyl derivative **237** in 24% yield; the acetate of (**237**) has a structure partially identical to isofumigaclavine A, and is therefore regarded as the depyrrole analog of the alkaloid and established a synthetic route to the alkaloid (*56*).

The crucial step in this synthetic route is the use of the furoyl moiety in the enamide that facilitated reduction of the number of stereoisomers expected from the photocyclization of enamide and that also was suited for the following modification of the resultant dihydrofuran ring (Scheme 88).

Analogously, the key intermediates in the above synthesis, the hydroxyaldehyde **236** and glycol **235**, were also converted to the depyrrole analogs of methyl lysergate **240** and isolysergate **241** (*56*). An unstable and inseparable mixture of epimeric hydroxyaldehydes **236** was dissolved in methanol–acetone and treated with chromic trioxide in sulfuric acid to give a 1 : 1 mixture

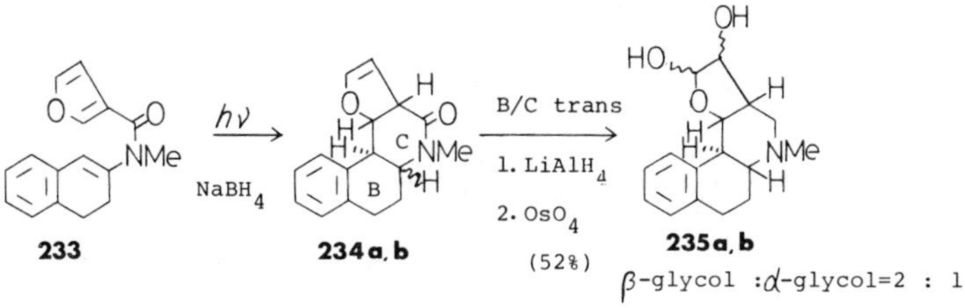

B/C	cis	:	trans	
C_6H_6-MeOH	1	:	1	(82%)
MeCN-MeOH	1	:	4	(72%)

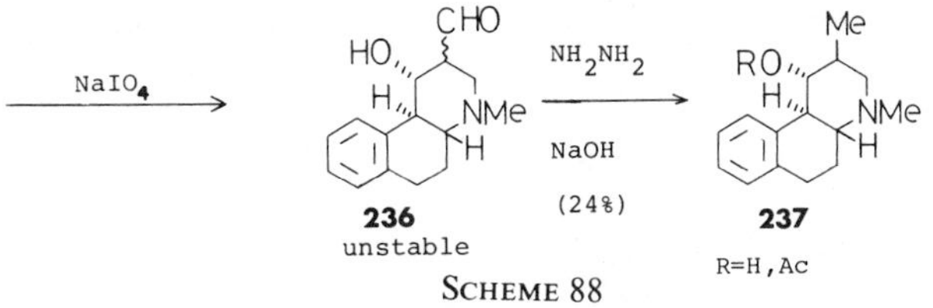

SCHEME 88

of the corresponding hydroxy esters **238** and **239** both in a yield of 20%. According to Horii's procedure (*131*), dehydration of the hydroxy esters (**238** and **239**) was achieved by heating with phosphorous oxychloride–phosphoric acid in pyridine to afford a mixture of epimeric unsaturated esters **241** and **240**, with almost identical ratios of 1:2 in yields of about 35%, respectively, from both hydroxy esters. Each was assigned as the depyrrole analogs of methyl lysergate and isolysergate, thereby establishing a potential synthetic route to lysergic acid and also demonstrating the potential of this synthetic route as a general method for the synthesis of many other ergoline groups of alkaloids (*56*) (Scheme 89).

2. Total Synthesis of Alkaloids

Costaclavine, Epicostaclavine, and Festuclavine. In the period before introduction of reductive photocyclization in 1981, Ninomiya *et al.* (*128,129*) had synthesized three epimeric 6,8-dimethylergolines costaclavine (**247**), epicostaclavine (**248**), and festuclavine (**249**) via the lactam that was prepared by the reaction of the tricyclic imine **242** with methacryloyl chloride or methacrylamide.

Although acylation of the imine with methacryloyl chloride in the presence of triethylamine at 0°C gave a mixture of the enamide **243** and the cyclized lactam **244** in the ratio of 2:1 with 52% yield, the reaction in the absence of base under refluxing temperature in benzene afforded the same mixture but with a different ratio, 1:4 in favor of the lactam, with 66% yield.

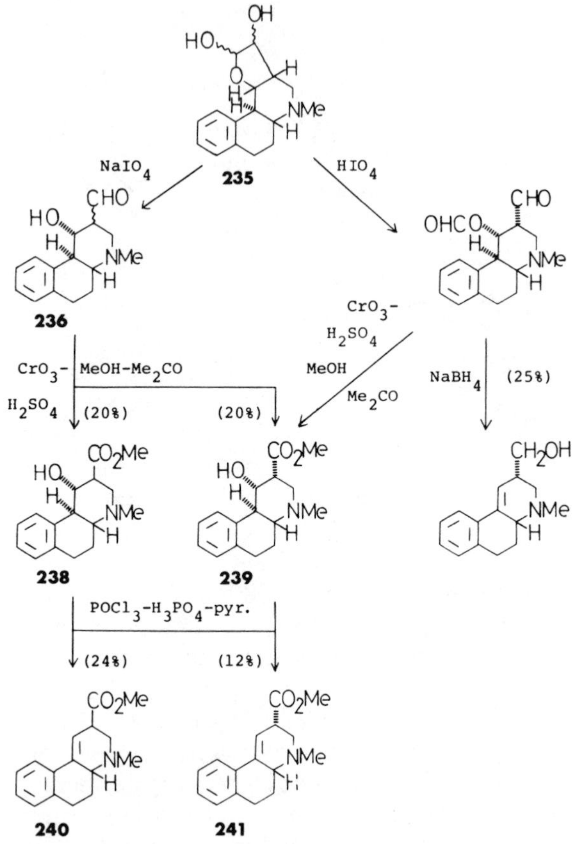

SCHEME 89

The lactam (244) thus obtained was also prepared by aza-annelation of the imine 242 with acrylamide by simple heating, followed by methylation with methyl iodide. Catalytic hydrogenation of the lactam 244 on platinum dioxide, followed by reduction with LAH and finally dehydrogenation, furnished costaclavine (247) and epicostaclavine (248) in yields of 16 and 4%, respectively. The latter (248) is the fourth epimer of 6,8-dimethylergoline, not found in nature and designated as such (128,129).

On the other hand, the lactam 244 was subjected to LAH reduction, followed by dehydrogenation to give festuclavine (249) in 22% yield in addition to small amount of costaclavine (247) and epicostaclavine (248) (128,129) (Scheme 90).

Epicostaclavine (248) and pyroclavine (250) were also synthesized by applying reductive photocyclization of the enamide (243) as shown in Scheme 91 (53).

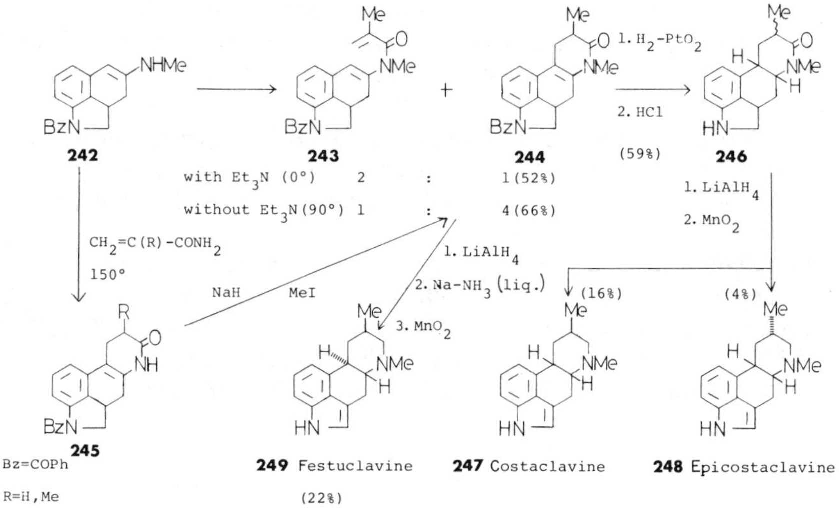

242

with Et$_3$N (0°)   2   :   1 (52%)

without Et$_3$N (90°)   1   :   4 (66%)

CH$_2$=C(R)-CONH$_2$

150°

NaH   MeI

245

Bz=COPh

R=H, Me

243

244   (59%)

246

1. LiAlH$_4$

2. MnO$_2$

1. LiAlH$_4$

2. Na-NH$_3$ (liq.)   (16%)

3. MnO$_2$ (4%)

249 Festuclavine

(22%)

247 Costaclavine

248 Epicostaclavine

SCHEME 90

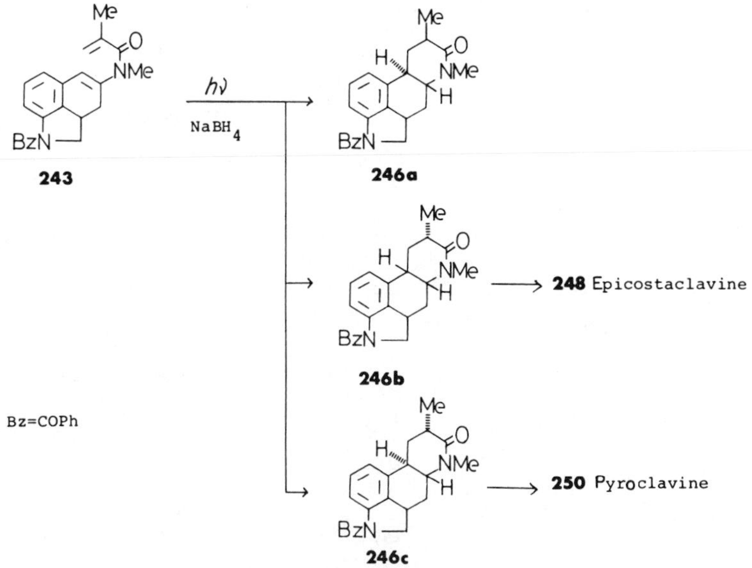

243

$h\nu$

NaBH$_4$

246a

246b → **248** Epicostaclavine

Bz=COPh

246c → **250** Pyroclavine

SCHEME 91

Lysergic Acid. Based on the results of the synthetic route using benzo[ƒ]quinoline derivatives (55,56), Ninomiya et al. (57) applied reductive photocyclization to the enamide **251**, prepared from the tricyclic imine **242** and 3-furoyl chloride.

Reductive photocyclization of the enamide **251** afforded a mixture of two hydrogenated lactams **252a** and **b** in 81% combined yield as a mixture of two stereoisomers with respect to C/D ring juncture. Lithium aluminum hydride reduction of the mixed lactams, followed by rebenzoylation, afforded the corresponding amines **253** and **254** in yields of 61 and 9%, respectively, on separation.

Ring opening of the dihydrofuran ring and the following modification of the partial structure was carried out according to the preliminary result (*55,56*) obtained on the synthesis of the depyrrole analogs of benzo[*f*]quinoline derivatives.

Osmylation afforded a mixture of two cis glycols **255**, which was cleaved into the corresponding hydroxyaldehydes **256**. Subsequent oxidation with chromic trioxide in methanol afforded a mixture of two hydroxy esters **257** and **258** that were separated. Contrary to the result in the model compounds (*56*), dehydration of both hydroxy esters **257** and **258** afforded the identical

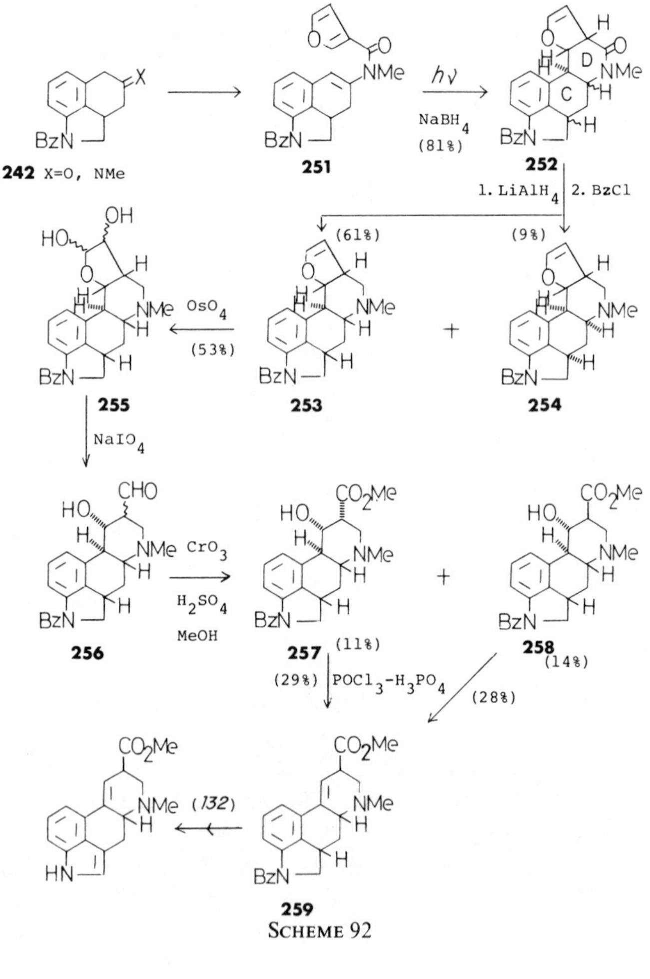

242 X=O, NMe   **251**   **252**

255   **253**   **254**

256   **257**   **258**

259

SCHEME 92

unsaturated ester **259** in 29% yield from **257** and 28% yield from **258**. This unsaturated ester **259** was identical to the authentic sample (*132*) prepared previously by Ramage *et al.* in their total synthesis of (±)-lysergic acid (Scheme 92).

F. MISCELLANEOUS ALKALOIDS

Alkaloids having a spiro-fused ring system, such as sesbanine and gramine, were also the targets of synthesis by the use of enamide photocyclization. Gramain *et al.* (*49,50*) succeeded in synthesizing various types of the spiro compounds **261** and **262** by applying nonoxidative photocyclization of the enamide **260** prepared from cyclohexanecarboxaldehyde (Scheme 93).

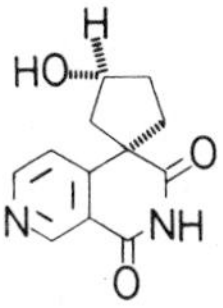

Sesbanine

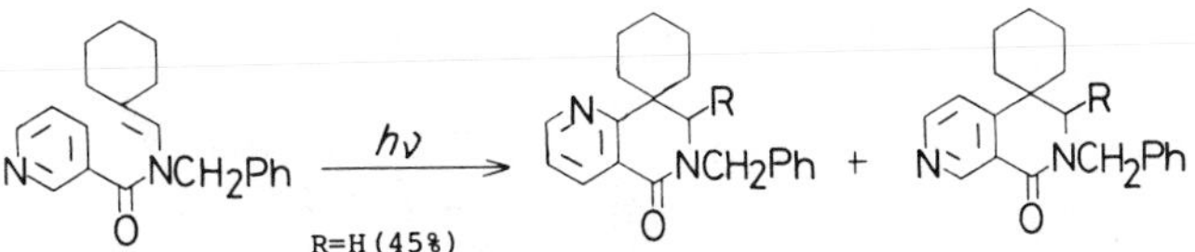

R=H (45%)   **261**   **262**
R=OMe (55%)   (1   :   1)

260

260 b   **260 c**

SCHEME 93

V. Alkaloid Syntheses: Cyclization of Related Systems

This chapter reviews the synthesis of alkaloids and related compounds by the use of photochemical or thermal cyclization reactions of conjugated systems, unconjugated systems structurally related to or derived from what we call enamide systems. As an extension of works that attempt to develop new cyclophiles devised from conjugated systems such as enamides, as well as an extension of works originating from stilbene-type compounds, a number of analogous conjugated systems and some unconjugated systems are shown to act as cyclophiles for the formation of six-membered ring systems involving nitrogen, and thus are applicable to the synthesis of alkaloids.

A. PHOTOCYCLIZATION OF 2-ACYL-1-BENZYLIDENEISOQUINOLINES

Photochemical conversion of stilbenes to phenanthrenes via a six π-electron conrotatory cyclization according to an electrocyclic mechanism to the dihydrophenanthrenes and subsequent dehydrogenation is a very famous and useful synthetic reaction (103).

Photocyclization of 1-benzylideneisoquinolines to the dibenzo[d,e,g]-quinolines is a well-studied example of the above stilbene–phenanthrene cyclization and therefore has been extended to the development of new photochemical methods for the synthesis of aporphine alkaloids (11–14). In this section, photocyclization of 2-acyl-1-benzylideneisoquinolines to dehydroaporphines and their subsequent conversion to aporphines and oxo-aporphines are summarized.

Irradiation of the enamide **118** with no substituent in the ortho position of the benzylidene group was carried out in the presence of an oxidizing agent. On the other hand, the o-halogenoenamide **263** was photocyclized in the presence of a base as an acid scavenger (Scheme 94).

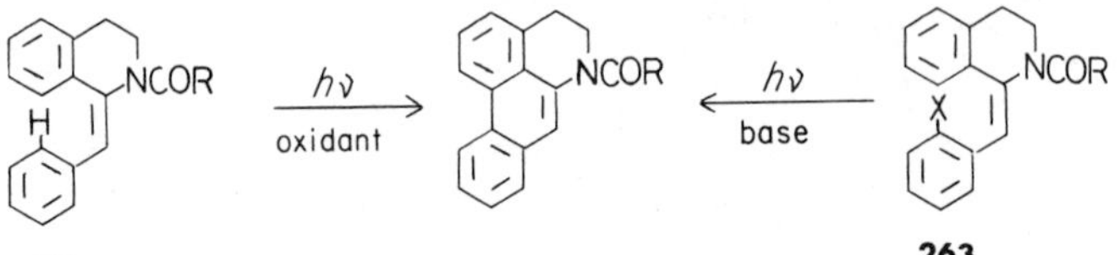

118   **263**

SCHEME 94

1. Synthesis of the Basic Skeleton

As described in Section IV,C, irradiation of a mixture of two geometrical isomers of enamides **118** and **119** that were prepared by acylation of the 1-benzylisoquinoline in the presence of iodine afforded the aporphine (**121**) in 65% yield (104,105) (Scheme 95).

Cava and Havlicek (108) also obtained the 7-phenylaporphine **132** as the major product in addition to small amount of the 13-phenylprotoberberine **130** by oxidative photocyclization of the enamide **131** (Scheme 96).

As an extension of enamide photocyclization as a general synthetic

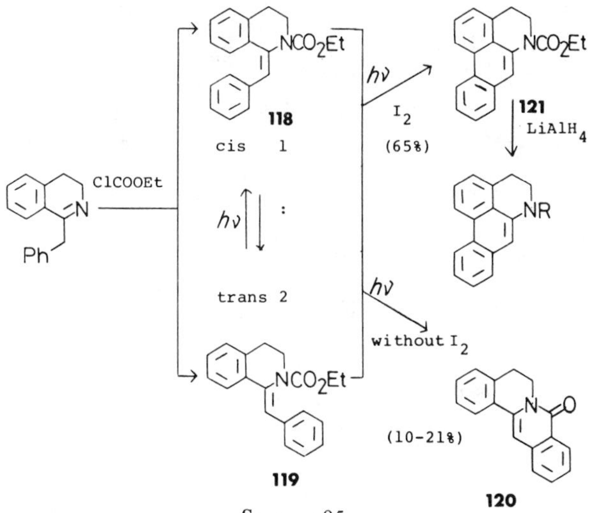
ClCOOEt
cis 1
trans 2
118
119
120
121
I2
(65%)
LiAlH4
without I2
(10-21%)

SCHEME 95

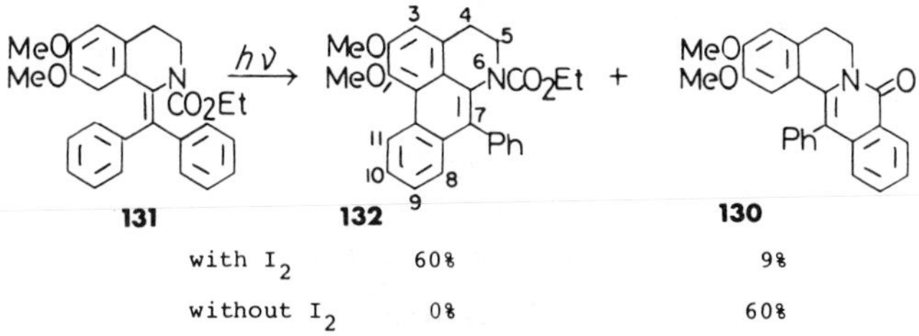
131
132
130
with I2 60% 9%
without I2 0% 60%

SCHEME 96

method, Ninomiya *et al.* (*133,134*) investigated oxidative photocyclization of the stilbene type of enamides **264** that contain the 4-oxazolin-2-one moiety and afforded the photocyclized compounds **265** in 50–89% yields. However, attempts to prepare an aporphine skeleton by photocyclization of the enamides **266a** and **b** were unsuccessful under various conditions, presumably due to steric hindrance arising from ring contraction (*133*) (Scheme 97).

On the other hand, the B-homo analog of the enamide **267** underwent smooth photocyclization under the same conditions as above to afford the B-homoaporphine skeleton, probably because of reduced steric congestion arising from involvement of the seven-membered ring in the enamide (*134*) (Scheme 98).

2. Total Synthesis of Alkaloids

Nuciferine. Cava *et al.* (*135,136*) synthesized (±)-nuciferine (**272**) by applying two types of cyclizations: one (*135*) is the nonoxidative photocycli-

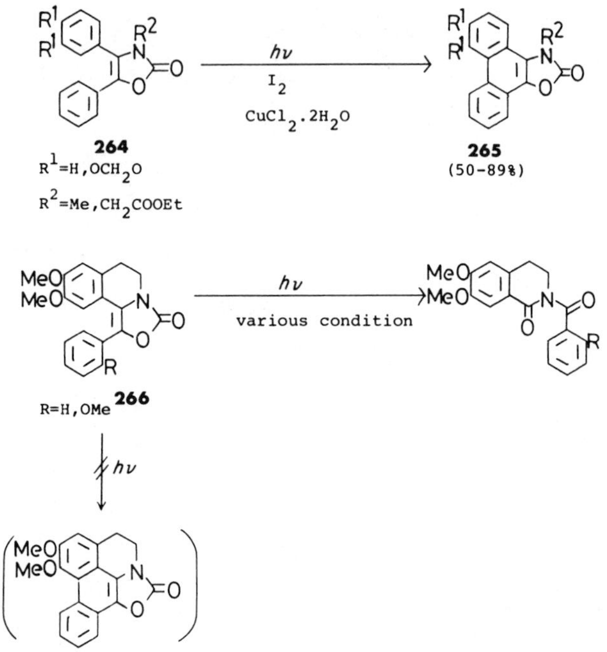
hν
I_2
$CuCl_2 \cdot 2H_2O$
264
R^1=H,OCH_2O
R^2=Me,CH_2COOEt
265
(50–89%)
hν
various condition
R=H,OMe
266
hν

SCHEME 97

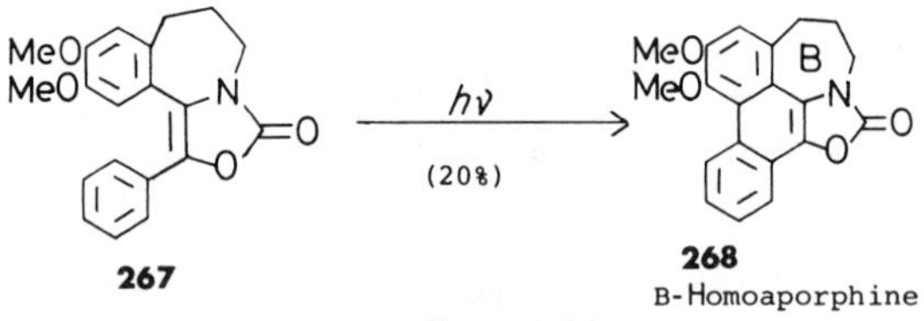
hν
(20%)
267
268
B-Homoaporphine

SCHEME 98

zation of the *o*-chloroenamide and the other (*136*) is oxidative photocyclization of the unsubstituted enamide.

Photocyclization of the *o*-chloroenamide **269** in the presence of either calcium carbonate or potassium *tert*-butoxide gave the aporphine **271** in 24 or 59% yield, respectively; this aporphine is considered formed via the intermediate **270** (*135*) (Scheme 99).

Oxidative photocyclization of the enamide **273** gave the same aporphine (**271**) in 15% yield; this was then converted to (±)-nuciferine by LAH reduction of the carbamate, followed by catalytic hydrogenation of ring C (*136*) (Scheme 100).

Glaucine. As in the synthesis of nuciferine (*135*), Yang *et al.* (*104*) synthesized *N*-carbethoxydehydro-L-norglaucine (**275**) by oxidative photocyclization of the enamide **274**. Independently, Cava *et al.* (*135*) completed

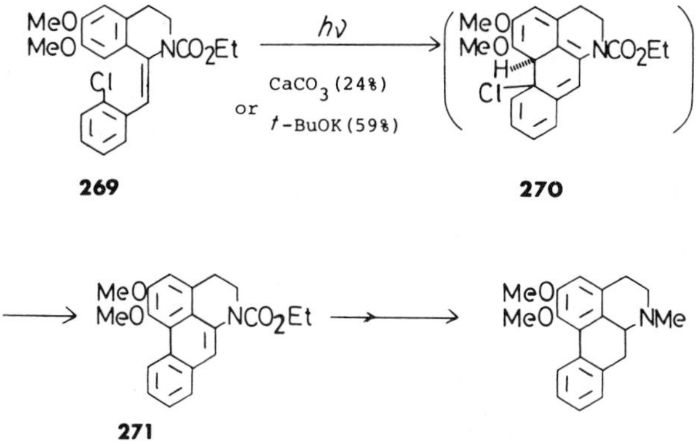

269   **270**

271

272 (±)-Nuciferine

SCHEME 99

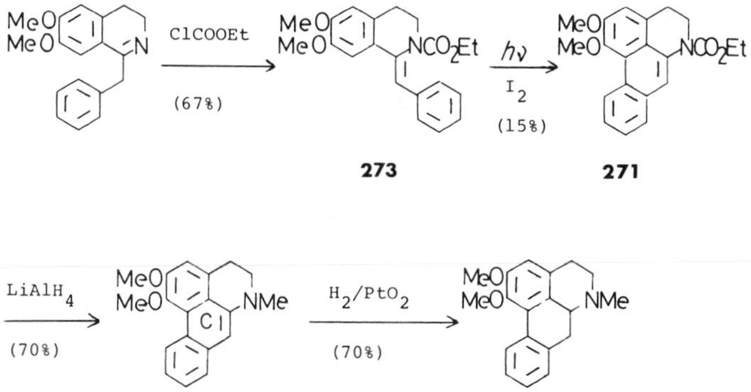

273   **271**

272 (±)-Nuciferine

SCHEME 100

total synthesis of (±)-glaucine via the same photoproduct (**275**), which was however prepared by nonoxidative photocyclization of the *o*-bromoenamide **276** in the presence of either calcium carbonate or potassium *tert*-butoxide (Scheme 101).

Recently, Castedo *et al.* (*137*) studied enamide photocyclization for aporphine synthesis and modified Cava's procedure (*135–137*) after consideration of the reaction course and structural requirement for the exclusive formation of aporphine. Castedo's group concluded that an enamide should have a forced *cis*-stilbene geometry, in order to avoid an undesired mode of cyclization leading to protoberberine-type compounds and also to avoid a facile conversion to oxoaporphines after cyclization.

Photocyclization of the enamide **278** in the presence of either *tert*-butylamine or triethylamine gave the aporphine **279** in 60% yield or the photoreduced aporphine **280** in 82% yield. The role of the amine in photocycliza-

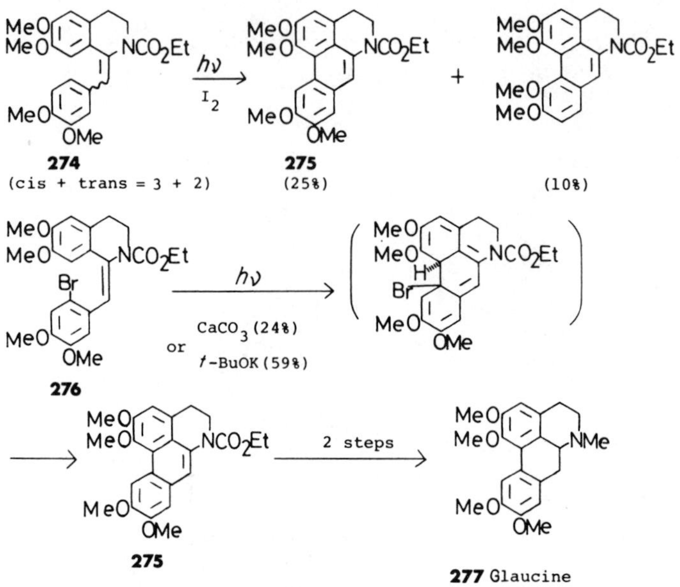
MeO
MeO
NCO2Et
MeO
OMe
hν
I2
+
274
(cis + trans = 3 + 2)
275
(25%)
(10%)
Br
CaCO3 (24%)
or
t-BuOK (59%)
276
2 steps
NMe
275
277 Glaucine

SCHEME 101

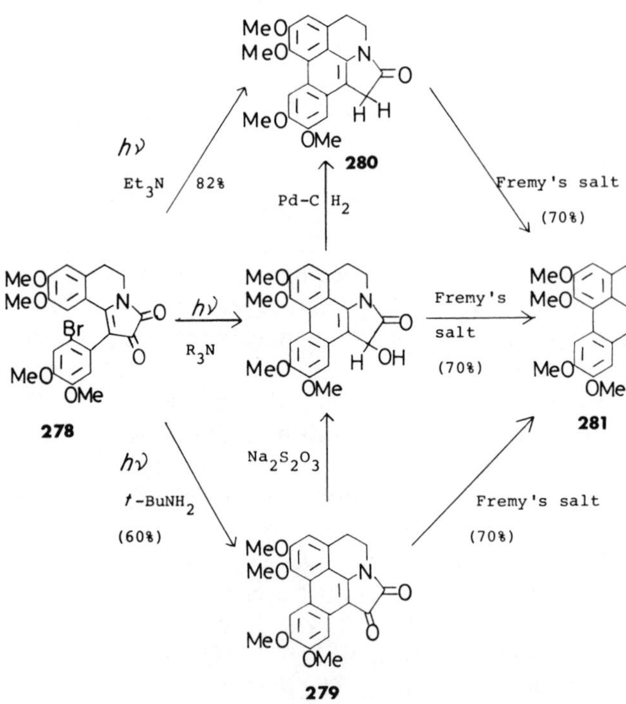
hν
Et3N
82%
280
Pd-C H2
Fremy's salt
(70%)
Br
hν
R3N
Fremy's salt
(70%)
278
281
hν
t-BuNH2
(60%)
Na2S2O3
Fremy's salt
(70%)
279

SCHEME 102

tion and photoreduction was also discussed (137). Oxoglaucine (281) was synthesized by oxidation with Fremy's salt from the above products (279 and 280) (137) (Scheme 102).

Dicentrine and Dicentrinone. Dicentrine (283) was synthesized by photocyclization of the o-bromo- and chloroenamides 282a and b by Cava et al. (138); the former enamide (282a) was best suited for the synthesis of the alkaloids in view of yield and reaction time (Scheme 103).

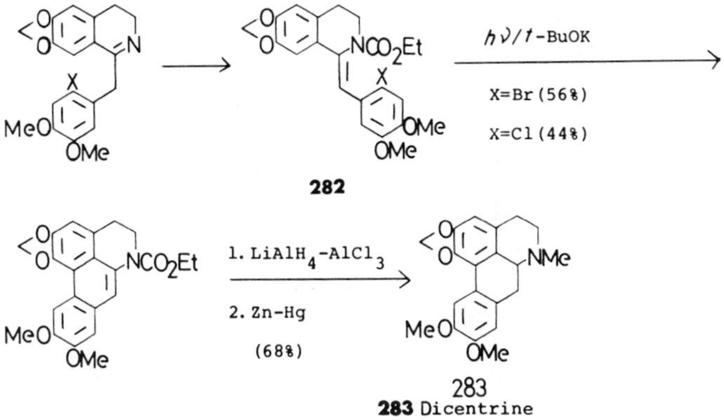

282

283
283 Dicentrine

SCHEME 103

Castedo et al. (137) completed an efficient synthesis of dicentrinone (284) in 20% overall yield following the methodology in the case of oxoglaucine (137) (Scheme 104).

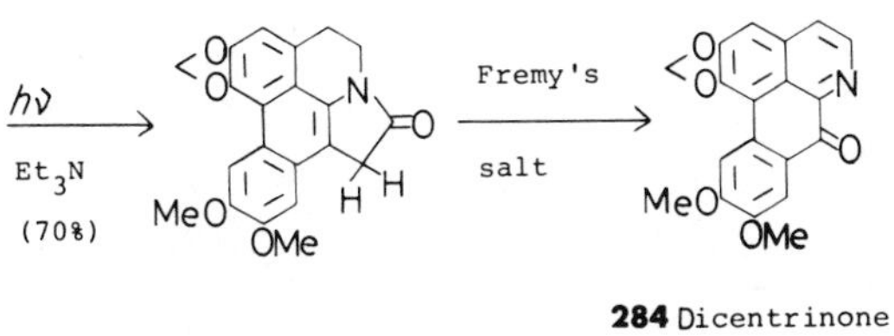

284 Dicentrinone

SCHEME 104

Cassameridine, Neolitsine, and Pontevedrine. Syntheses of cassameridine (139), a rare oxoaporphine alkaloid, by Cava et al. (138); pontevedrine

(*139*), a dioxoaporphine alkaloid; and lysicamine (*37*), an oxoaporphine alkaloid, by Castedo *et al.* (*137,139*); and neolitsine, an aporphine alkaloid, by Moltrasio *et al.* (*140*) have been reported, all via routes involving enamide photocyclization (Schemes 105–108).

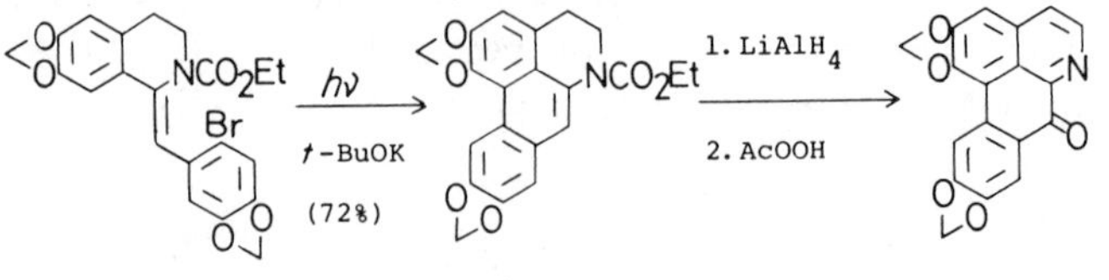

285 Cassameridine

SCHEME 105

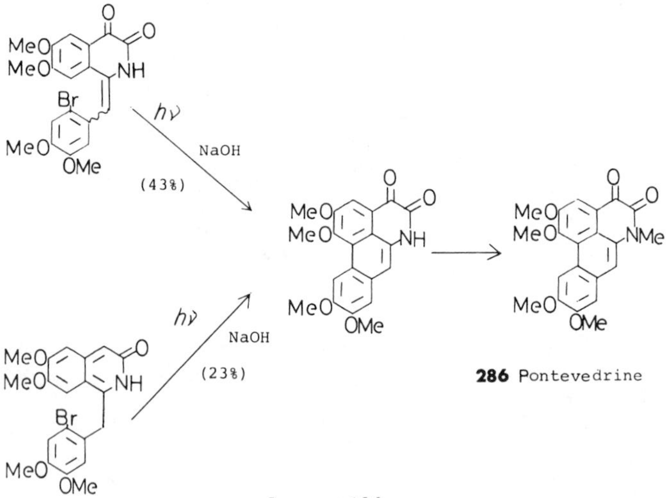

286 Pontevedrine

SCHEME 106

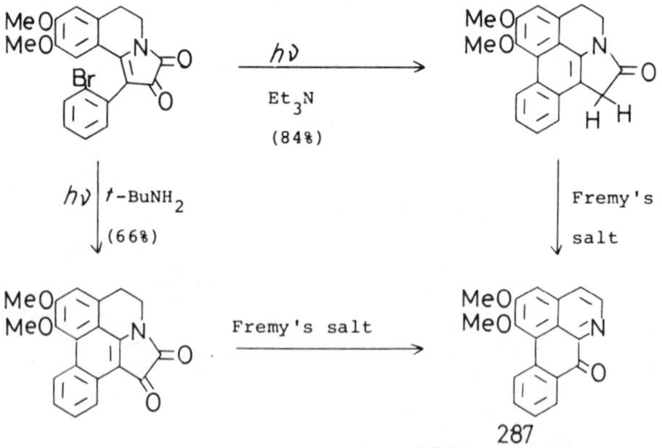

287 Lysicamine

SCHEME 107

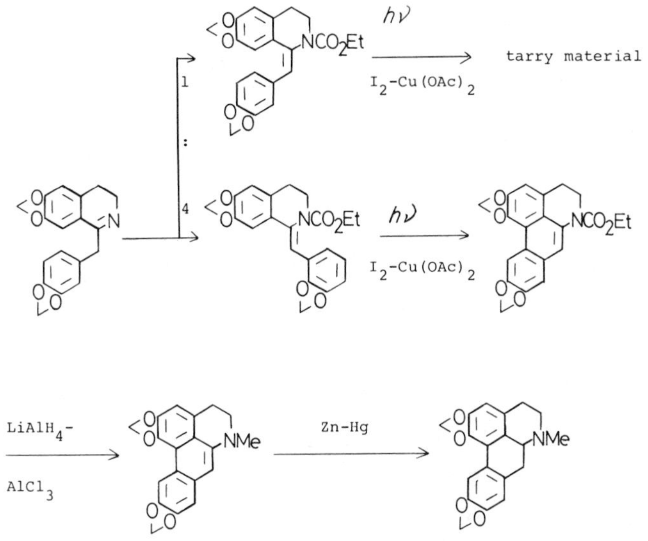

SCHEME 108

Corunnine, Nandazurine, (±)-Caaverine, (±)-Isoboldine, (±)-Thalicmidine, and (±)-Domesticine. Relative to the aporphine synthesis by enamide photocyclization, Kupchan and O'Brien (*141*) have developed the oxidative photochemical synthesis of aporphine alkaloids.

The Reissert compound **289** was condensed with the 1-iodobenzaldehyde **290** to give the 1-benzyl compounds **291**, which were then irradiated. The photocyclized products **292** were the oxoaporphines, which were then converted to the respective alkaloid: corunnine, nandazurine, (±)-caaverine, (±)-isoboldine, (±)-thalicmidine, and (±)-domesticine in good yields (Scheme 109).

B. INTRAMOLECULAR DIELS–ALDER CYCLIZATION

The intramolecular Diels–Alder reaction promises to become widely used in the synthesis of natural products from reports of recent research (*142–144*). In natural alkaloid synthesis by a Diels–Alder reaction, an enamide was utilized as either a suitable dienophile or diene group. There exist numerous examples of bimolecular and intramolecular Diels–Alder reactions of enamides with various dienophiles as well as bimolecular (4 + 2) cycloadditions of enamines with electron-deficient dienes. On the other hand, there appear to be only few reports of the (4 + 2) cycloaddition reactions of enamides with unactivated dienes.

In this section, we summarized some examples of alkaloid synthesis by

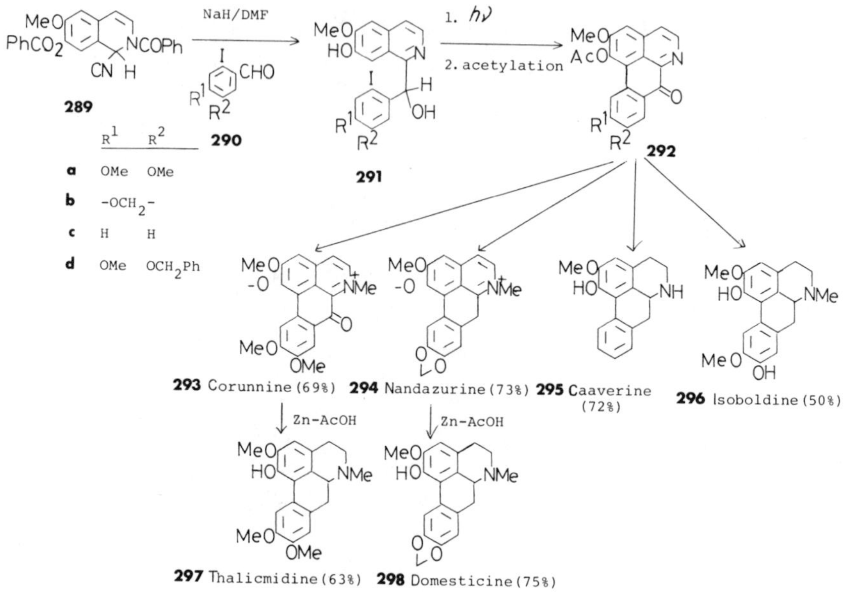

266   ICHIYA NINOMIYA AND TAKEAKI NAITO

	R^1	R^2
a	OMe	OMe
b	-OCH$_2$-	
c	H	H
d	OMe	OCH$_2$Ph

293 Corunnine (69%) 294 Nandazurine (73%) 295 Caaverine (72%) 296 Isoboldine (50%)

297 Thalicmidine (63%) 298 Domesticine (75%)

SCHEME 109

using an enamide function as the starting component of either diene or dienophile.

Chelidonine. In 1971 Oppolzer and Keller (*145*) discovered and developed an intramolecular Diels–Alder reaction using *o*-quinodimethane for the synthesis of polycyclic natural products.

Chelidonine, a representative benzo[*c*]phenanthridine alkaloid (*1,2*), was synthesized as the first application of this reaction in natural product synthesis (*145*). Initial thermal opening of the four-membered ring in **299** led to the formation of a transient *E-o*-quinodimethane (**301**), which has a dienamide structure in the diene part and is then trapped by the suitably positioned multiple bond in the same molecule. They applied this intramolecular reaction to the acetylenic cyclobutene **300** for the synthesis of (±)-chelidonine (Scheme 110).

After investigating the reaction in order to control its stereochemistry, they succeeded in improving this approach considerably, as shown below. Introduction of a nitro group into the known styrene **302** by reaction with silver nitrite in the presence of iodine, gave the nitrostyrene **303** in 72% yield. Heating of **303** at 120°C gave the cis-fused adduct **304** stereoselectively in 97% yield. Treatment of the nitro compound **304** thus obtained with titanium trichloride furnished the sensitive ketone **305**. Concomitant reduction of the carbonyl and urethane groups of the crude product with LAH afforded (±)-chelidonine in 54% yield from **304** (*146*) (Scheme 111).

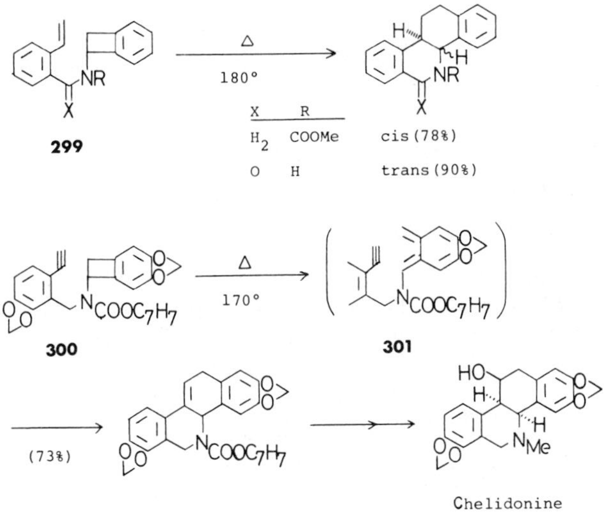
Δ
180°
X R
H2 COOMe cis (78%)
O H trans (90%)
299
Δ
170°
300
301
(73%)
Chelidonine

SCHEME 110

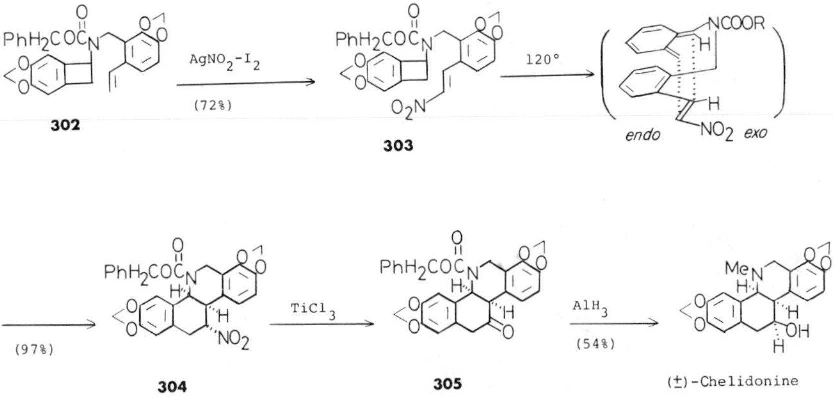
AgNO2-I2
(72%)
120°
302
303
endo
exo
(97%)
TiCl3
AlH3
(54%)
304
305
(±)-Chelidonine

SCHEME 111

(+)-, (−)-, and (±)-Pumiliotoxin-C. Attempts to synthesize pumiliotoxin-C in a stereoselective manner led to the study of an intramolecular Diels–Alder reaction of two types of enamides **306** and **307** (*147–149*). One enamide (**306:** X = H_2, Y = O, R = Me) was converted at 200°C to the almost pure cis-fused octahydroquinolone **308**, whereas the other enamide (**307:** X = O, Y = H_2, R = Et) afforded a mixture of the trans lactam **309** and the cis isomer **308** in the ratio of 3:2, thus showing less and reverse stereoselectivity (*147*) (Scheme 112).

Based on these results, the readily available amine **310** was converted to the enamide **311**, which, when heated at 215°C for 24 hr, underwent

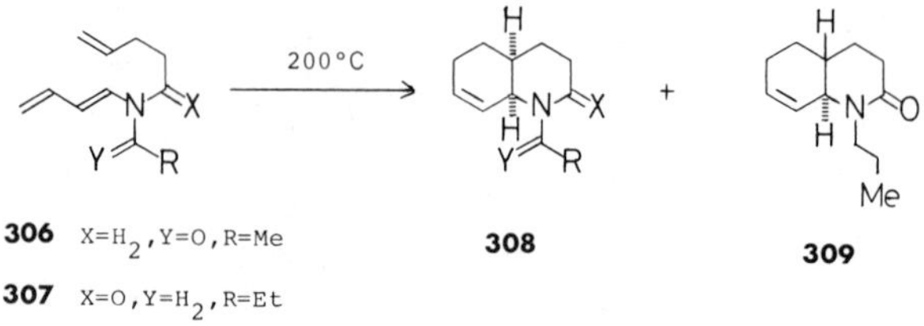

306 X=H₂,Y=O,R=Me

307 X=O,Y=H₂,R=Et

308

309

SCHEME 112

intramolecular cycloaddition to afford the corresponding adduct in 60–90% yield, and the subsequent reactions involving hydrogenation and hydrolysis afforded racemic pumiliotoxin-C in 60% yield (*148*) (Scheme 113).

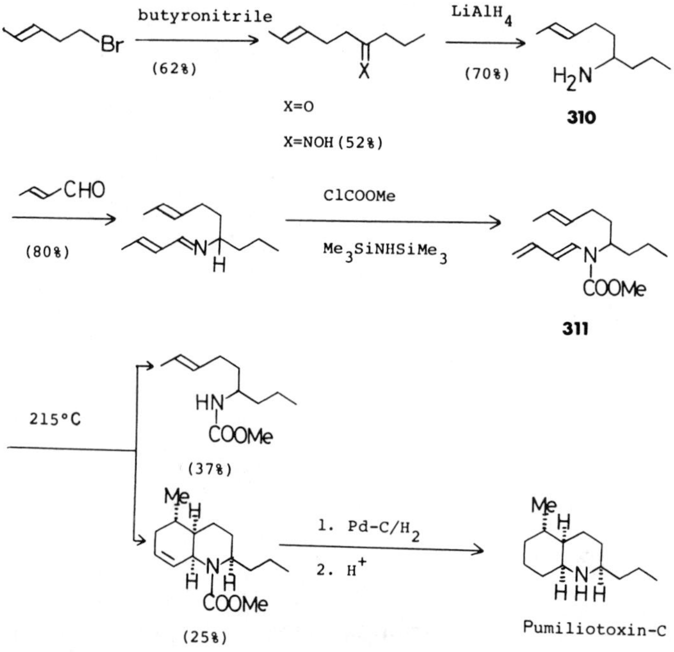

SCHEME 113

Similarly, an enantioselective synthesis of (+)-(2*R*)- and (−)-(2*S*)-pumi-liotoxin-C was completed from the optically active (*R*)- and (*S*)-norvalines (*149*) (Scheme 114).

(±)-**Lycorane and Galanthane.** Stork and Morgans (*150,151*) investigated the intramolecular Diels–Alder reaction of two types of enamides **312** and **319**, in which an enamide group is positioned at the dienophile in the former (**312**) (*150*) and at the diene part in the latter (**319**) (*151*), respectively.

The enamide **312** was prepared from 2-carboxybenzaldehyde and 3-car-

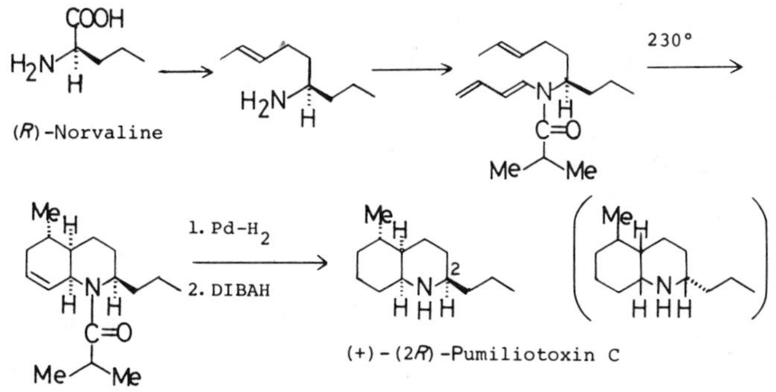

SCHEME 114

bomethoxypyrrolidine and then heated in *o*-dichlorobenzene under reflux for 25 hr to give a mixture of the cyclized lactams **313** and **314** in 84% yield, with a ratio of 0.84:1; the former (**313**) constitutes the galanthane ring

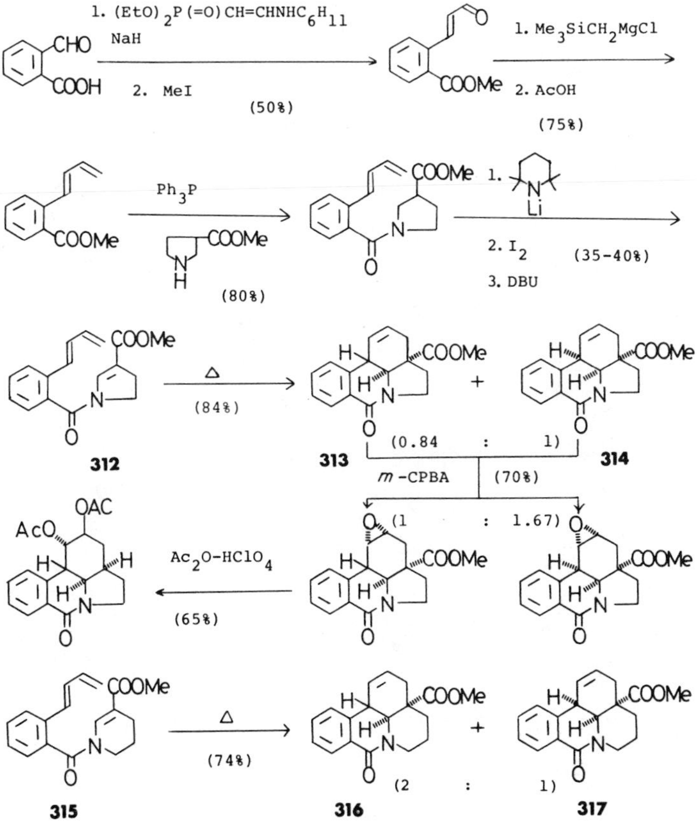

SCHEME 115

system (*150*). On the other hand, the enamide **315** including a six-mem-bered ring, afforded a mixture of two lactams, **316** and **317** in the ratio of 2 : 1. These subtle differences in the stereochemical course of the reaction caused by the structure of enamides illustrate the delicate nature of the interactions controlling the stereochemistry of an intramolecular Diels–Alder reaction (*150*) (Scheme 115).

Enamide **319**, prepared from the lactone **318** and 3-pyrrolidinol, in several steps, was heated at 140°C to give the homogeneously cyclized lactam **320** in 51% yield; this was then hydrogenated to afford 7-oxo-α-ly-corane (**321**) (*151*) (Scheme 116).

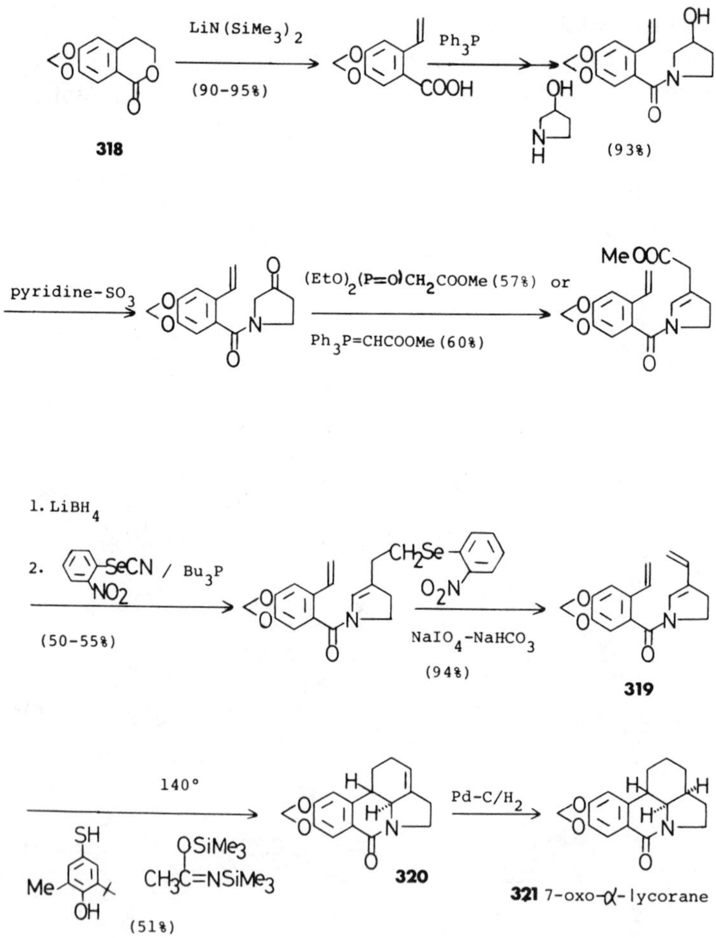

SCHEME 116

Aspidospermine and Aspidospermidine. The substituted hydroindole and hydroquinoline rings, common to a diverse array of alkaloids found in nature, were synthesized by applying intramolecular Diels–Alder reaction of the endocyclic enamide as dienophile (*152*).

The enamides **323** were prepared from tetrahydropyridine with 3,5-hexadienoyl chloride. Heating of a 1% solution of **322** in toluene resulted in the subsequential cheletropic expulsion of sulfur dioxide, followed by an intramolecular Diels–Alder reaction to give the hydroindoles **324** in 45–67% yields. Then **324a** was converted to hydrolulolidine (**325**) and **324c** to the ketolactam **326**, thereby furnishing a new formal total synthesis of aspidospermine (**327**) (*152*) (Scheme 117).

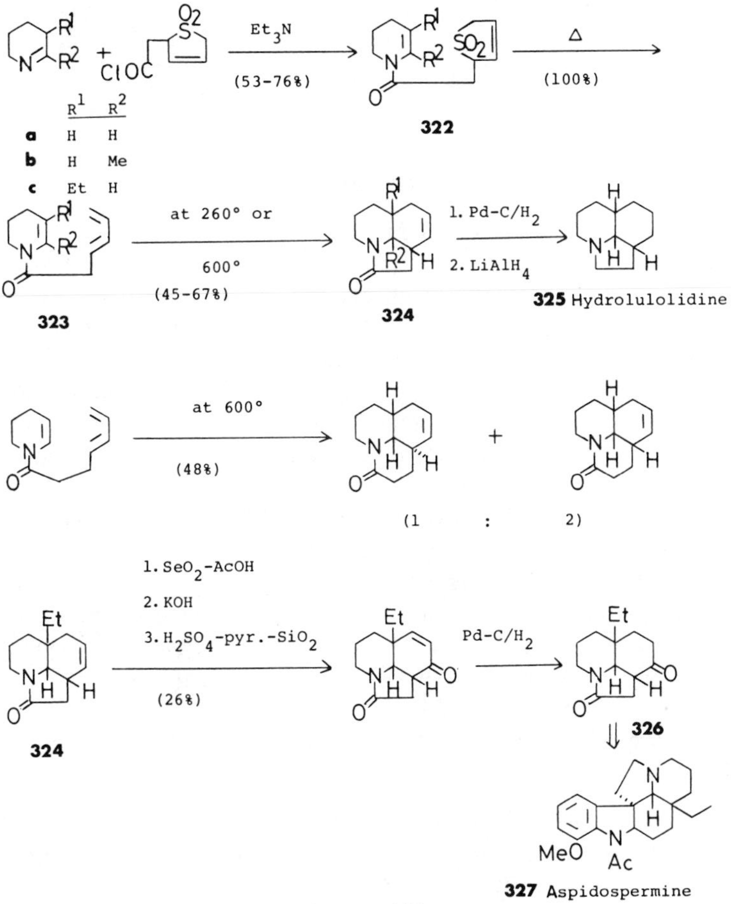

SCHEME 117

Magnus *et al.* (*153,154*) reported total synthesis of (±)-aspidospermidine (**330**) by applying the intramolecular Diels–Alder reaction to the enamide **329** prepared from the 2-methylindole derivative **328** (Scheme 118).

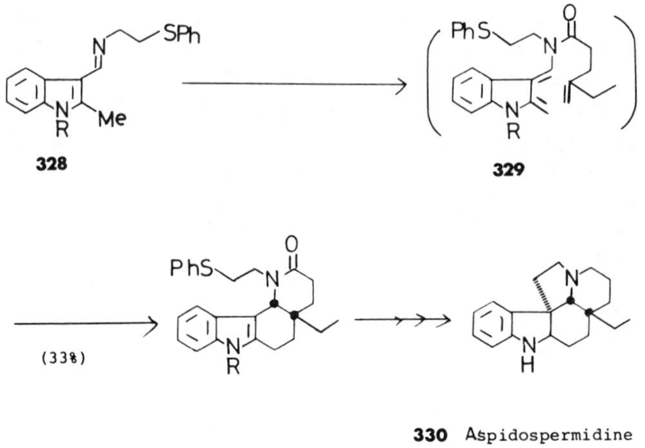

SCHEME 118

Attempted Synthesis of *Erythrina* Alkaloids. During the course of investigating the intramolecular (4 + 2) cycloaddition reaction of enamides with unactivated butadiene moieties, efforts were directed toward the construction of the spirocyclic compounds (**332**) that possess the molecular framework characteristic to the *Erythrina* alkaloids (*155,156*).

Acylation of 1-methyl-3,4-dihydroisoquinolines with 3,5-hexadienoyl chloride, in which the dienoic moiety is expeditiously masked as a sulfolene derivative, gave the enamide **331** in 67% yield. Interestingly, thermolysis of **331** in refluxing xylene did not afford the expected spirocyclic compounds (**332**) but rather the bridged cycloadducts (**333**) in 97% yields (*155,156*).

Thermolysis of **331** in refluxing xylene for only 3.5 min produced a mixture of the enamidodiene **334** and vinylcyclobutane **335**. When either **334** or the enamide **331** was heated in xylene at refluxing temperature for 15 min, the vinylcyclobutane (**335**) was obtained in a quantitative yield. The vinylcyclobutane then underwent facile thermal rearrangement to the 2-azabicyclo[3.3.1]nonenone (**333**).

This is the first example of a cycloaddition reaction of enamide with unactivated butadiene (*155,156*) (Scheme 119).

cis-Dihydrolycoricidine. The tricyclic skeleton of lycoricidine was prepared by a route involving an intramolecular (4 + 2) cycloaddition of a trimethylsilyloxyenamide **338** (*157*). The amide **336** was converted to the key aldehyde **337** which was exposed to triethylamine and trimethylchloro-

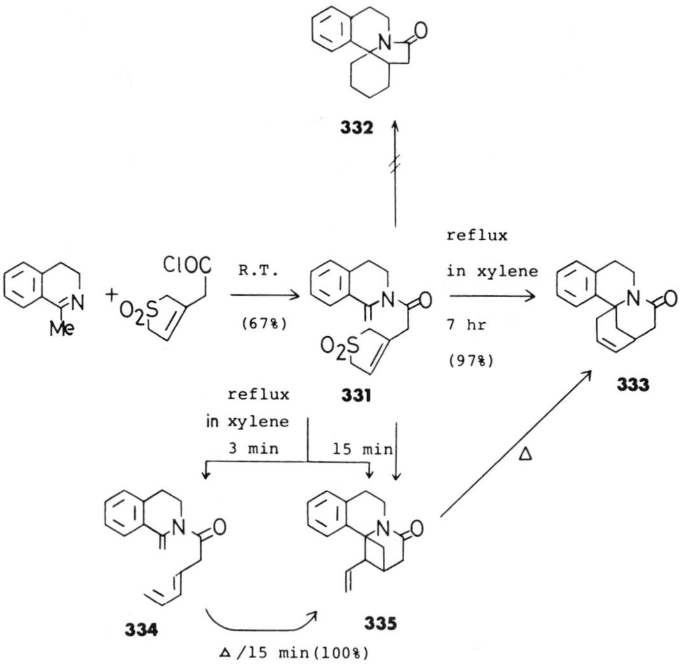

SCHEME 119

silane to give the cyclized lactam **339** in 60–65% yield. This simple *in situ* process for generating the requisite enamide (**338**) with concomitant (4 + 2) cycloaddition proved superior to the alternative procedures involving a low-temperature generation of **338**, followed by isolation and thermolysis to effect cyclization. The lactam **339** was acetylated and debenzylated to give the *cis*-dihydrolycoricidine triacetate **340**.

This approach demonstrated the utility of 4-trimethylsilyloxyenamide (**338**) as a diene component for the Diels–Alder reaction (*157*) (Scheme 120).

C. INTRAMOLECULAR PHOTOINDUCED REACTION BETWEEN ARYL HALIDES AND ENAMIDES

Snieckus *et al.* (*158,159*) demonstrated the usefullness of photocyclization of the *o*-halogenophenethylenamide **341** in alkaloid synthesis. The bromoenamide **341** was irradiated with 2537-nm light in benzene–triethylamine solution to give the seven-membered product **342** in 20–21.5% yield. This (**342**) was then converted to the Schöpf–Schweickert amine **343** on reduction (*158*); **343** has a structure closely related to the skeleton of the rhoeadine group of alkaloids (Scheme 121).

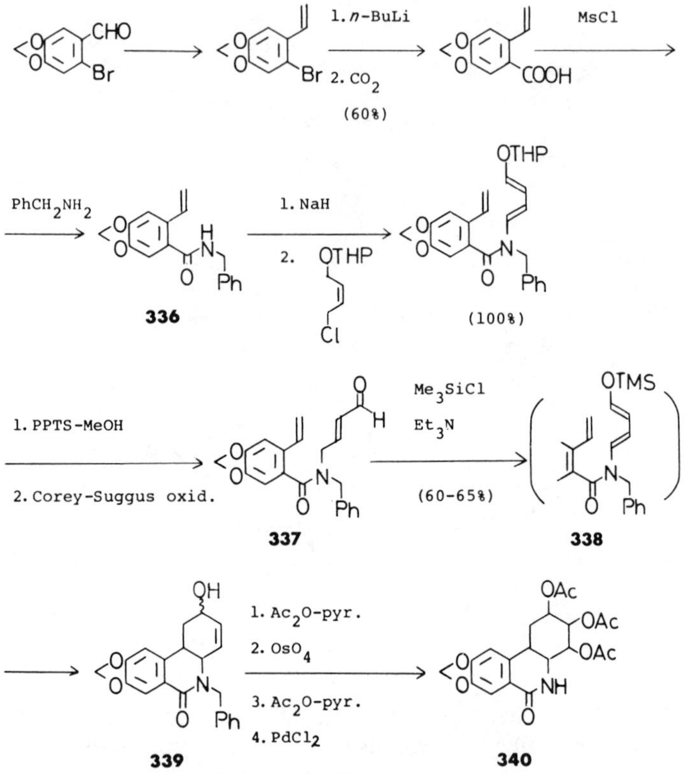

SCHEME 120

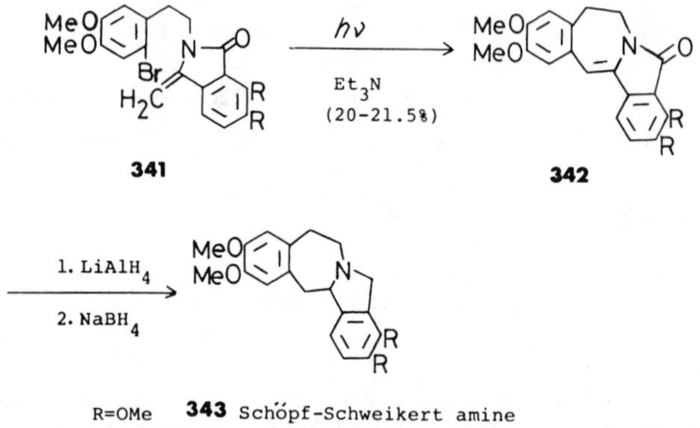

R=OMe **343** Schöpf-Schweikert amine

SCHEME 121

Cephalotaxine. Later, Tse and Snieckus (*159*) also carried out the same cyclization on the iodoenamide **344** and obtained the tricyclic compound **345** in 46% yield. This lactam **345** was then reduced first with platinum dioxide in hydrogen stream and then with LAH to give the corresponding amine **346**, which served as a key intermediate in the synthesis of cephalotaxine (**347**) (*160*) (Scheme 122).

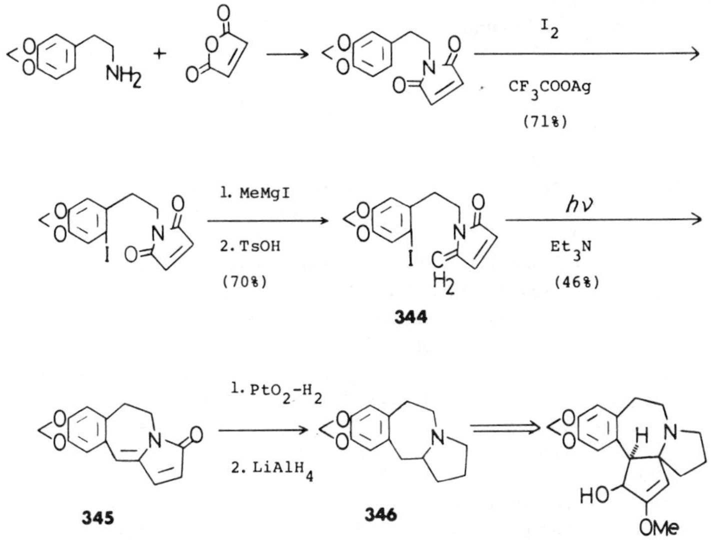

344

345   **346**

347 Cephalotaxine

SCHEME 122

Acknowledgments

The authors greatly appreciate the devotion and collaboration of their colleagues, particularly Drs. Toshiko Kiguchi and Okiko Miyata, Miss Yukiko Tada, and all other members of the Medicinal Chemistry Laboratory in Kobe.

REFERENCES

1. M. Shamma, "The Isoquinoline Alkaloids." Academic Press, New York, 1972.
2. M. Shamma and J. L. Moniot, "Isoquinoline Alkaloid Research, 1972–1977," Plenum, New York, 1978.
3. A. G. Cook, "Enamines: Synthesis, Structure and Reactions." Dekker, New York, 1969.
4. E. Bertele, H. Boos, J. D. Dunitz, F. Elsinger, A. Eschenmoser, I. Felner, H. P. Gribi, G. Gschwend, E. F. Meyer, M. Pesaro, and R. Scheffold, *Angew. Chem.* **76**, 393 (1964).
5. N. C. Yang and G. R. Lenz, *Tetrahedron Lett.* 4897 (1967).
6. P. T. Izzo and A. S. Kende, *Tetrahedron Lett.* 5731 (1966).
7. R. W. Hoffmann and K. R. Eicken, *Tetrahedron Lett.* 1759 (1968).
8. R. W. Hoffmann and K. R. Eicken, *Chem. Ber.* **102**, 2987 (1969).

9. I. Ninomiya, T. Naito, and T. Mori, *Tetrahedron. Lett.* 2259 (1969).
10. I. Ninomiya, T. Naito, and T. Mori, *J. Chem. Soc., Perkin Trans. 1* 505 (1973).
11. I. Ninomiya and T. Naito, *Heterocycles* **15,** 1433 (1981).
12. I. Ninomiya, *Heterocycles* **2,** 105 (1974).
13. G. R. Lenz, *Synthesis* 489 (1978).
14. T. Kametani and K. Fukumoto, *Acc. Chem. Res.* **5,** 212 (1972).
15. T. Naito, Y. Tada, Y. Nishiguchi, and I. Ninomiya, *Hetercycles* **16,** 1137 (1981).
16a. T. Naito, Y. Tada, and I. Ninomiya, *Heterocycles* **16,** 1141 (1981).
16b. T. Naito, K. Katsumi, Y. Tada, and I. Ninomiya, *Heterocycles* **20,** 779 (1983).
17. I. Ninomiya, T. Naito, and T. Kiguchi, *Tetrahedron Lett.* 4451 (1970).
18. I. Ninomiya, T. Naito, and T. Kiguchi, *J. Chem. Soc., Perkin Trans. 1* 2257 (1973).
19. I. Ninomiya, T. Naito, and T. Mori, *Tetrahedron Lett.* 3643 (1969).
20. I. Ninomiya, T. Naito, T. Kiguchi, and T. Mori, *J. Chem. Soc., Perkin Trans. 1* 696 (1973).
21. I. Ninomiya and T. Naito, *J. Chem. Soc., Chem. Commun.* 137 (1973).
22. I. Ninomiya, H. Takasugi, and T. Naito, *Heterocycles* **1,** 17 (1973).
23. I. Ninomiya, T. Naito, and H. Takasugi, *J.Chem. Soc., Perkin Trans. 1* 1720 (1975).
24. I. Ninomiya, T. Naito, and H. Takasugi, *J. Chem. Soc., Perkin Trans. 1* 1791 (1975).
25. I. Ninomiya, H. Takasugi, and T. Naito, *J. Chem. Soc., Chem. Commun.,* 732 (1973).
26. I. Ninomiya, T. Naito, and H. Takasugi, *J.Chem. Soc., Perkin Trans. 1* 1865 (1976).
27. I. Ninomiya, T. Kiguchi, and T. Naito, *J. Chem. Soc., Chem. Commun.,* 81 (1974).
28. I. Ninomiya, T. Kiguchi, O. Yamamoto, and T. Naito, *J. Chem. Soc., Perkin Trans. 1* 1723 (1979).
29. T. Kametani, T. Honda, T. Sugai, and K. Fukumoto, *Heterocycles* **4,** 927 (1976).
30. T. Kametani, T. Sugai, Y. Shoji, T. Honda, F. Satoh, and K. Fukumoto, *J. Chem. Soc., Perkin Trans. 1* 1151 (1977).
31. G. R. Lenz, *Tetrahedron Lett.* 1963 (1973).
32. G. R. Lenz, *J. Org. Chem.* **39,** 2839 (1974).
33. J. Cornelisse and E. Havinga, *Chem. Rev.* **75,** 353 (1975).
34. I. Ninomiya, S. Yamauchi, T. Kiguchi, A. Shinohara, and T. Naito, *J. Chem. Soc., Perkin Trans. 1* 1747 (1974).
35. I. Ninomiya, T. Kiguchi, S. Yamauchi, and T. Naito, *J. Chem. Soc., Perkin Trans. 1* 197 (1980).
36. O. L. Chapman and W. R. Adams, *J. Am. Chem. Soc.* **90,** 2333 (1968).
37. P. G. Cleveland and O. L. Chapman, *J. Chem. Soc., Chem. Commun.* 1064 (1967).
38. Y. Ogata, K. Takagi, and I. Ishino, *J. Org. Chem.* **36,** 3975 (1971).
39. Y. Kanaoka, *Kagaku no Ryoiki, Zokan* **123,** 41 (1979).
40. B. S. Thyagarajan, N. Kharasch, H. B. Lewis, and W. Wolf, *Chem. Commun.* 614 (1967).
41. D. H. Hey, G. H. Jones, and M. J. Perkins, *J. Chem. Soc. C* 116 (1971).
42. A. Mondon and K. Krohn, *Chem. Ber.* **105,** 3726 (1972).
43. K. A. Muszkat and E. Fischer, *J. Chem.Soc. B* 662 (1967).
44. G. R. Lenz, *J. Org. Chem.* **39,** 2846 (1974).
45. T. Naito, Y. Tada, K. Katsumi, and I. Ninomiya, *Heterocycles* **20,** 775 (1983).
46. J. Grimshaw and A. P. de Silva, *J. Chem. Soc., Chem. Commun.* 301 (1980).
47. J. Grimshaw and A. P. de Silva, *J. Chem. Soc., Chem. Commun.* 302 (1980.
48. G. R. Lenz, *J. Heterocycl. Chem.* **16,** 433 (1979).
49. J. C. Gramain, Y. Troin, and D. Vallée, *Bull. Chem. Soc. Fr.* Part 2, Suppl., 22 (1980).
50. J. C. Gramain, Y. Troin, and D. Vallée, *J. Chem. Soc., Chem. Commun.* 832 (1981).
51. A.R. Martin, V. M. Paradkar, G. W. Peng, R. C. Speth, H. I. Yamamura, and A. S. Horn, *J. Med. Chem.* **23,** 865 (1980).

52. S. Prabhakar, A. M. Lobo, and M. R. Tavares, *J. Chem. Soc., Chem. Commun.*, 884 (1978).
53. I. Ninomiya and T. Naito, unpublished results.
54. T. Naito, Y. Tada, Y. Nishiguchi, and I. Ninomiya, *Heterocycles* **18,** 213 (1982).
55. T. Kiguchi, C. Hashimoto, T. Naito, and I. Ninomiya, *Heterocycles* **18,** 217 (1982).
56. T. Kiguchi, C. Hashimoto, T. Naito, and I. Ninomiya, *Heterocycles* **19,** 1873 (1982).
57. T. Kiguchi, C. Hashimoto, T. Naito, and I. Ninomiya, *Heterocycles* **19,** 2279 (1982).
58. J. D. Morrison and H. S. Mosher, "Asymmetric Organic Reactions," p. 160. Prentice-Hall, Englewood Cliffs, New Jersey, 1971.
59. R. Noyori, I. Tomino, and Y. Tanimoto, *J. Am. Chem. Soc.* **101,** 3129 (1979).
60. R. Noyori, I. Tomino, and M. Nishizawa, *J. Am. Chem. Soc.* **101,** 5843 (1979).
61. A. I. Meyers and M. E. Ford, *Tetrahedron Lett.* 1337 (1974).
62. M. Asami, H. Ohno, S. Kobayashi, and T. Mukaiyama, *Bull. Chem. Soc. Jpn.* **51,** 1869 (1978).
63. M. Asami and T. Mukaiyama, *Heterocycles* **12,** 499 (1979).
64. S. Terashima, N. Tanno, and K. Koga, *Chem. Lett.* 981 (1980).
65. S. Terashima, N. Tanno, and K. Koga, *J. Chem. Soc., Chem. Commun.* 1026 (1980).
66. T. Kametani, N. Takagi, M. Toyota, T. Honda, and K. Fukumoto, *Heterocycles* **16,** 591 (1981).
67. T. Naito and I. Ninomiya, *Heterocycles* **14,** 959 (1980).
68. T. Naito, O. Miyata, and I. Ninomiya, *J. Chem. Soc., Chem. Commun.* 517 (1979).
69. M. Sainsbury and N. L. Uttley, *J. Chem. Soc., Chem. Commun.* 2109 (1977).
70. T. Naito and I. Ninomiya, *Heterocycles* **15,** 735 (1981).
71. F. Hotellier, P. Delaveau, and J.-L. Pousset, *Phytochemistry* **14,** 1407 (1975).
72. M. Sainsbury and N. L. Uttley, *J. Chem. Soc., Chem. Commun.* 319 (1977).
73. T. Naito, O. Miyata, I. Ninomiya, and S. C. Pakrashi, *Heterocycles* **16,** 725 (1981).
74. R. B. Mujumdar and A. R. Martin, *Tetrachedron Lett.* **23,** 1455 (1982).
75. Atta-ur-Rahman and M. Ghazala, *Synthesis* 372 (1980).
76. W. Carruthers and N. Evans, *J. Chem. Soc., Perkin Trans. 1* 1523 (1974).
77. H. Hara, O. Hoshino, and B. Umezawa, *Tetrahedron Lett.* 5031 (1972).
78. H. Iida, S. Aoyagi, and C. Kibayashi, *J. Chem. Soc., Chem. Commun.* 499 (1974).
79. I. Ninomiya, T. Naito, and T. Kiguchi, *Chem. Commun.* 1669 (1970).
80. I. Ninomiya, T. Naito, and T. Kiguchi, *J. Chem. Soc., Perkin Trans 1* 2261 (1973).
81. J. B. Hendrickson, T. L. Bogard, and M. E. Fisch, *J.Am. Chem. Soc.* **92,** 5538 (1970).
82. W. M. Messer, M. Tin-Wa, H. H. S. Fong, C. Bevelle, N. R. Farnworth, D. J. Abraham, and J. Trojanek, *J. Pharm. Sci.* **61,** 1858 (1972).
83. M. E. Wall, M. C. Wani, and H. L. Taylor, *162nd Natl. Meet., Am. Chem. Soc.* Med. I-34 (1971).
84. D. H. Hey, G. H. Jones, and M. J. Perkins, *J. Chem. Soc., Perkin Trans. 1* 105 (1972).
85. I. Ninomiya, O. Yamamoto, T. Kiguchi, and T. Naito, *J. Chem. Soc., Perkin Trans. 1* 203 (1980).
86. I. Ninomiya, O. Yamamoto, and T. Naito, *Heterocycles* **4,** 743 (1976).
87. I. Ninomiya, T. Kiguchi, O. Yamamoto, and T. Naito, *Heterocycles* **4,** 467 (1976).
88. S. V. Kessar, G. Singh, and P. Balakrishnan, *Tetrahedron Lett.* 2269 (1974).
89a. I. Ninomiya, O. Yamamoto, and T. Naito, *Heterocycles* **7,** 137 (1977).
89b. I. Ninomiya, O. Yamamoto, and T. Naito, *J. Chem. Soc. Perkin Trans. 1,* in press (1983).
90. H. Ishii and T. Ishikawa, *Yakugaku Zasshi* **101,** 663 (1981).
91. H. Ishii, K. Harada, T. Ishida, E. Ueda, K. Nakajima, I. Ninomiya, T. Naito, and T. Kiguchi, *Tetrahedron Lett* 319 (1975).

92. H. Ishii, E. Ueda, K. Nakajima, T. Ishida, T. Ishikawa, K. Harada, I. Ninomiya, T. Naito, and T. Kiguchi, *Chem. Pharm. Bull.* **26,** 864 (1978).
93. I. Ninomiya, T. Naito, H. Ishii, T. Ishida, M. Ueda, and K. Harada, *J. Chem. Soc., Perkin Trans. 1* 762 (1975).
94. W. J. Begley and J. Grimshaw, *J. Chem. Soc., Perkin Trans. 1* 2324 (1977).
95. I. Ninomiya, T. Naito, and H. Ishii, *Heterocycles* **3,** 307 (1975).
96a. I. Ninomiya, O. Yamamoto, and T. Naito, *Heterocycles* **7,** 131 (1977).
96b. I. Ninomiya, O. Yamamoto, and T. Naito, *J. Chem. Soc., Perkin Trans. 1,* in press (1983).
97. G. Nonaka, H. Okabe, I. Nishioka, and N. Takao, *J. Pharm. Soc. Jpn.* **93,** 87 (1973).
98. G. Nonaka and I. Nishioka, *Chem. Pharm. Bull.* **23,** 521 (1975).
99. N. Takao, *Chem. Pharm. Bull.* **19,** 247 (1971).
100. I. Ninomiya, O. Yamamoto, and T. Naito, *J. Chem. Soc., Chem. Commun.,* 437 (1976).
101. I. Ninomiya, O. Yamamoto, and T. Naito, *J. Chem. Soc., Perkin Trans. 1* 212 (1980).
102. I. Ninomiya, O. Yamamoto, and T. Naito, *Heterocycles* **5,** 67 (1976).
103. E. V. Blackburn and C. J. Timmons, *Q. Rev., Chem. Soc.* **23,** 482 (1969).
104. N. C. Yang, G. R. Lenz, and A. Shani, *Tetrahedron Lett.* 2941 (1966).
105. N. C. Yang, A. Shani, and G. R. Lenz, *J. Am. Chem. Soc.* **88,** 5369 (1966).
106. G. R. Lenz and N. C. Yang, *Chem. Commun.* 1136 (1967).
107. G. R. Lenz, *J. Org. Chem.* **42,** 1117 (1977).
108. M. P. Cava and S. C. Havlicek, *Tetrahedron Lett.* 2625 (1967).
109. G. R. Lenz, *J. Org. Chem.* **41,** 2201 (1976).
110. K.-Y. Zee-Cheng, K. D. Paull, and C. C. Cheng, *J. Med. Chem.* **17,** 347 (1974).
111. T. Kametani, A. Ujiie, M. Ihara, K. Fukumoto, and S.-T. Lu, *J. Chem. Soc., Perkin Trans 1* 1218 (1976).
112. S. Sasagawa, K. Kametani, and S. Kiyofugi, *Chem. Pharm. Bull.* **17,** 1 (1969).
113. M. Konda, T. Shioiri, and S. Yamada, *Chem. Pharm. Bull.* **23,** 1063 (1975).
114. H. Taguchi and I. Imazeki, *Yakugaku Zasshi* **84,** 955 (1964).
115. S. C. Pakrashi, B. Achari, E. Ali, P. P. Ghosh Dastidar, and R. R. Sinha, *Tetrahedron Lett* **21,** 2667 (1980).
116. I. Ninomiya, T. Kiguchi, and Y. Tada, *Heterocycles* **6,** 1799 (1977).
117. I. Ninomiya, Y. Tada, T. Kiguchi, O. Yamamoto, and T. Naito, *Heterocycles* **9,** 1527 (1978).
118. I. Ninomiya, Y. Tada, O. Miyata, and T. Naito, *Heterocycles* **14,** 631 (1980).
119. T. Kametani, Y. Hirai, M. Kajiwara, T. Takahashi, and K. Fukumoto, *Chem. Pharm. Bull.* **23,** 2634 (1975).
120. T. Y. Au, H. T. Cheung, and S. Sternhell, *J. Chem. Soc., Perkin Trans. 1* 13 (1973).
121. J. D. Phillipson, S. R. Hemingway, N. G. Bisset, P. J. Houghton, and E. J. Shellard, *Phytochemistry* **13,** 973 (1974).
122. M. Sainsbury and B. Webb, *Phytochemistry* **14,** 2691 (1975).
123. M. Sainsbury and N. L. Uttley, *J. Chem. Soc., Perkin Trans. 1* 2416 (1976).
124. I. Ninomiya and T. Naito, *Heterocycles* **2,** 607 (1974).
125. A. Shafiee and A. Rashidbaigi, *J. Heterocycl. Chem.* **14,** 1317 (1977).
126. B. Danieli, G. Lesma, and G. Palmisano, *J. Chem. Soc., Chem. Commun.* 109 (1980).
127. B. Danieli, G. Lesma, and G. Palmisano, *Gazz. Chim. Ital.* **111,** 257 (1980).
128. I. Ninomiya and T. Kiguchi, *J. Chem. Soc., Chem. Commun.* 624, (1976).
129. I. Ninomiya, T. Kiguchi, and T. Naito, *J. Chem. Soc., Perkin Trans. 1* 208 (1980).
130. I. Ninomiya, T. Kiguchi, and T. Naito, *Heterocycles* **4,** 973 (1976).
131. Z. Horii, T. Watanabe, T. Kurihara, and Y. Tamura, *Chem. Pharm. Bull.* **13,** 420 (1965).
132. R. Ramage, V. M. Armstrong, and S. Coulton, *Tetrahedron* **37,** Suppl. 1, 157 (1981).

133. I. Ninomiya, I. Furutani, O. Yamamoto, T. Kiguchi, and T. Naito, *Heterocycles* **9**, 853 (1978).
134. I. Ninomiya and T. Naito, *Heterocycles* **10**, 237 (1978).
135. M. P. Cava, M. J. Mitchell, S. C. Havlicek, A. Lindert, and R. J. Spangler, *J. Org. Chem.* **35**, 175 (1970).
136. M. P. Cava, S. C. Havlicek, A. Lindert, and R. J. Spangler, *Tetrahedron Lett.* 2937 (1966).
137. L. Castedo, C. Soá, J. M. Soá, and R. Suau, *J. Org. Chem.* **47**, 513 (1982).
138. M. P. Cava, P. Stern, and K. Wakisaka, *Tetrahedron* **29**, 2245 (1973).
139. L. Castedo, R. Estevez, J. M. Soá, and R. Suau, *Tetrahedron Lett.* 2179 (1978).
140. G. H. Moltrasio, R. M. Sotelo, and D. Giacopello, *J. Chem. Soc., Perkin Trans. 1* 349 (1973).
141. S. M. Kupchan and P. F. O'Brien, *J. Chem. Soc., Chem. Commun.* 915 (1973).
142. W. Oppolzer, *Angew. Chem., Int. Ed. Engl.* **16**, 10 (1977).
143. W. Oppolzer, *Angew. Chem.* **89**, 10 (1977).
144. G. Brieger and J. N. Bennett, *Chem. Rev.* **80**, 63 (1980).
145. W. Oppolzer and K. Keller, *J. Am. Chem. Soc.* **93**, 3836 (1971).
146. W. Oppolzer, *Pure Appl. Chem.* **53**, 1181 (1981).
147. W. Oppolzer and W. Fröstl, *Helv. Chim. Acta* **58**, 590 (1975).
148. W. Oppolzer, W. Fröstl, and H. P. Weber, *Helv. Chim. Acta* **58**, 593 (1975).
149. W. Oppolzer and E. Flaskamp, *Helv. Chim. Acta* **60**, 204 (1977).
150. D. J. Morgans, Jr. and G. Stork, *Tetrahedron Lett.* 1959 (1979).
151. G. Stork and D. J. Morgans, Jr., *J. Am. Chem. Soc.* **101**, 7110 (1979).
152. S. F. Martin, S. R. Desai, G. W. Phillips, and A. C. Miller, *J. Am. Chem. Soc.* **102**, 3294 (1980).
153. T. Gallagher and P. Magnus, *Tetrahedron* **37**, 3889 (1981).
154. T. Gallagher, P. Magnus, and J. C. Huffman, *J. Am. Chem. Soc.* **104**, 1140 (1982).
155. S. F. Martin, T. Chou, and C. Tu, *Tetrahedron Lett.* 3823 (1979).
156. S. F. Martin, C. Tu, and T. Chou, *J. Am. Chem. Soc.* **102**, 5274 (1980).
157. G. E. Keck, E. Boden, and U. Sonnewald, *Tetrahedron Lett.* **22**, 2615 (1981).
158. H. O. Bernhard and V. Snieckus, *Tetrahedron Lett.* 4867 (1971).
159. I. Tse and V. Snieckus, *J. Chem. Soc., Chem. Commun.* 505 (1976).
160. B. Weinstein and A. R. Craig, *J. Org. Chem.* **41**, 875 (1976).

—— CHAPTER 5 ——

THE IMIDAZOLE ALKALOIDS

L. MAAT AND H. C. BEYERMAN

Laboratory of Organic Chemistry
Delft University of Technology, Delft, The Netherlands

THE ALKALOIDS, VOL. XXII
ISBN 0-12-469522-1

I. Introduction

Thirty years have passed since Battersby and Openshaw (*1*) discussed the imidazole alkaloids in this series. The group of alkaloids was small then and dealt mainly with pilocarpine and related alkaloids. Meanwhile, the relative and absolute stereochemistries of pilocarpine have been elucidated and some different syntheses of this medicinally important compound have been reported. However, the mechanism of many biochemical transformations still remains obscure.

One could remark that histidine and its derivative histamine belong to the most widely distributed compounds containing the imidazole nucleus, but in spite of this broad occurrence, only a few alkaloids are known whose biogenesis can be attributed to histidine.

Since 1953, quite a number of alkaloids containing the imidazole nucleus have been isolated that can be added to the originally small group of imidazole alkaloids. Some of the new members have been mentioned, if only briefly, in previous volumes of this treatise in the chapter "Alkaloids Unclassified and of Unknown Structure." Here we divide these new compounds in Alkaloids Related to Pilocarpine (Section IV) and Other Imidazole Alkaloids (Section V). The last category comprises now over 80 alkaloids isolated from different sources. These imidazole alkaloids proved to be of strongly divergant structures, some of which have not yet been completely clarified. Therefore, we classify them in alphabetical order according to their name or the natural source. As in other alkaloid fields, the term (plant) alkaloids is not restricted to plant material only, so we discuss also imidazole alkaloids isolated from mussels, sea urchins, and so forth. We have mentioned a few references on the pharmacology of imidazole alkaloids, mainly as a guide to general aspects.

Short reports on the progress in the chemistry of imidazole alkaloids have appeared in annual reviews (*2–11*), and many recent monographs on alkaloids contain a chapter on imidazole alkaloids (*12–17*).

II. Bases Related to Histidine

Relatively few simple derivatives of imidazole occur naturally; mainly products of fungi have been found, such as ergothioneine (**1**) and hercynine, the betaine of histidine (**2**). This is rather astonishing because the amino acid histidine is a widespread constituent of many proteins. The base histamine (**3**) is derived from histidine and plays an important role in allergic reactions in man and in animal tissues.

1 Ergothioneine   **2** Histidine   **3** Histamine

Histamine and two other unidentified alkaloids have been found in the alcohol extract of the foliage of *Capsella bursapastoris* (shepherd's purse) (*18*). The perennial ryegrass *Lolium perenne* L. has yielded histamine among a fascinating variety of other types of alkaloids (*19*). Also the N^π,N^τ-dimethylimidazolium derivative of histidine has been found, besides histidine, in the seaweed *Gracilaria secundata* Harvey (*20*). Imidazole alkylamines were isolated from the skin of some amphibians from Australia and Papua New Guinea (*21*) and from South America (*22*). N^α,N^α-Dimethylhistamine has been found in the plant *Casimiroa edulis* (*23*); the compound has also been isolated from the sponge species *Geodia gigas* (*24*) and *Ianthella* (*25*). N^α,N^α-Dimethylhistamine was quite recently isolated from *Echinocereus blanckii*. The amine was shown to be responsible for the cardioactivity of extracts of the plant (*26a*). One synthesis (*26b*) makes use of a method by which 4-(2-chloroethyl)imidazole reacts with dimethylamine in propanol. From the fresh fruiting bodies of the mushroom *Clitocybe acromelalga* the amino acid betaine clithioneine was isolated. By spectral analysis and chemical degradations it has been identified as 2(*S*)-*S*-[3(*S*)-amino-3-carboxy-2(*S*)-hydroxypropyl]ergothioneine (*27*).

In 1982 Palumbo *et al.* (*28*) reported the isolation of a relatively large amount of 5-mercapto-1-methyl-L-histidine (**4**) from unfertilized sea urchin eggs (*Paracentrotus lividus*), in addition to glutathione. In preliminary experiments, both **4** and the corresponding disulfide **5** were found in aqueous extracts. Therefore, an isolation procedure was developed that involved, in an early stage, the conversion of the thiol to the more stable

4 5-Mercapto-1-methyl-L-histidine   **5**

6 N^α-*trans*-Cinnamoylhistamine

disulfide by aerial oxidation of the lipid-free aqueous extracts in the presence of a few crystals of iodine. The structure elucidation was based on the usual spectrographic evidence. The position of the NMe group and the stereochemistry of the chiral center followed from desulfuration of 5 with Raney nickel, affording N^τ-methyl-L-histidine, identified by comparison to an authentic sample (29)

N^α-trans-Cinnamoylhistamine (6) has been isolated from different plants, for example, as the major alkaloid from the leaves and stems of *Argyrodendrum peralatum* (30) and from both fresh roots and leafy stems of *Acacia spirorbis* Labill. (31).

III. Pilocarpine: A Jaborandi Alkaloid

A. OCCURRENCE

The Jaborandi alkaloids comprise a group of closely related alkaloids that occur in South American plants belonging to the family of the Rutaceae. The first samples of the shrubs were investigated *inter alia* for their pharmacological properties in the third quarter of the nineteenth century (32). These samples came from *Pilocarpus jaborandi* Holmes. Later, several species and varieties became known; these were reviewed in 1953 (1). Pilocarpine was and still is the only naturally occurring imidazole alkaloid that is used in clinical medicine. In spite of many references to *Pilocarpus jaborandi* Holmes in ethnological and botanical sources and suggestions of its employment for a variety of diseases, it has not been possible to pin down the use of its leaves to any particular purpose among South American Indians. The medicinally important "Jaborandis" are species of *Pilocarpus*, but it should be remarked that this vernacular name is commonly applied to other rutaceous and numerous piperaceous plants as well. The reader is referred to an extensive interdisciplinary appraisal by Holmstedt *et al.* (33).

B. ISOLATION

The pilocarpine used in medicine is of natural origin, which means it is being produced by extraction only; as far as known, the leaves of *Pilocarpus microphyllus* are used at present. The syntheses reported (Section III,F) apparently cannot compete with the cheaper natural product. This is probably explained by the intricate stereochemical problems involved (Sections III,D and F). In 1928 Chemnitius described a typical large-scale extraction procedure (34). It is believed that this publication is the basis for later industrial manufacturing processes. Some older modifications or otherwise improvements were recorded in 1953 (*1*). The isolation of pilocarpine by silica gel chromatographic (*35*) and ion-exchange countercurrent (*36*) methods has since been reported.

In order to avoid the undesirable isomerization to isopilocarpine (Section III,D,3) and, probably, to avoid enzymatic degradation, the Jaborandi leaves are processed as rapidly as possible. Acidification converts the alkaloids to their water-soluble salts, which allows defatting of the leaves with, for example, petroleum ether. Alkalinization and extraction with a suitable lipid-extracting solvent or solvent mixture gives the crude alkaloid mixture. (+)-Pilocarpine (mp 34°C) crystallizes only with difficulty; it can be obtained readily as the nitrate salt (mp 178°C) in the absence of isopilocarpine. The separation of isopilocarpine nitrate from pilocarpine nitrate by crystallization is often difficult because mixtures may give products of constant melting points; the same holds for the hydrochlorides.

C. Structure Elucidation

The structure elucidation, mainly by chemical degradation methods, has been extensively dealt with by Battersby and Openshaw (*1*). There remained the problems of stereochemistry that were solved mostly in the years following the review just mentioned. They are described in Section III,D.

D. Stereochemistry

1. Relative Configuration

It was found long ago that pilocarpine (**7**) isomerizes to isopilocarpine (**8**) under a wide variety of conditions. In 1880 it already was observed that pilocarpine undergoes a change when heated alone or with hydrochloric acid (*37*). Other investigators noted a change when the alkaloid was heated with aqueous sodium hydroxide. This isomerization is discussed in Section III,D,3.

In 1900 Jowett (*38*) suggested that the differences between pilocarpine and isopilocarpine had their origin in the steric structures. This was in contrast to the opinion of Pinner *et al.* who assumed the differences to be caused by N-methylation at different sites in the imidazole ring (*39*). Langenbeck in 1924 assumed on the basis of chemical reactions, such as the formation of quaternary salts and ozonolysis, that the differences were of a steric nature (*40,41*). Preobrazhenski *et al.* in 1936 (*42*) supposed on the basis of different stability and optical rotation that pilocarpine and isopilocarpine should possess the cis and trans configuration, respectively. An identical conclusion was drawn by Zavyalov (*43*) in 1952, on the basis of

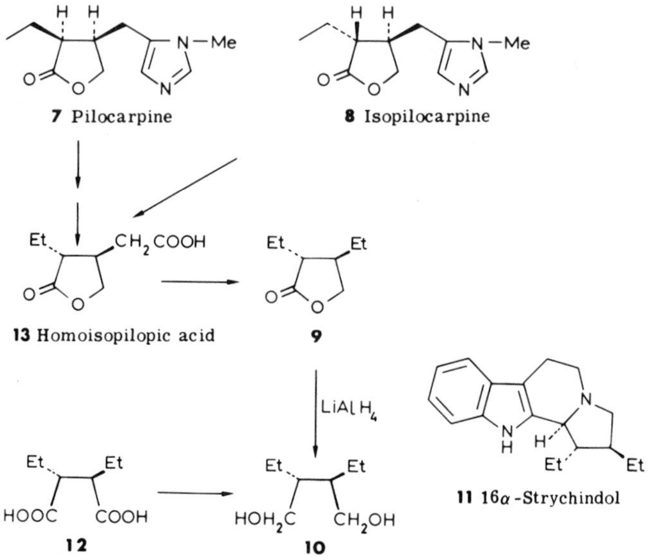

7 Pilocarpine   **8** Isopilocarpine

13 Homoisopilopic acid   **9**

LiAlH₄

11 16α-Strychindol

HOOC   COOH   HOH₂C   CH₂OH
12   **10**

chemical reactions of the butyrolactone ring that contains the two chiral centers.

2. Absolute Configuration

Nagarajan et al. (44) in 1963 determined the absolute configuration of (+)-trans-2,3-diethyl-4-butanolide (**9**) and of (+)-2,3-diethyl-1,4-butanediol (**10**). Starting with (−)-**9**, they prepared (−)-16α-strychindol (**11**), a degradation product of strychnine. The absolute configuration of strychnine had been determined in 1956 by Bijvoet and Peerdeman by X-ray diffraction studies (45,46), and, therefore, the absolute configuration of **9** may be adjudged. Moreover, it appeared that the (+)-2,3-diethylsuccinic acid (**12**), the starting material for the preparation of the (+)-diol **10** by Nagarajan et al., possessed the absolute configuration 2R,3R just as **9** and **11**. Thereafter, in 1966, Hill and Barcza (47) prepared the butyrolactone **9** by an electrolytic procedure according to Kolbe from homoisopilopic acid (**13**), a degradation product of isopilocarpine (**8**). The compound **13** could also be prepared, with epimerization, from pilocarpine (**7**) by ozonolysis followed by acid hydrolysis of the amide formed. It leads to the structure of (2S,3R)-2-ethyl-3-[(1-methylimidazol-5-yl)methyl]-4-butanolide for natural (+)-pilocarpine.*

*CA nomenclature: (3S-cis)-3-ethyldihydro-4-[(1-methyl-1H-imidazol-5-yl)methyl]-2(3H)-furanone.

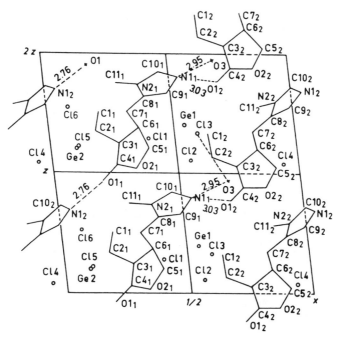

FIG. 1. The *xz* projection of half a unit cell indicating packing and hydrogen bond scheme of pilocarpine trichlorogermanate(II), according to Fregerslev and Rasmussen (*48*).

The preceding conclusions were confirmed by Fregerslev and Rasmussen (*48*) in 1968 by X-ray diffraction measurements of pilocarpine trichlorogermanate(II) hemihydrate. Figure 1 shows the clinographic projection with the cis position of the two substituents of the lactone ring. It also transpired that in the crystalline state, hydrogen bonds are present between the nitrogen atom N-1 and the oxygen atom O-1 of a second molecule of pilocarpine (Fig. 2). Haase and Kussäther (*49*) reported further crystallographic data on pilocarpine.

Chumachenko *et al.* (*50,51*) confirmed the structure of pilocarpine by their synthesis of homopilopic acid, a degradation product of pilocarpine (Section III,F). Inch and Lewis (*52*) also gave structural evidence by their synthesis of (+)-(R)-1-acetoxy-2,3-(bisacetoxymethyl)pentane (**14**), which

14

15 Isopilopic acid

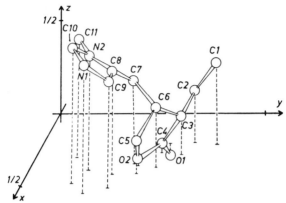

FIG. 2. Clinographic projection of a pilocarpine molecule (*48*).

could also be obtained by reduction of (+)-isopilopic acid (**15**), an oxidation product of isopilocarpine.

Quantum chemical calculations on the conformation of pilocarpine were performed by Kang (*53*). It was found that the calculated minimum-energy conformation of pilocarpine agreed well with the conformation found by the X-ray diffraction studies.

3. Pilocarpine – Isopilocarpine Isomerization

The phenomenon of isomerization of pilocarpine to isopilocarpine was already discussed in 1880. The change takes place under a wide variety of conditions, which were reviewed earlier (*1*). Döpke and d'Heureuse (*54*) in 1968 proved that the isomerization (*inter alia* an undesirable side reaction occurring during the isolation of pilocarpine from plant material) is not caused by an opening of the lactone ring, but proceeds via epimerization at

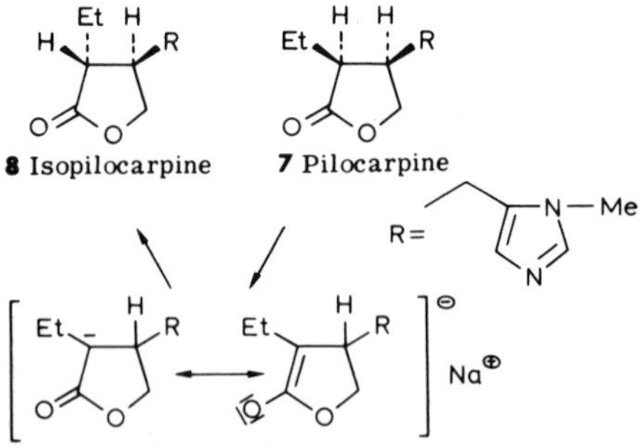

SCHEME 1. Pilocarpine (**7**)–isopilocarpine (**8**) isomerization according to Döpke and d'Heureuse (*54*).

C-2. This isomerization proceeds via an enol salt, stabilized by mesomerism, which was isolated by the investigators (Scheme 1).

Nunes and Brochmann-Hanssen *et al.* (*55,56*) proposed a reaction scheme for the hydroxide ion-catalyzed hydrolysis of pilocarpine (Scheme 2) based on the theory of Döpke and d'Heureuse. Two main pathways may be distinguished: (1) hydrolysis of pilocarpine (**7**) to the pilocarpate anion (**7**$^\ominus$) and (2) epimerization of pilocarpine (**7**) to isopilocarpine (**8**), following to the isopilocarpate anion (**8**$^\ominus$). These two pathways appeared to be competing pseudo-first-order reactions.

The degree of isomerization of pilocarpine has since been determined by several investigators. Ryan (*57*) used infrared spectroscopy. Weber (*58*) used high-performance liquid chromatography. Neville *et al.* (*59*) studied the epimerization of pilocarpine in deuterium oxide with the aid of nuclear magnetic resonance spectroscopy. They concluded that in the hydroxide ion-catalyzed hydrolysis of pilocarpine (**7**), the pilocarpate (**7**$^\ominus$) may be formed both directly from pilocarpine and indirectly via the enolate. From the ratio of the products formed, it appears that the most important reaction is the enolization and not the opening of the lactone ring. Neville *et al.* give an explanation for the nonepimerization of isopilocarpine in alkaline medium. Examination of Dreiding stereomodels of pilocarpine shows a severe steric hindrance between the ethyl and the imidazole substituents. This steric hindrance can be relieved through dissociation of the α proton of the lactone to give the enolate with a nearly planar sp^2 α carbon. In isopilocarpine there is less, if any, steric interaction between the ethyl and imidazole substituents. Thus, the α proton is probably less acidic and not prone to enolization under aqueous conditions. This interpretation is supported by the observed absence of deuterium exchange at the α carbon with isopilocarpine. Neville *et al.* found in contrast to the results of Nunes, that with decreasing temperatures, the process of alkaline hydrolysis decreases even more than that of the competing enolization, which precedes the epimerization.

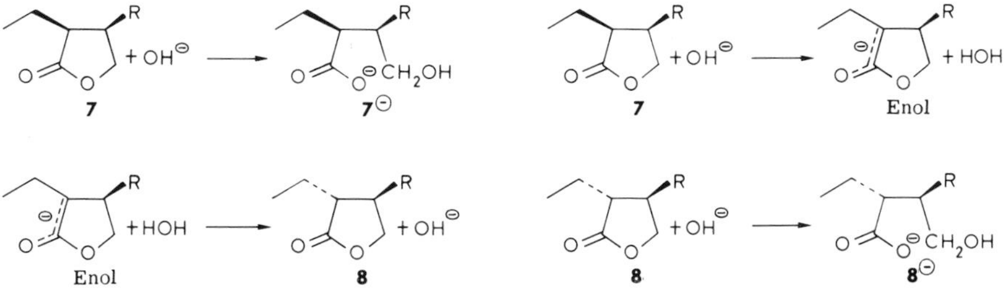

SCHEME 2. Pilocarpine (**7**)–isopilocarpine (**8**) isomerization according to Nunes and Brochmann-Hanssen (*55,56*).

E. Analytical Methods

The decomposition of pilocarpine in aqueous solution — to be avoided as far as possible in pharmaceutical formulations — has been studied by Baeschlin *et al.* (*60*). In aqueous medium, pilocarpine (**7**) can hydrolyze to pilocarpic acid (**16**) and epimerize to isopilocarpine (**8**), which, in turn, can hydrolyze to isopilocarpic acid (**17**). The kinetics of hydrolysis of pilocarpine have been determined by Chung *et al.* (*61*). A paper describing the thin-layer chromatographic analysis of pilocarpine appeared in 1969 (*62*). A thin-layer electrophoretic method for separation of pilocarpine from drug sources has been developed (*63*). Other analytical methods used for the determination of pilocarpine include TLC (*64*), potentiometry with ion-selective membrane electrodes (*65*), photodensitometry (*66*), and gas–liquid chromatography (*67–69*). The pilocarpine content of drug preparations was estimated by colorimetry (*70*). A review has been published on analyses of commercial eye drops containing pilocarpine (*71*). Quantitative analysis of degradation products in ophthalmic formulations containing pilocarpine has been described (*72*). Ultraviolet spectrophotometry–polarimetry was used for the anlaysis of pilocarpine and isopilocarpine in ophthalmic solutions (*73*). A spectrophotometric assay for pilocarpine has appeared (*74*). A recommendation of official methods for the analysis of pilocarpine was published in 1972 (*75*).

The kinetics of the hydroxide ion-catalyzed epimerization of pilocarpine to isopilocarpine and of its hydrolysis to pilocarpic acid have been studied (*56*). Both forms of degradation lead to loss of pharmacological activity. The importance of possible inactivation by epimerization during thermal sterilization of ophthalmic preparations of pilocarpine was pointed out. It was also considered that some epimerization would always occur during the extraction of pilocarpine from Jaborandi leaves, and that isopilocarpine might, therefore, be an artifact and not a genuine plant alkaloid (*55*).

A suitable method for the determination of pilocarpine (**7**) and its degradation products (**8, 16, 17**) requires a technique with a high specificity and sensitivity that can operate under mild conditions. Many methods lack at least one of these requirements. High-performance liquid chromatography (HPLC) on a cation-exchange resin still suffers from some drawbacks (*58,76,77*). The use of these resins implies operating conditions (pH 9) where pilocarpine is known to undergo alkali-catalyzed hydrolysis, and the

16 Pilocarpic acid   **17** Isopilocarpic acid

method has the disadvantage of poor resolution and a variable retention time. Noordam *et al.* reported on a rapid and convenient reversed-phase HPLC procedure (*78*) that could be used for the quantitative determination of pilocarpine and its degradation products in ophthalmic formulations (*79*). The method was also applied in a modified form (*80,81*).

The ^{13}C-NMR spectra of pilocarpine, isopilocarpine, and their methiodides have been discussed (*82*). The spectral assignments reflect the differences between the cis and trans isomers.

A polarographic study on complexes formed between pilocarpine and various imidazoles and copper was carried out. The half-wave potentials of reduction waves of pilocarpine in a copper citrate medium were recorded and found to be a function of the imidazole concentration at low copper concentrations (*83*).

The complexation reactions of pilocarpine with cobalt(II), nickel(II), copper(II), and zinc(II) ions were investigated (*84*). The complexes were characterized by spectrophotometry, conductivity, ESR, NMR, and magnetic susceptibility measurements. In each case, the metal ion was pseudotetrahedrally coordinated. Pilocarpine interacts with the metal ion through the imidazole ring.

F. Synthesis

The syntheses of pilocarpine via homopilopic acid by Preobrazhenski *et al.* and Dey have been discussed previously (*1*). In 1968, Chumachenko *et al.* (*50,51*) reported the synthesis of homopilopic acid, starting from furfural. A few years later, DeGraw (*85*) also reported on a synthesis of homopilopic acid; his approach to the problem was similar to Chumachenko's work.

Furfural (**18**) was oxidized to give the butenolide **19**, which on Michael condensation with diethyl ethylmalonate afforded lactone **20**. Hydrolysis of **20** yielded the dilactone **21**. This was treated with ethanol in the presence of sulfuric acid. The isomer mixture obtained was separated by preparative thin-layer chromatography, yielding 4-ethoxy-3-ethoxycarbonylmethyl-2-ethyl-4-butanolide (**22**). Elimination of ethanol, with the aid of *p*-aminobenzenesulfonic acid, gave 2-(3*H*)-furanone **23**; the 2-(5*H*)-furanone isomer **24** was obtained with the aid of orthophosphoric acid. Both isomeric esters gave the acid **25** after hydrolysis, which on catalytic hydrogenation afforded *cis*-3-carboxymethyl-2-ethyl-4-butanolide **26**, identical to homopilopic acid. This synthesis of homopilopic acid differs from earlier syntheses because the less stable *cis*-2,3-disubstituted butanolide (**26**) is formed in the last step.

The same principle, namely, the cis hydrogenation of the butenolide ring in the last step, has been discussed by Chumachenko (*86*) in a paper on the synthesis of pilocarpine starting from 4-ethoxycarbonylmethyl-3-ethyl-2-(5H)-furanone (24). Conversion of 24 with phthaloylglycine hydrochloride in the presence of sodium ethanolate yielded 3-ethyl-4-(phthalimidoaceto)(ethoxycarbonyl)methyl-2-(5H)-furanone (27). The latter compound gave, after hydrolysis with diluted hydrochloric acid, aminomethyl didehydrohomopilopyl ketone (28). Analogous to a known method (87), Chumachenko described a way to synthesize pilocarpine via the 2,3-didehydro derivative of pilocarpine (86).

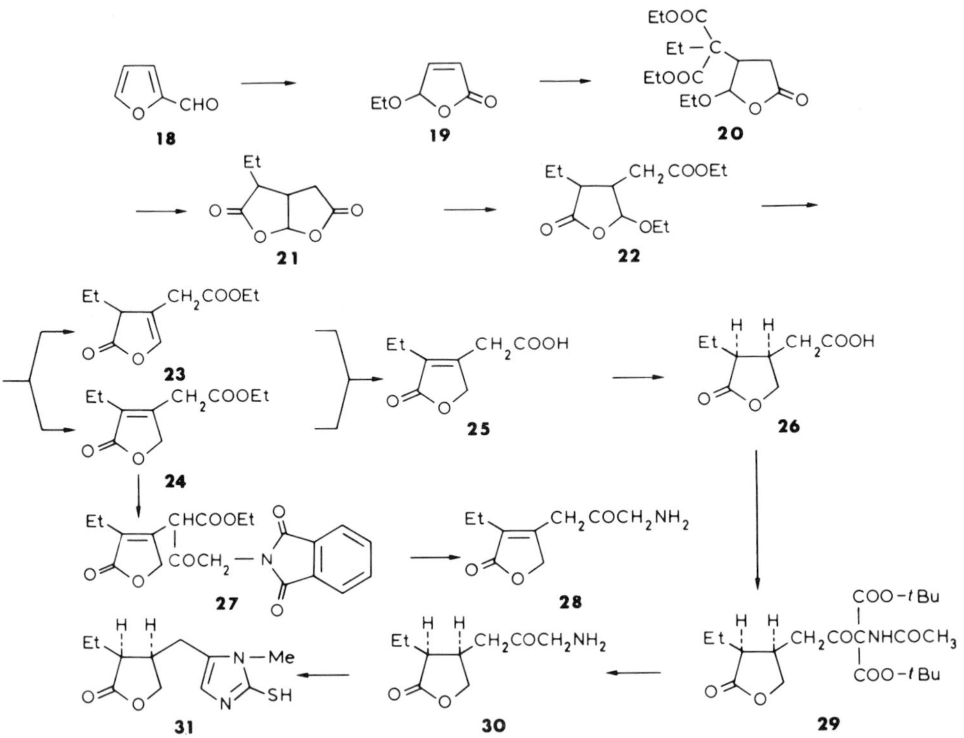

A total synthesis of optically active pilocarpine specifically labeled at nitrogen [N-¹⁴CH₃] via homopilopic acid was published by DeGraw et al. in 1975 (77). Optical resolution of racemic homopilopic acid (26) was achieved with the aid of (+)-α-methylbenzylamine. The acid chloride of 26 was treated with the sodium salt of di-tert-butyl acetamidomalonate, yielding 29. Hydrolysis and decarboxylation afforded (+)-aminomethyl homopilopyl ketone (30). This ketone was condensed with [¹⁴C]methyl isothiocyanate to give 2-mercaptopilocarpine (31) according to Preobrazhenski (87). Removal of the 2-mercapto group with aqueous hydrogen peroxide yielded (+)-[N-¹⁴CH₃]pilocarpine, containing about 5% isopilocarpine.

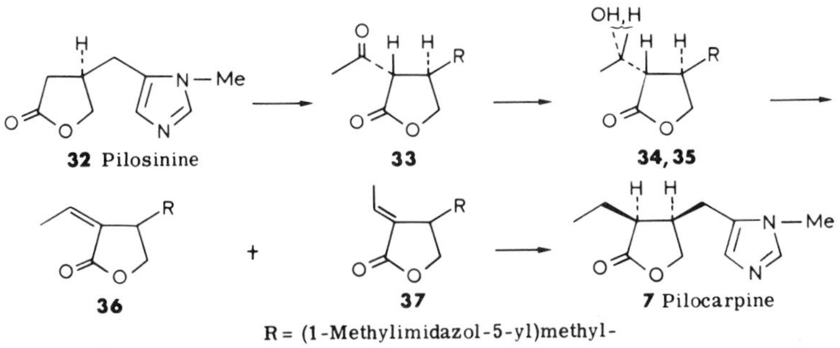

R = (1-Methylimidazol-5-yl)methyl-

A synthesis of (+)-pilocarpine, based on a starting material containing the imidazole ring, was reported by Link and Bernauer in 1972 (*69*). They first prepared (+)-pilosinine (**32**) (Section IV,B), which was converted to the keto lactone **33** with ethyl acetate in the presence of potassium *tert*-butanolate in *tert*-butanol. Hydrogenation of **33** yielded a mixture of the epimeric alcohols **34** and **35**. This mixture was treated with acetic anhydride and gave, via the acetates at 130°C, a mixture of the two isomeric ethylidene compounds **36** and **37**. Hydrogenation over platinum in methanol afforded (+)-pilocarpine (**7**) containing about 7% (+)-isopilocarpine (**8**).

A stereoselective synthesis of (+)-pilocarpine (**7**) starting from L-histidine (**2**) has been worked out by Noordam *et al.* (*88–90*). Use was made of the *S* configuration of the amino acid, which is the same as that of C-3 of the lactone ring in both (+)-pilocarpine and (+)-isopilocarpine. Furthermore, regioselective N-alkylation reactions of the imidazole nucleus of histidine had been developed by Beyerman *et al.* (*29,91*). Schemes 3 and 4 depict the different ways of the regioselective alkylations. For the synthesis of pilocarpine, the N^π-methylation has been performed via N^τ-protection with the 4-nitrobenzenesulfonyl group, instead of the benzoyl group (*29*).

L-Histidine (**2**) was converted to (*S*)-2-hydroxy-3-(imidazol-5-yl)propionic acid (**38**), with retention of configuration using silver nitrite and orthophosphoric acid. The acid (**38**) was esterified and the ester (**39**) treated with 4-nitrobenzenesulfonyl chloride. With the latter treatment, the hydroxyl group was esterified and the N^τ atom was sulfonated. Walden inversion of the reaction product (**40**), using lithium bromide in 2-butan-

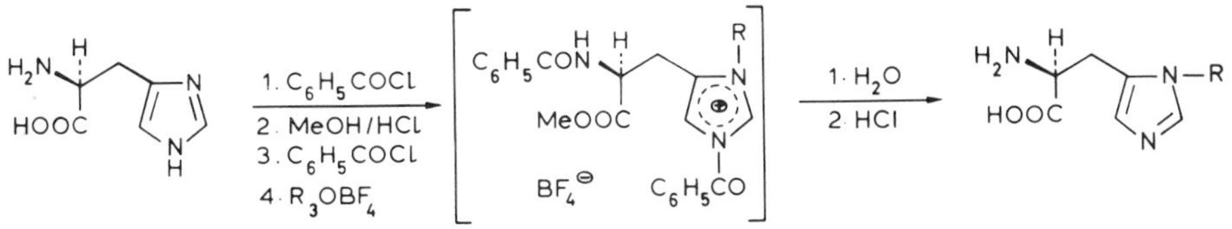

SCHEME 3. Regioselective N^π-alkylation of L-histidine (*29*).

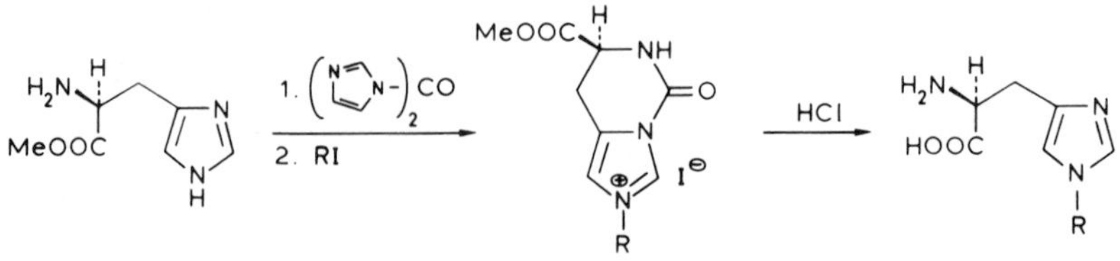

SCHEME 4. Regioselective N$^\tau$-alkylation of L-histidine (91).

one, afforded **41**. Compound **41**, after methylation with trimethyloxonium tetrafluoroborate, gave methyl (R)-2-bromo-3-(1-methylimidazol-5-yl) propionate (**42**). Walden inversion of **42**, using the sodium salt of dibenzyl ethylmalonate, yielded the ester **43**, which was hydrogenolyzed to 2-methyl dihydrogen (R)-1-(1-methylimidazol-5-yl)-2,3,3-pentanetricarboxylate (**44**). Compound **44** was decarboxylated at 140°C, yielding a mixture of diastereoisomers **45**. The ester group of **45** was reduced selectively to the alcohol function, affording **46**. Lactonization of **46** under acidic conditions yielded a 1:1 mixture of (+)-pilocarpine (**7**) and (+)-isopilocarpine (**8**) (88–90). The separation of the isomers is well established and the conversion of (+)-isopilocarpine to (+)-pilocarpine has been described (92).

From comparison with 2-bromo- and 2-chloro-3-(1-methylimidazol-5-yl)propionic acid intermediates (93–95) and the optical rotation of the final product, it followed that about 35% racemization took place at the C-3

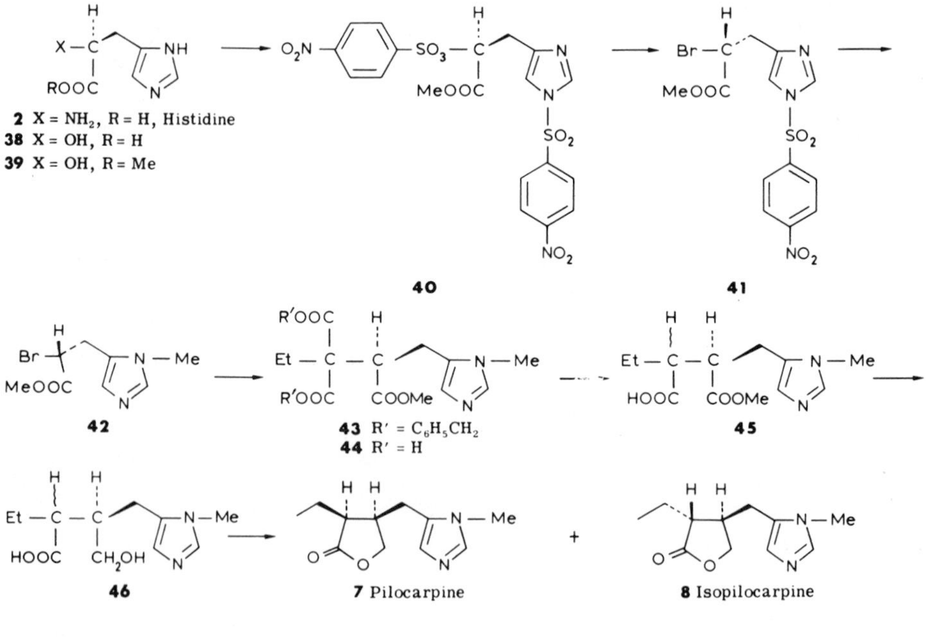

2 X = NH$_2$, R = H, Histidine
38 X = OH, R = H
39 X = OH, R = Me

40

41

42

43 R' = C$_6$H$_5$CH$_2$
44 R' = H

45

46

7 Pilocarpine

+

8 Isopilocarpine

chiral center of the lactone ring during the alkylation with the sodium salt of dibenzyl ethylmalonate (89,90).

G. Biogenesis and Biosynthesis*

Boit (97) and Leete (98) suggested that in the biogenesis of pilocarpine, the imidazole compound **47** reacts with acetylacetic acid (**48**). It should be remarked that the phosphate of **47** is assumed to be an intermediate in the biogenesis of histidine. The reaction between the methylene group in **48** and

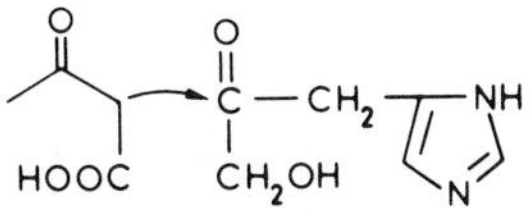

the carbonyl function of **47** is a type of aldol condensation that has been shown to occur in the plant biosynthesis of the butenolide ring of digitoxigenin (99). The condensation is followed by lactone ring closure by esterification of the alcoholic function with the acid group. In the scheme proposed by Boit and which has been further elaborated by Nunes (55,100), esterification takes place first, followed by formation of the lactone ring by an aldol condensation (Scheme 5).

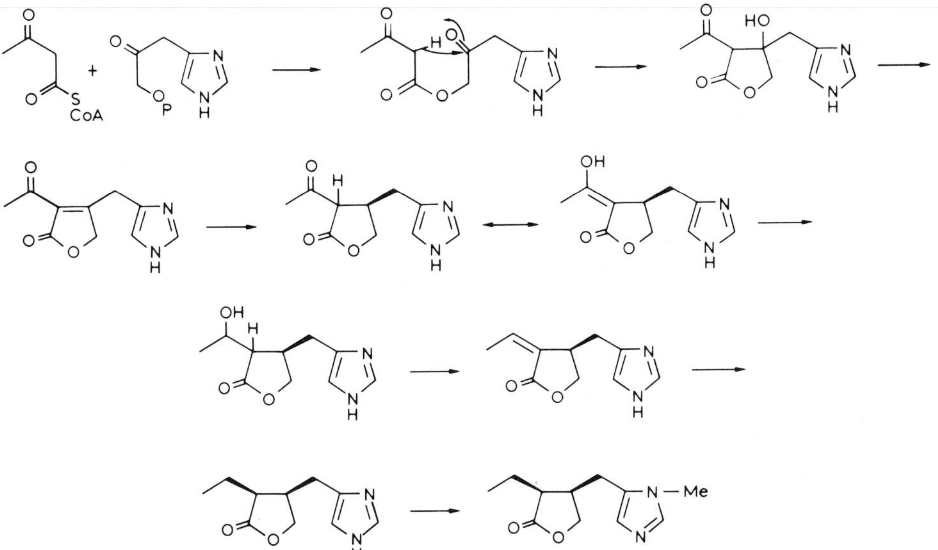

SCHEME 5. First biosynthetic pathway for pilocarpine, proposed by Nunes (55).

*These terms are used here as defined by Rapoport and by Leete (96).

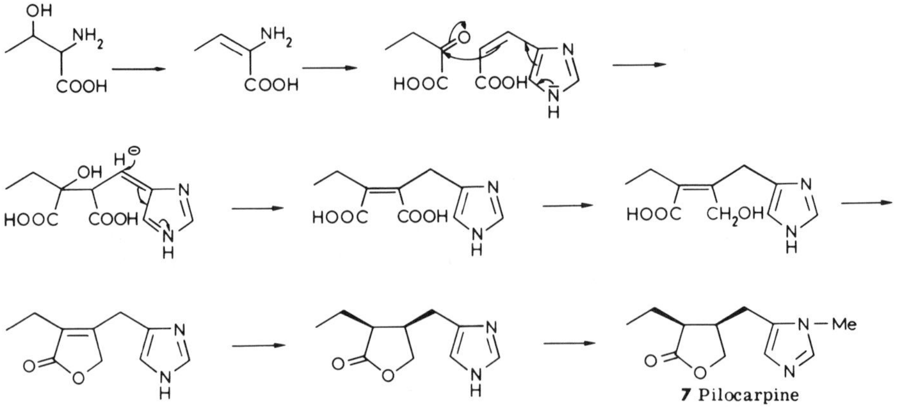

SCHEME 6. Second biosynthetic pathway for pilocarpine, proposed by Nunes (55).

Nunes proposed also a second mode of biogenesis, based on L-threonine (Scheme 6). This scheme makes first use of an aldol condensation, analogous to the early proposal of Leete, followed by esterification to the lactone. In both schemes the last step is a selective methylation of N^π by S-adenosyl-methionine.

The biosynthesis of pilocarpine in *Pilocarpus pennatifolius* was studied by the administration of radiolabeled precursors. Radioactive sodium acetate, histidine, histidinol, methionine, and threonine were administered by the cut-stem method. Histidine, methionine, and threonine were administered together by a wick inserted through the stem of an intact plant. Sodium acetate and histidine were fed to root cuttings by suspending the roots in aqueous solutions of the precursors. After 64–75 h, the roots were harvested and total alkaloid extracts made. These extracts were then fed to stem cuttings.

Of the cut-stem feedings, only L-[*methyl*-^{14}C]methionine showed a significant incorporation of 0.06% into pilocarpine. Subsequent degradation showed 98% of the radioactivity located in the N-methyl group. The wick feeding showed no incorporation of radioactivity into pilocarpine after 7 days. The total alkaloid extracts from the root feedings showed significant radioactivity. However, the pilocarpine isolated after feeding of the extracts to the stems was not radioactive. Many of the feedings could not be performed under ideal conditions. Thus, a lack of incorporation from these feedings is not definitive evidence that the precursor is not a reactant in the biogenetic scheme.

The methyl group of methionine has been found to be the biological source of the N-methyl group of pilocarpine. This portion of the biosynthesis occurred within the leaves. At present, it seems reasonable to suppose that pilocarpidine is biosynthesized in the roots and translocated to the leaves where it is N-methylated to pilocarpine.

H. Pharmacology

Pilocarpine is a widely studied peripheral stimulant of the parasympathetic system (101). It is used topically as a myotic to counteract the mydriatic effect of atropine and other parasympatholytic drugs. It has clinical value in the treatment of glaucoma when used as eye drop solutions ranging from 0.5 to 10% in concentration (102). Pilocarpine is reported to stimulate the growth of hair and therefore was employed in hair lotions (1). Internally, it was used as a diaphoretic in the treatment of nephritis (103).

Pilocarpine is unique among the cholinergic agents in that it is a tertiary amine derivative, whereas the majority of the members of this class are quaternary ammonium compounds. It has been demonstrated (104,105) that pilocarpine loses its activity when it is measured at pH 9, whereas acetylcholine does not, suggesting that pilocarpine acts in its ionized form, i.e., its activity decreases with increasing pH.

Although the pharmacological properties of pilocarpine have been studied extensively (106), little is known of the structural requirements for its parasympathomimetic activity. In 1982 Aboul-Enein and Al-Badr published an extensive review on the structure–activity relationship of compounds that are structurally related to pilocarpine (102); this review also gives a prognosis of receptor sites. A general conclusion is that any slight structural modification of the pilocarpine molecule causes a drastic reduction in or complete loss of its biological activity.

Formulations of pilocarpine in clinical usage — mainly for the treatment of glaucoma — are given in a pocket handbook issued jointly by the British Medical Association and The Pharmaceutical Society of Great Britain (107).

IV. Alkaloids Related to Pilocarpine

A. Pilocarpidine

The Jaborandi alkaloid pilocarpidine (**49**) is the N-nor lower homolog of pilocarpine. No new reports on the occurrence of pilocarpidine have been published since the last review in this series (1). The synthesis of pilocarpidine that terminates with the N^{π}-methylation of the imidazole nucleus is likewise the synthesis of pilocarpidine (Section III,F).

B. Pilosine, Isopilosine, and Epiisopilosine

The minor alkaloid pilosine, isolated in 1912 by Pyman and by Léger and Roques (*1*) from *Pilocarpus microphyllus,* was named isopilosine (**50**) by Voigtländer and Rosenberg in 1959 (*108*). The latter authors isolated from the same plant another alkaloid and named it pilosine (**51**). Repetition of this work by Löwe and Pouk in 1973 (*109*) showed that the pilosine described by Voigtländer and Rosenberg was found to be a 1 : 1 mixture of pilosine and isopilosine. The two bases were separated by crystallization of pilosine as the salicylate; isopilosine was isolated from the mother liquors.

A further alkaloid from *P. microphyllus* was detected and identified as (−)-epiisopilosine (**52**). Revised physical data of (+)-pilosine and (+)-isopilosine are given (*109*).

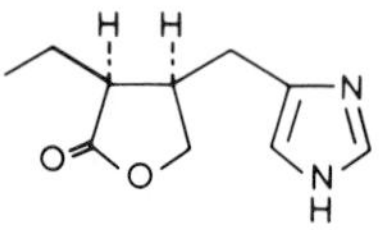

49 Pilocarpidine

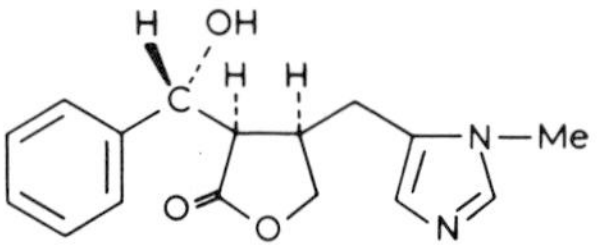

51 Pilosine

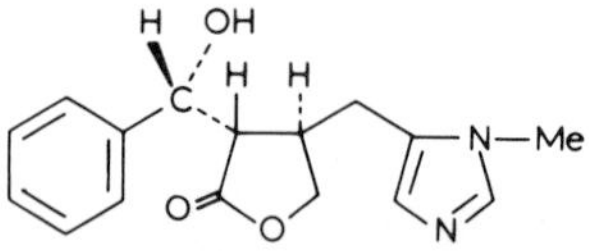

50 Isopilosine

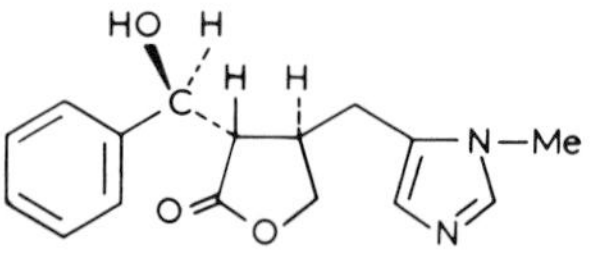

52 Epiisopilosine

In 1972 Link and Bernauer (*69*) published a synthesis of (+)-isopilosine and of (+)-pilocarpine, and then obtained (−)-epiisopilosine as a by-product. The readily available ester **53** was converted in two steps to the aldehyde (**54**), which on Stobbe condensation with succinic ester gave the half-ester acid salt **55**. Lithium borohydride reduction followed by prolonged acid treatment gave (±)-pilosinine [(±)-**32**], together with 2,3-dehydropilosin-

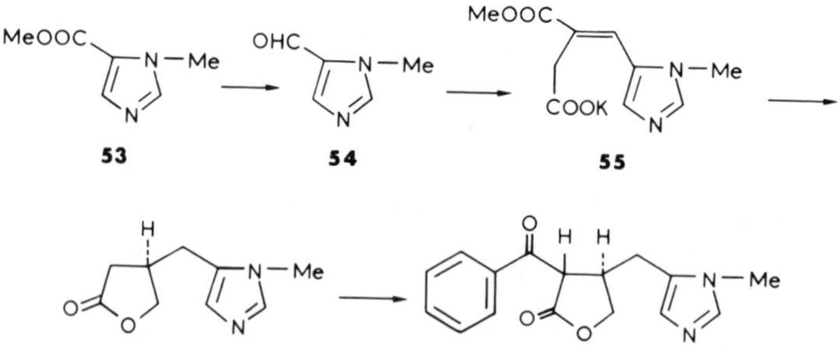

53   **54**   **55**

32 Pilosinine

56

ine. Resolution with (−)-di-O,O'-p-toluyl-L-tartaric acid afforded (−)-pilo-sinine (32) in low yield, which was converted with methyl benzoate in the presence of potassium *tert*-butanolate to the ketone 56. Platinum-catalyzed hydrogenation provided a 1 : 1 mixture of (+)-isopilosine (50) and (−)-epi-isopilosine (52). Correlation with (+)-pilocarpine via (+)-pilosinine gave the absolute configuration (69). This was confirmed by X-ray analysis (110).

A synthesis of (±)-pilosinine (32) via aminomethyl homopilosinyl ketone, starting from ethyl ω-phthalimidoacetylacetate and 3-cyano-1-ethoxypro-pene was described by Mehrotra and Dey (111).

The correctness of the absolute configuration of (+)-isopilosine and (−)-epiisopilosine and of (+)-pilosine, whose absolute configuration also rests on the work of Link and Bernauer (69), has been questioned by Sarel *et al.* (112). The latter authors proposed for the pilosines structures epimeric at the phenyl carbinol carbon atom on the basis of comparison of their circular dichroism (CD) spectra to those of aromatic amino acids of known absolute configuration. (+)-Pilosine (51) and (+)-isopilosine (50) exhibit a positive CD at 220 nm, as do aromatic amino acids with *S* configuration. (−)-Epiiso-pilosine (52) shows a negative CD in the low-wavelength region. Link and Bernauer expressed their doubt about the applicability of the "aromatic amino acid chirality rule" (113). By summarizing the data on X-ray analysis and the transformation of (+)-isopilosine to (+)-pilocarpine, they concluded that the earlier published structures of the pilosines were correct. Finally, this was also confirmed by Sarel *et al.* (114) on the basis of high-dilution IR studies.

Sarel *et al.* (112,115) found also that the pilosine originally isolated from *P. jaborandi* was a 1 : 1 molecular compound composed of (+)-pilosine and (+)-isopilosine. The mass spectra of 50, 51, and 52 have been studied, and a detailed investigation of the aldol condensation between pilosinine and benzaldehyde and its 2- and 4-fluoro derivatives has been described (112).

Dermatological preparations containing pilosine, isopilosine, and their salts have been described in a patent (116).

C. EPIISOPILOTURINE

Epiisopiloturine (57), in addition to pilocarpine, was isolated in 1978 by Voigtländer *et al.* (117) from the leaves of a variety of *Pilocarpus micro-*

57 Epiisopiloturine   58

phyllus Stapf. growing in the Alto Turi region of Brazil. Structure elucidation followed from IR, ^{1}H- and ^{13}C-NMR, and MS data and from comparison with the known stereoisomers of the pilosine series that have been found
as secondary alkaloids in other varieties of *P. microphyllus.* The lactone
moiety is not linked with the imidazole ring by the methylene bridge at C-5,
as in the case with the pilosines, but rather at C-4; compound **57** is an
N^τ-methylepiisopilosine isomer. The methiodide of **57** was found to be
identical to that of epiisopilosine, thus establishing the stereochemistry.

Alkaline hydrolysis of **57** gave benzaldehyde and neopilosinine. After
treatment with acetic acid anhydride and pyridine, anhydroepiisopiloturine
was obtained, isolated as its nitrate. Biogenetic aspects were discussed (*117*).

D. METAPILOCARPINE

Metapilocarpine was first reported by Pinner; he obtained it by heating
pilocarpine hydrochloride at 225–235°C (*118*). Polonovski proposed a
betaine structure (**58**) for the compound (*119*). On reinvestigation, metapilocarpine was shown to be a racemic mixture of isopilocarpine (*120*). The
structure was proved by spectral analysis and by GLC comparison with
authentic isopilocarpine. The pharmacological activity of the racemic product was compared to that of (+)-pilocarpine and (+)-isopilocarpine (*120*).

E. SYNTHETIC ANALOGS RELATED TO PILOCARPINE

Structural analogs of pilocarpine have been synthesized mainly for pharmacological investigations. In 1961 Mehrotra and Dey (*111*) reported the
synthesis of a structural isomer of pilocarpine with the ethyl substituent at
the methylene bridge instead of at the lactone ring. The condensation
product of 3-cyano-1-ethoxy-1-propene and diethyl ethylmalonate was
reacted with phthalimidoacetyl chloride to give **59**. Hydrolysis and conversion with potassium thiocyanate yielded **60**, which gave (±)-6-ethylpilosin-
ine (**61**) after removal of the thiol group and methylation.

In 1973 Döpke and Mücke studied the reaction between the oxolactone
62 with 5-chloro-1-methylimidazol-2-yl lithium and prepared 3-(5-chloro-
1-methylimidazol-2-yl)-2,2-diethyl-3-hydroxy-4-butanolide   (**63**).   They
found that the carbonyl activity of the lactone ring thus substituted had been

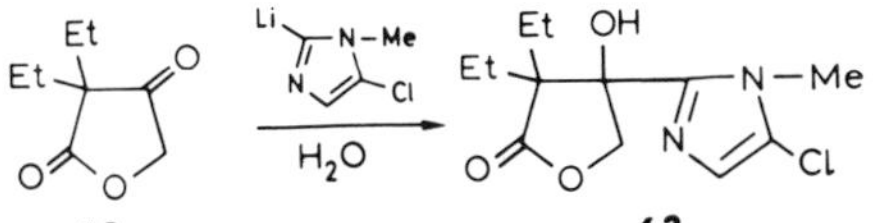

changed enough to prevent ring opening on addition with organometallic compounds. The structure of this alkaloid analog was determined with the aid of spectral data (*121*).

In 1973 Koda *et al.* (*122*) exposed pilocarpine to aminolysis and lithium aluminum hydride reduction, and then obtained several analogs. Treatment with ammonia or with aqueous methylamine or isopropylamine at room temperature gave the hydroxy amides **64–66**. Reaction of pilocarpine with ammonia at 200–210°C yielded a lactam, the pilocarpine analog **67**. Similarly, the *N*-methyllactam **68** was prepared by reaction with liquid methylamine at 225°C. Reduction of pilocarpine with lithium aluminum hydride (LAH) in tetrahydrofuran yielded the tetrahydrofuran analog **69**. Preliminary pharmacological studies indicated interesting cholinergic activity.

Borne *et al.* (*123*) synthesized ten analogs of pilocarpine in which the imidazole ring had been replaced by a pyridine, pyrimidine, or pyrazine ring. In some of the compounds, the ethyl group of the lactone moiety was also replaced by either a hydrogen atom or a methyl group. Finally, four compounds were prepared in which the lactone ring oxygen atom had been replaced by a methylene group, affording cyclopentanone analogs. The synthesis was based on the Michael addition of the anions derived from 2-methylpyridine, 4-methylpyridine, 4-methylpyrimidine, and 2-methyl-

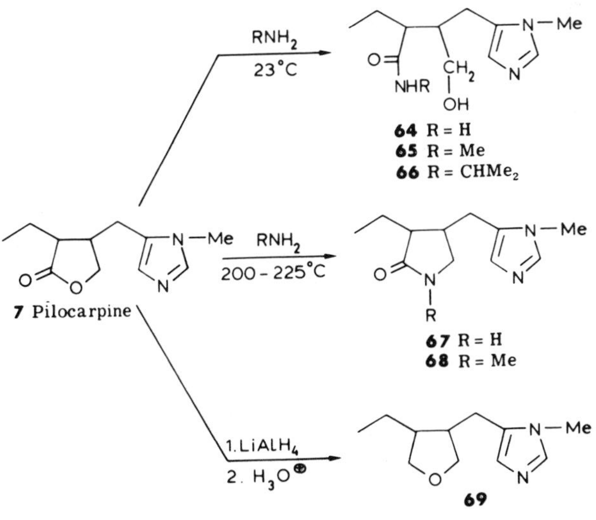

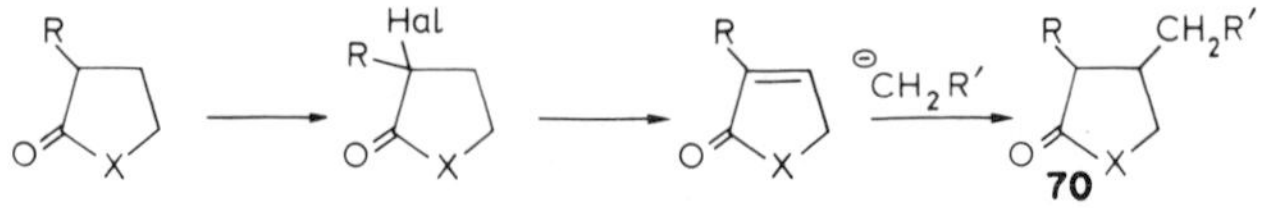

R = H, Me, or Et   R' = 2- or 4-pyridyl, 4-pyrimidyl,
X = O or CH₂   or 2-pyrazinyl

SCHEME 7. Synthesis of pilocarpine analogs by Borne *et al.* (*123*).

pyrazine to the appropriate but-2-en-4-olides or cyclopent-2-enones. The butenolides were prepared by dehydrohalogenation of the corresponding 2-bromo derivatives that, in turn, were prepared by bromination of the butanolides (Scheme 7).

In the same way, the sulfur analogs of pilocarpine, dihydro-4-(4-pyrimidylmethyl)-2-(3*H*)-thiophenone (Scheme 7, **70**: X = S, R = H, R' = 4-pyrimidyl) and 3-ethyldihydro-4-(4-pyrimidylmethyl)-2-(3*H*)-thiophenone (**70**, X = S, R = C₂H₅, R' = 4-pyrimidyl), were synthesized (*124*)

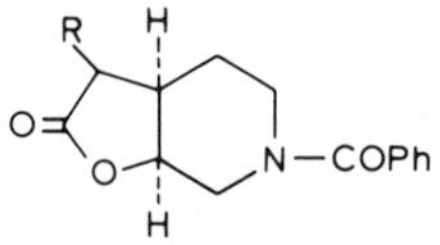

Finally, mention must be made of the preparation of the perhydro-furo[2,3-*c*]pyridin-2-ones **71** and **72**. These compounds are conformationally restricted analogs of pilocarpine in which the imidazole ring has been replaced by a simple alkylamine moiety fused to the lactone ring (*125*).

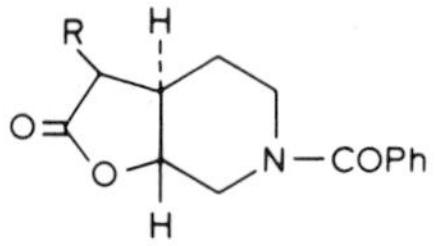

V. Other Imidazole Alkaloids

A. ALKALOIDS OF *Alchornea* SPECIES

1. Alchornine and Alchornidine

Alchornine (**73**) and alchornidine (**74a** or **74b**) have been isolated by Hart *et al.* (*126*) from the bark and leaves of *Alchornea javanensis* (Bl.) Muell.-Arg. (Euphorbiaceae), a small tree of the New Guinea rain forest. From spectroscopic data and chemical evidence, the major alkaloid alchornine was shown to have the hexahydroimidazo[1,2-*a*]pyrimidine structure **73**. Hydrolysis of a minor alkaloid, alchornidine, with ethanolic potassium hydroxide gave alchornine and 2,2-dimethylacrylic acid. But alchornidine is not simply an *N*-acyl derivative of alchornine, because mild hydrolysis with dilute acetic acid gives an isomer of **73**. This isoalchornine can be shown to be **75**, and it is readily converted to alchornine by reaction with alkali. This conversion occurs by opening of the lactam ring at N-4 and recyclization to N-1; it indicates that alchornine is the thermodynamically more favored form. Although structure **74a** is favored for alchornidine, the spectroscopic data do not enable a firm choice to be made between **74a** and **74b** (*127*). The mass spectra of alchornine (**73**), alchornidine (**74a** or **74b**), and their derivatives are consistent with the proposed structures and provide an excellent method for determining the position of the isopropenyl substituent on the imidazole ring.

73 Alchornine   **74a**   **74b**   **75** Isoalchornine

Alchornidine

2. Alchorneine, Isoalchorneine, and Alchorneinone

Alchorneine (**76**) was isolated in 1970 by Khuong-Huu *et al.* (*128*) from the leaves of *Alchornea floribunda* Muell.-Arg. and *Alchornea hirtella* Benth. of the Euphorbiaceae family, native to Congo and Cameroon. Spectroscopic data and chemical degradation indicated the alkaloid to be the tetrahydroimidazo[1,2-*a*]pyrimidine **76**. The monomethiodide and monomethiobromide were prepared, and treatment with sodium methoxide opened the pyrimidine ring. Acid hydrolysis gave 4,5-dihydro-4-isopropenyl-3-methoxy-2-imidazolone (**77**). The structure of the methiobromide of **76** was confirmed by X-ray analysis (*129,130*). The tartrate of **76** showed strong vagolytic action and inhibition of intestinal peristalsis in dogs, and exhibited ganglioplegic parasympathy. Therefore, alchorneine has been proposed as a potential spasmolytic agent in dogs (*131*).

In 1972 Khuong-Huu *et al.* (*132*) isolated isoalchorneine (**78**) and alchorneinone (**79**), along with alchorneine from the leaves and roots of *Alchornea floribunda* Muell.-Arg. Infrared, nuclear magnetic resonance, and mass spectra, as well as chemical correlation reactions, were used to elucidate the structures. The absolute configuration of the chiral centers of **78** was determined by synthesis of (*S*)-4,5-dihydro-4-isopropyl-2-imidazolone (**80**) from (*S*)-valinamide. The latter compound was reduced with LAH to give 1,2-diamino-3-methylbutane. This was reacted with phosgene, affording **80**. The optical rotation of **80** was equal but of opposite sign compared to that of the imidazolone derived from alchorneine via **77**. Chemical degradation of

76 Alchorneine   **77**   **78** Isoalchorneine

79 Alchorneinone   **80**

78 yielded also (R)-4,5-dihydro-4-isopropyl-2-imidazolone; thus isoalchorneine possesses the same absolute configuration at position 2 as alchorneine. The S configuration has been attributed to the second chiral center of isoalchorneine at position 6 because no rotation was observed. The R configuration would introduce a much greater assymetry in the molecule and it is therefore unlikely.

The imidazolone alkaloid alchorneinone (**79**), which might be either a precursor of **76** and **77** or a degradation product, possesses the same absolute configuration as **76** because it could be correlated also with **80**.

B. ALKALOIDS OF *Casimiroa edulis*

1. Casimiroedine

Casimiroedine (**81**) is the principal alkaloid of the seeds of the fruit of *Casimiroa edulis* La Llave et Lejarza, a tree of the Rutaceae family widely distributed throughout Mexico and Central America. It was isolated in 1911 by Power and Callan (*133*). Casimiroedine was found to be the cinnamic acid amide of casimidine (**82**). It contains the N^α-cinnamoyl-N^α-methylhistamine moiety (*134*). Acetylation of casimiroedine, followed by reaction with hydrogen bromide/glacial acetic acid/dichloromethane, hydrolysis with silver carbonate, and again an acetylation afforded β-D-glucose pentaacetate and confirmed the presence of a carbohydrate fragment (*135*). X-Ray analysis established the basic structure of casimiroedine, but left the unanswered question of whether it was the *cis*- or *trans*-cinnamic acid amide of **82** (*136*). The stereochemistry of **81** was confirmed by total synthesis from the chloromercury derivative **83** of 4-(2-chloroethyl)imidazole hydrochloride. Glycosylation with 2,3,4,6-tetra-O-acetyl-α-D-glucosyl bromide (**84**) and treatment with methanolic methylamine gave casimidine (**82**). When **82** was reacted with *trans*-cinnamic acid in the presence of

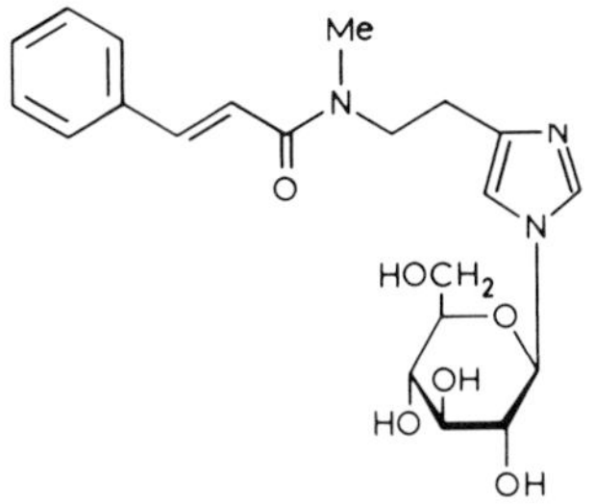

81 Casimiroedine

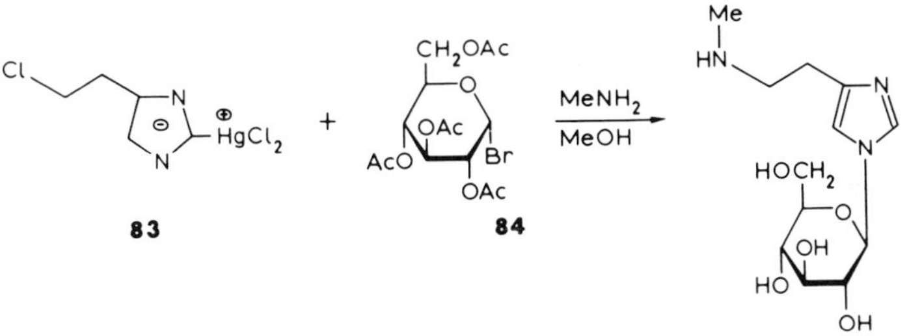

83 + **84** $\xrightarrow[\text{MeOH}]{\text{MeNH}_2}$

82 Casimidine

N-ethoxycarbonyl-2-ethoxy-1,2-dihydroquinoline (EEDQ), only one nucleoside product was formed. A comparison to authentic casimiroedine showed it to be identical (*137*). The cis analog of the alkaloid was obtained similarly from **82** and *cis*-cinnamic acid. Cytostatic activity of casimiroedine has been reported (*137*).

2. Zapotidine

Zapotidine (**85**) was obtained in 1956 by Kincl *et al.* from the seeds of *Casimiroa edulis* Llave et Lex., apparently the same notation as La Llave et Lejarza mentioned in other publications (*138*). In the Pharmacopoeia of Mexico, the fruit and the seed are recognized under the title of "*Zapota blanco*" (literally, white fruit). Lithium aluminum hydride reduction of zapotidine gave N^{α},N^{α}-dimethylhistamine, and boiling with 20% potassium hydroxide in ethanol yielded N^{α}-methylhistamine (**86**). Spectroscopic data, together with other chemical findings, show zapotidine to possess the structure of 6-methyl-5,6,7,8-tetrahydroimidazo[1,5-*c*]pyrimidine-5-thione (**85**), one of the few sulfur-containing alkaloids (*139*)

Zapotidine has been synthesized from histamine (**3**). Histamine was converted to 5,6,7,8-tetrahydroimidazo[1,5-*c*]-pyrimidin-5-one (**87**) with *N,N'*-carbonyldiimidazole. Reduction with LAH in tetrahydrofuran gave

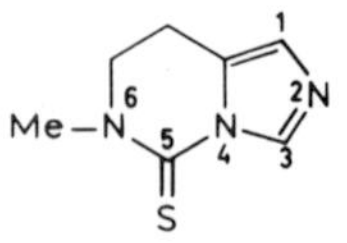

85 Zapotidine

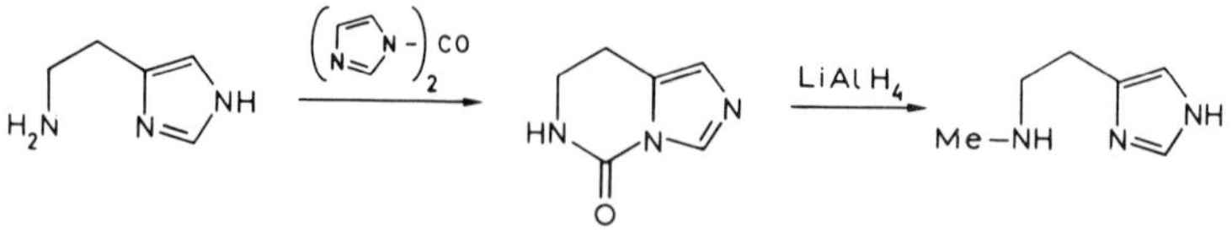

3 Histamine   **87**   **86** N^{α}-Methylhistamine

88

90   **89** Goitrine

N^{α}-methylhistamine (**86**), which then afforded zapotidine by reaction with N,N'-thiocarbonyldiimidazole in an overall yield of 22% (*140*). This synthesis made preliminary pharmacological tests possible, the results of which are given. A synthesis of zapotidine reported earlier produced only the oxo analog (**87**) (*141*).

Speculations on the biosynthesis of zapotidine start from the isothiocyanate derivative of histamine (**88**) that must undergo a ring closure followed by a methylation at N-6. Although not known in biosyntheses, a similar conversion is to be found in the synthesis of goitrine (**89**) from the isothiocyanate **90** (*142*).

C. CHAKSINE

Chaksine (**91**) was isolated in 1935 by Siddiqui and Ahmad (*143*) from the seeds of *Cassia absus* L. (Caesalpiniaceae), a plant used in folk medicine. On the basis of degradation reaction products, Wiesner *et al.* proposed structure **91** (*144*), but Singh *et al.* arrived at a different structure (**92**) that would satisfy the experimental data (*145*). Final proof of structure **91** was obtained

91 Chaksine   **92**   **93**

both from NMR data of **91** and the corresponding ureido acid **93** (*146*) and from X-ray analysis (*147*).

An isomer of **91**, isochaksine, has also been found in the seeds of *C. absus* L. It does not appear to be an artifact of chaksine formed during isolation (*143*). Neither the structure of isochaksine nor that of an intermediate isomer, neochaksine, are described (*148*). Perhaps the existence of these isomers can be explained by stereoisomerism at the chiral carbon atoms, as has also been suggested for an isomer of the ureido acid **93** (*146*). In 1979, the isolation of chaksine and isochaksine was reported also from the leaves and the roots of *C. absus* L. (*149*).

Pharmacological actions of chaksine chloride in rabbits (*150*) and in mice and rats (*151*) have been studied.

D. ALKALOIDS OF *Cynometra* SPECIES

In 1973, Khuong-Huu *et al.* isolated three alkaloids anantine (**94**), cyno-metrine (**95**), and cynodine (**96**), from the leaves of *Cynometra ananta,* a plant of the Leguminoseae family native to Zaire. The structures were deduced on the basis of spectral and chemical evidence (*152*). The structures bear some similarity to that of pilocarpine (**7**), which contains a lactone ring instead of a lactam ring. Treatment of cynometrine with polyphosphoric acid yielded the N-methylated anantine (**97**). The latter compound could not be prepared by a simple N-methylation of anantine. Hydrolysis of cynodine afforded cynometrine and benzoic acid.

Besides anantine (**94**) and cynometrine (**95**), six other imidazole alkaloids of the same type have been isolated from *Cynometra lujae* growing in Congo–Brazzaville (*153–155*). These alkaloids are isoanantine (**98**), iso-cynometrine (**99**), isocynodine (**100**), noranantine (**101**), hydroxyanantine (**102**), and cynolujine (**103**).

^{13}C-NMR spectroscopy indicated that the isomerism between anantine and isoanantine and between cynometrine and isocynometrine can be explained by the position of the methyl group in the imidazole nucleus. Anantine and cynometrine are the N^τ-methyl derivatives, and the isomers have the methyl substituent at the N^π-nitrogen. The same holds for cynodine

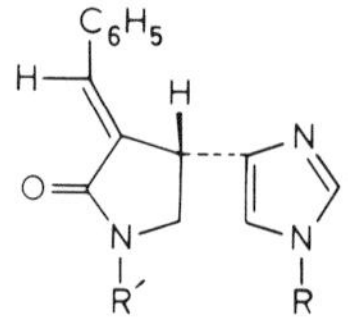

94 R = Me, R′ = H Anantine
97 R = R′ = Me
101 R = R′ = H Noranantine

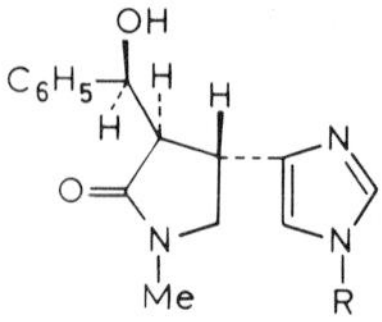

95 R = Me Cynometrine
104 R = H

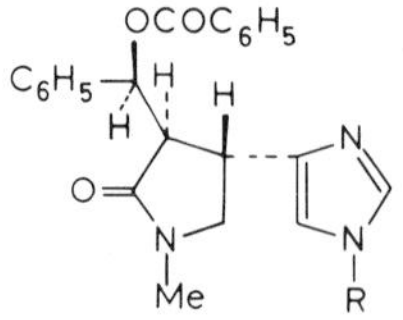

96 R = Me Cynodine
105 R = H

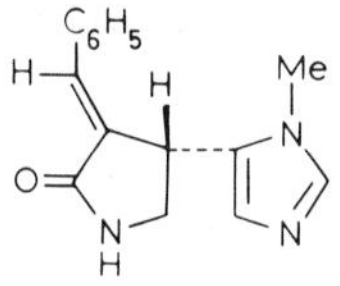

98 Isoanantine

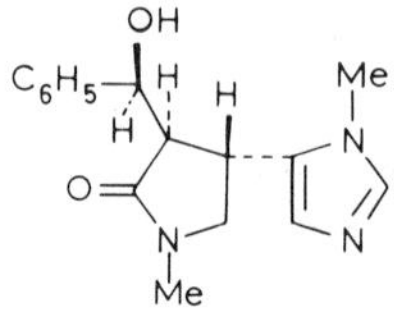

99 Isocynometrine

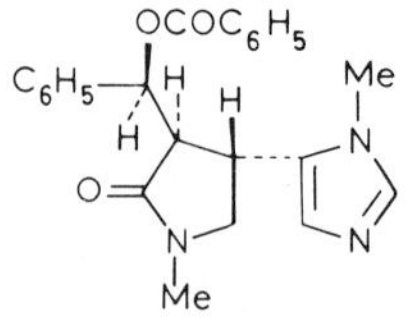

100 Isocynodine

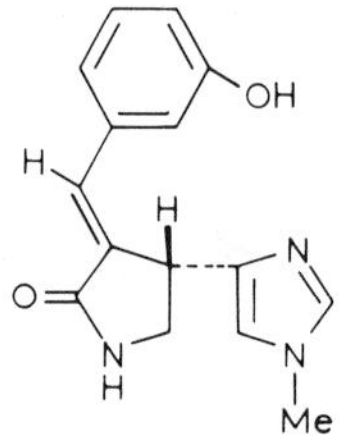

102 Hydroxyanantine

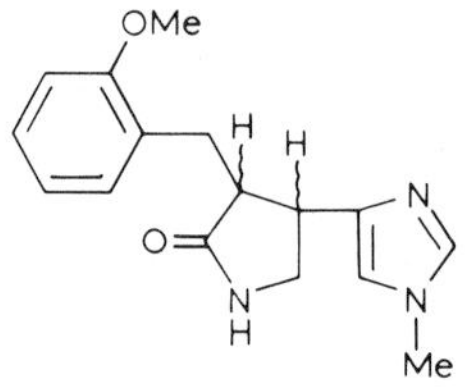

103 Cynolujine

and isocynodine, because the latter compound proved to be the benzoyl ester of isocynometrine. N-Methylation of noranantine yielded anantine as the main product and isoanantine. Hydroxyanantine (**102**), which contains a phenolic hydroxyl group, afforded *m*-hydroxybenzoic acid on oxidation with permanganate. Cynolujine proved to possess the *o*-methoxydihydroanantine structure **103**.

The relative configuration of cynometrine and isocynometrine were determined by X-ray diffraction analysis. The absolute configuration of the phenyl carbinol chiral center was determined by the method of Horeau and Kagan (*154*). From UV spectroscopic data and the rotations of the different alkaloids and their derivatives, the steric structures **94–102** have been deduced as depicted in the illustrations (*155*).

From *Cynometra hankei* Harms., cynometrine (**95**) and cynodine (**96**) and their *N*-demethyl derivatives **104** and **105** have been isolated by Waterman and Faulkner. The structures were elucidated by spectral analysis and by methylation of **104** and **105** to the parent compounds **95** and **96**, respectively (*156*).

Racemic isoanantine (**98**) has been synthesized starting from 5-formyl-1-methylimidazole. This was converted via a Wittig–Horner reaction to ethyl

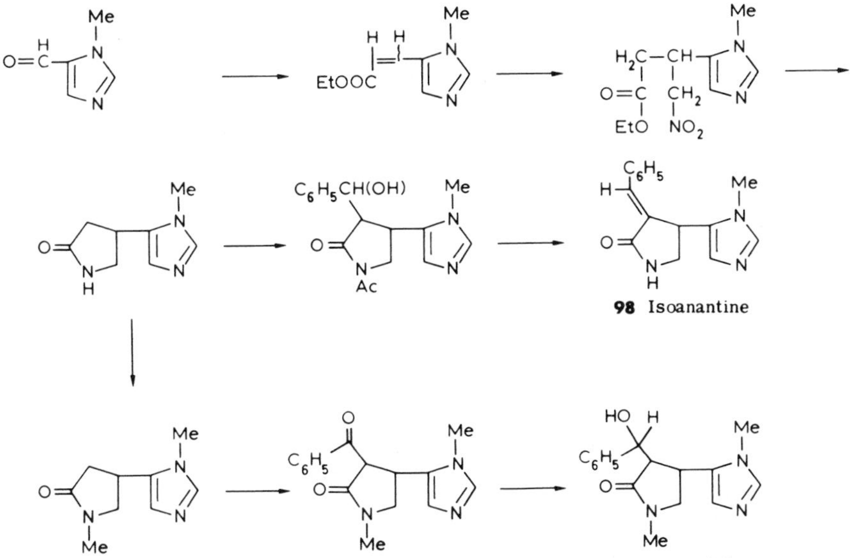

98 Isoanantine

99 Isocynometrine

SCHEME 8. Synthesis of the *Cynometra* alkaloids isoanantine (**98**) and isocynometrine (**99**) (*153,155*).

β-(1-methylimidazol-5-yl)acrylate. Treatment with nitromethane in tetramethylguanidine, followed by hydrogenation over palladium on carbon, afforded the lactam ring. Acetylation of the lactam nitrogen and condensation with benzaldehyde in dimethylformamide and sodium hydride yielded the phenyl carbinol, which gave racemic isoanantine after dehydration and deacetylation with polyphosphoric acid or phosphoryl chloride in pyridine (*153*).

Methylation of the lactam nitrogen and condensation with methyl benzoate in the presence of lithium diethylamide in THF afforded a ketone, which, on sodium borohydride reduction, yielded racemic isocynometrine (**99**) as the major component of a mixture of two epimeric phenyl carbinols (Scheme 8) (*155*). Racemic anantine (**94**) and cynometrine (**95**) have been synthesized analogously, starting from 4-formyl-1-methylimidazole (*155*).

E. ALKALOIDS OF *Cypholophus friesianus*

Cypholophine (**106**) and *O*-acetylcypholophine (**107**) were isolated in 1970 by Hart *et al.* (*157*) from *Cypholophus friesianus* (K. Schum.) H. Winkl. (Urticaceae), a shrub growing in Australia and New Guinea. Spectral data and degradation products led to the proposal of structure **106** for cypholophine. This structure was confirmed by the following synthesis.

106 R = H, Cypholophine
107 R = MeCO, O-Acetylcypholophine

108

109

The starting material in the synthesis of **106** (*157*) was 3-(3,4-dimethoxy-phenyl)propanoyl chloride, which with the aid of diazomethane at $-20°C$, was converted to the diazo ketone **108**. Treatment of **108** in ether with 48% hydrogen bromide gave 4-bromo-1-(3,4-dimethoxyphenyl)-3-pentanone (**109**). The bromo ketone **109** was heated in an autoclave with 2-iminotetra-hydropyran hydrochloride in methanolic ammonia and yielded the product **106**, identical to the alkaloid cypholophine, in an overall yield of 6%.

In a second synthetic attempt, the diazo ketone **108** was converted to the α-ketol 1-(3,4-dimethoxyphenyl)-4-hydroxy-3-pentanone via its acetate. The α-ketol was oxidized *in situ* with bismuth trioxide in the presence of ammonium acetate and 5-hydroxyvaleric anhydride (overall yield 0.2%).

A standard acylation of **106** with acetic anhydride in pyridine at room temperature gave **107**, identical with O-acetylcypholophine.

It has been suggested that these alkaloids apparently are derived biogenetically from histidine (*17*, p. 840), but biosynthetic evidence still has to be forthcoming.

F. ALKALOIDS OF *Dendrobium* SPECIES

From *Dendrobium anosmum* Lindl. and *Dendrobium parishii* Rchb. f. (Orchidaceae) octahydrodipyrido[1,2-*a*:1′,2′-*c*]imidazol-10-ium bromide (**110**) was isolated in 1968 by Leander and Lüning (*158*). It was stable toward catalytic hydrogenation, but afforded with LAH a saturated diacidic base $C_{11}H_{20}N_2$, establishing a tricyclic nature. The structure of **110** was further indicated on the basis of the NMR spectrum; the structure was then proved by synthesis. Heating of 2-bromomethylpyridine hydrobromide and 2-bromopyridine together yielded dipyrido[1,2-*a*:1′,2′-*c*]imidazol-10-ium bromide, which could be hydrogenated over platinum oxide in methanol to give **110** (*158*).

The crystal structure of **110** was obtained from an X-ray determination.

110

The atoms of the five-membered ring and the four neighbor atoms in position 1, 4, 6, and 9 are all close to a plane. The remaining carbon atoms deviate significantly from this plane, in such a way that the conformation of the ion may be described as a twisted chair (*159*).

G. ALKALOIDS OF *Dolichothele sphaerica*

Dolichotheline (**111**) is a histamine-derived alkaloid produced by the cactus *Dolichothele sphaerica* Britton and Rose (Cactaceae) native to southern Texas and northern Mexico. The alkaloid was first isolated in 1969 by Rosenberg and Paul (*160*). Spectroscopic data suggested structure **111**, 4(5)-(*N*-isovalerylaminoethyl)imidazole or N^α-isovalerylhistamine. The structure was proved by synthesis. Refluxing of histamine with isovaleric anhydride yielded **111**, identical to the natural product (*160*). In addition to the major alkaloid dolichotheline, five minor alkaloids have been isolated (*161*). These were identified as *N*-methylphenethylamine, *β-O*-methylsynephrine, *N*-methyltyramine, synephrine, and *β-O*-ethylsynephrine by IR, NMR, and comparison to authentic materials; *β-O*-ethylsynephrine was probably an artifact of synephrine, since it was not found in a second extraction attempt when no ethanol was used.

111 Dolichotheline

Biosynthetic studies were performed by Rosenberg *et al.* (*162–165*), and by O'Donovan *et al.* (*166,167*) (Scheme 9). A feeding experiment with racemic [2-^{14}C]histamine showed that the histamine residue arises specifically from this precursor. The isovaleryl unit incorporated activity efficiently from racemic [2-^{14}C]leucine and [1-^{14}C]isovalerate to label in each case the carbonyl group of **111**. Mevalonic acid also proved to be a specific but less efficient precursor of the residue of **111**.

It was possible also to incorporate unnatural precursors into the alkaloids of *Dolichothele sphaerica* (*164,165,168,169*). On feeding of 3-aminoethylpyrazole, the alkaloid 3-(*N*-isovalerylaminoethyl)pyrazole (**112**) was produced, and N^α-isocaproylhistamine was obtained by the simultaneous feeding of 4-methylpentanoic acid (isocaproic acid) and of 4(5)-aminoethy-

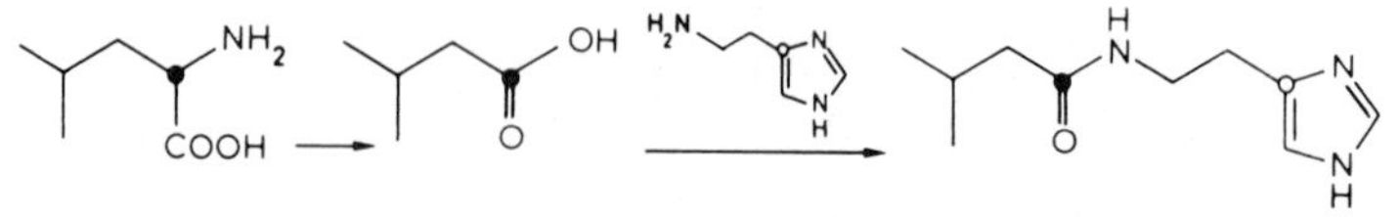

SCHEME 9. Biosynthetic pathway for dolichotheline (**111**) (*162–167*).

limidazole. [1-¹⁴C]Cinnamic acid when fed gave rise to labeled N^α-cinna-moylhistamine (**6**). The production of 4(5)-(*N*-isovalerylaminoethyl)imida-zole could be influenced positively by giving the plants a pretreatment with inhibitors of histidine decarboxylase, e.g., α-methylhistidine and α-hydra-zinohistidine. These inhibitors decreased the formation of histamine from histidine, by which the availability of the natural precursor decreases and the formation of the new alkaloid from the unnatural precursor increases (*165*).

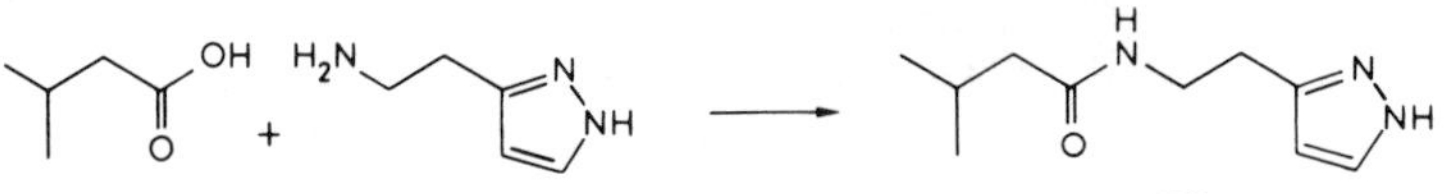

112

H. Alkaloids of *Glochidion philippicum*

From the leaves of *Glochidion philippicum* (Euphorbiaceae), a tree that grows in New Guinea, Johns and Lamberton in 1966 isolated four alkaloids (*170*): glochidine (**113**), glochidicine (**114**), N^α-(4-oxodecanoyl)histamine (**115**), and N^α-cinnamoylhistamine (**6**). The structures were deduced with the aid of IR, NMR, and mass spectroscopic methods.

Acid hydrolysis by concentrated hydrochloric acid at 100°C of **113** and **115** yielded histamine hydrochloride and 4-oxodecanoic acid, respectively. In contrast, **114** proved to be stable to this acidic condition. The relative instability of glochidine (**113**) in strong acid was attributed to the N–C–N system, which is not present in the isomeric glochidicine (**114**).

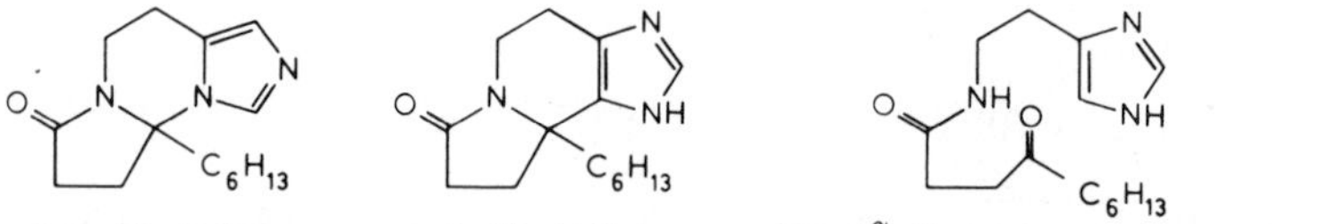

113 Glochidine   **114** Glochidicine   **115** N^α-(4-Oxodecanoyl)histamine

Acylation of histamine with 4-oxodecanoyl chloride in pyridine yielded a mixture of **113**, **114**, and **115**, which could be separated to yield compounds identical to the respective natural alkaloids. Heating of **113** in dilute acetic acid gave a mixture of **113**, **114**, and **115**; further heating finally yielded pure glochidicine (**114**). One may therefore assume **113** and **114** to be cyclization products of **115**.

By analogy, one could postulate **113** and **114** to arise biogenetically by condensation of the carbonyl of **115** with the NH— or the CH— position, respectively. Because the N – C – N group in **113** is unstable in acid medium, ring opening and recyclization occur to **114**, which is stable in acid surroundings (*171*).

I. ALKALOIDS OF *Macrorungia longistrobus*

A shrub that occurs in South Africa and is called *Macrorungia longistrobus* C.B. Cl. (Acanthaceae) contains a number of imidazole alkaloids. The following have been identified: macrorine (**116**), isomacrorine (**117**), macrorungine (**118**), normacrorine (**119**), longistrobine (**120**), isolongistrobine (**121**), and dehydroisolongistrobine (**122**).

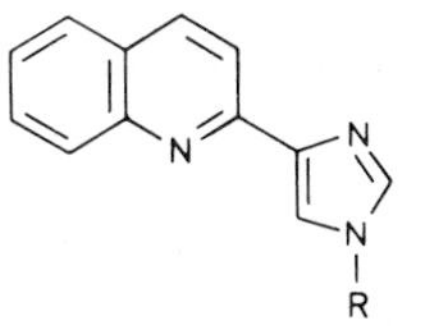

116 R = Me, Macrorine
119 R = H. Normacrorine

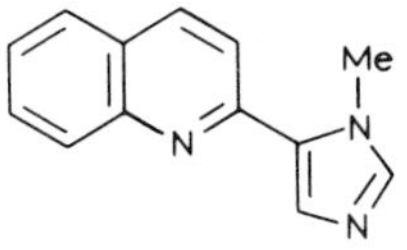

117 Isomacrorine

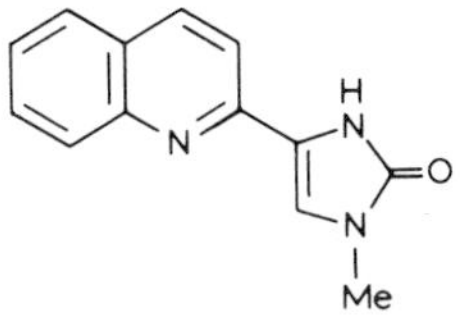

118 Macrorungine

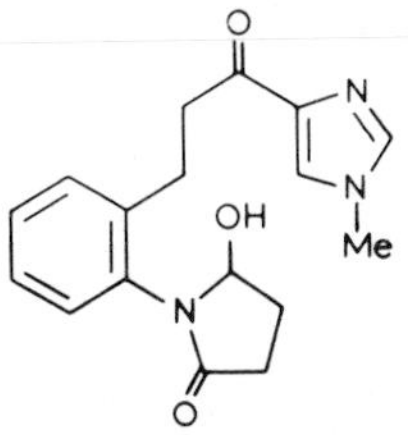

120 Longistrobine

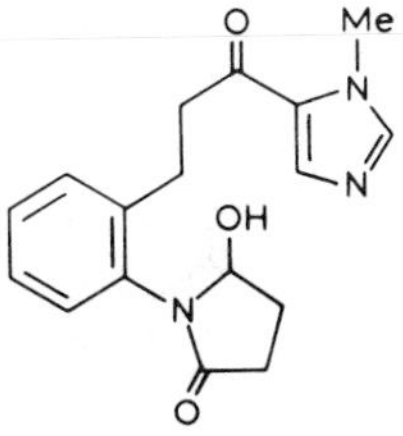

121 Isolongistrobine

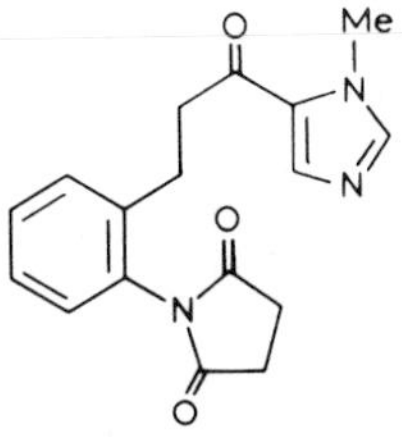

122 Dehydroisolongistrobine

The first three alkaloids were isolated in 1964 by Arndt *et al.* (*172*). Spectroscopic data of degradation products led to the structural formulations. A synthesis of isomacrorine (**117**) is described later and is further proof of the proposed structure (*173*).

From the residue that remained after the isolation of macrorine, isomacrorine, and macrorungine, a fourth alkaloid could be isolated. This was called normacrorine (*174*). The structure (**119**) was deduced mainly from IR and NMR spectra and confirmed by synthesis. This synthesis makes use of the method of Clemo for the preparation of pyridylimidazoles. The oxime of 2-acetylquinoline (**123**) was converted, via the *p*-toluenesulfonyl ester, to 2-ω-aminoacetylquinoline hydrochloride (**124**). Cyclization of **124** in the presence of potassium thiocyanate gave 4-quinol-2-yl-2-mercaptoimidazole

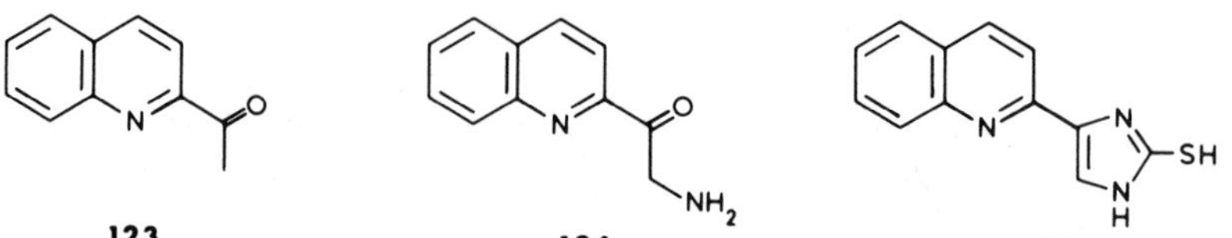

123 **124** **125**

hydrochloride (**125**). The latter compound, on oxidation with dilute nitric acid, gave 4-quinol-2-ylimidazole identical to natural normacrorine (**119**).

  Methylation of normacrorine with dimethyl sulfate gave a 2 : 1 mixture of **116** and **117**. On the basis of the negative inductive effect of the quinolyl substituent on the imidazole ring, one would expect on methylation in nonbasic medium isomacrorine to be the major product. The fact that, contrary to this expectation the main product is macrorine, can be explained by assuming a shielding of the most reactive imidazole nitrogen (N^π) by the quinolyl substitutent.

  In 1969 Arndt et al. (*175*) again isolated three imidazole alkaloids from

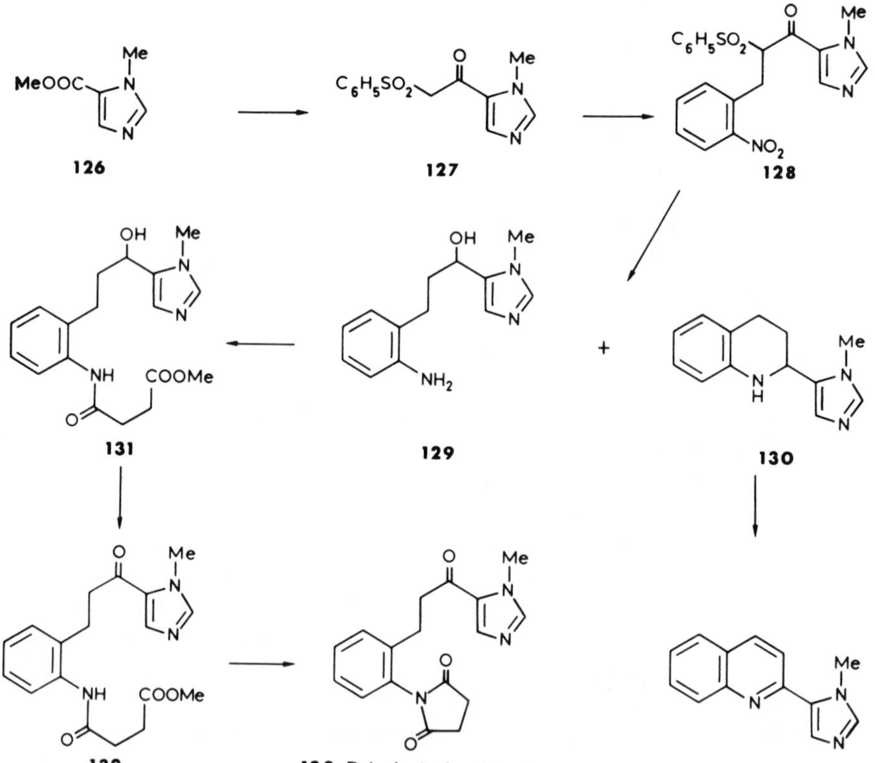

126 **127** **128**

131 **129** + **130**

132 **122** Dehydroisolongistrobine **117** Isomacrorine

M. longistrobus: longistrobine (**120**), isolongistrobine (**121**), and dehydro-isolongistrobine (**122**). Zinc-dust distillation of longistrobine and isolongistrobine yielded the known alkaloids macrorine and isomacrorine, respectively. This and spectroscopic data suggested that the two alkaloids were substituted tetrahydroisoquinolylimidazoles with a 4-hydroxy-1-oxobutano bridge over positions 1 and 4 of the quinoline moiety. Dehydroisolongistrobine would then be the compound where the bond between the nitrogen and C-2 of the quinoline part had been hydrolyzed.

These structures were refuted by Wuonola and Woodward (*173,176,177*) who arrived at a different conclusion with regard to the structures. Structures **121** and **122** were proposed for isolongistrobine and dehydroisolongistrobine, respectively, and the structures were confirmed by total synthesis.

The starting material for the synthesis of dehydroisolongistrobine was the imidazole ester **126**, which was converted to the β-ketosulfone **127**. Compound **127** was treated with potassium *tert*-butanolate and *tert*-butanol in tetrahydrofuran, and reacted with o-nitrobenzyl bromide in tetrahydrofuran. This gave the phenylsulfonyl nitroketone **128**. Reduction of **128** with aluminum amalgam in aqueous tetrahydrofuran gave a mixture of two compounds, **129** and **130**, which could be separated by preparative thin-layer chromatography. Dehydrogenation of **130**, either with palladium on carbon or with sulfur in xylene, gave isomacrorine (**117**) (overall yield of isomacrorine 15%). Compound **129**, mixed with pyridine in dichloromethane, was reacted with 3-methoxycarbonylpropanoyl chloride to give the ester **131**. Oxidation of **131** with chromium trioxide in aqueous pyridine yielded the ketone **132**. On heating of **132** for 5 min *in vacuo* above its melting point of 128°C, dehydroisolongistrobine (**122**) was formed, identical to the natural product. The overall yield in the synthesis of **122** was 10%.

129   **133** R = OH   **121** Isolongistrobine
           134 R = =O

The synthesis of isolongistrobine also started with the imidazole ester **126**, which in three steps was converted to the amino alcohol **129**. Acylation with 4-pentenoyl chloride gave the amide alcohol **133**. Oxidation of **133** with chromium trioxide in aqueous pyridine yielded the 4-pentene carboxamide **134**. Oxidative splitting of the vinyl group of **134** with sodium periodate aided by a catalytic amount of osmium tetroxide in dioxane/water, gave the aldehyde equivalent of **121**, identical to natural isolongistrobine. The overall yield in this synthesis was 12%. Because longistrobine and isolongistrobine differ only in the position of the N^{im}-methyl group, it must be concluded from the preceding that structure **120** must be allotted to longistrobine.

J. OROIDIN AND DIBROMOPHAKELLIN

Oroidin (**135**) was isolated in 1971 by Forenza *et al.* (*178*) from the sea sponge *Agelas oroides*. Structure **136** had first been proposed on the basis of spectral data and degradative products, but was later revised by Garcia *et al.* (*179*) because of its close resemblance to the isomeric alkaloid dibromophakellin. This weakly basic substance was isolated from the sponge *Phakellia flabellata* (*180,181*). The structure of dibromophakellin **137** was deduced from spectroscopic data and chemical evidence. It was confirmed by a single-crystal X-ray diffraction analysis of the monoacetyl derivative (*181*).

135 Oroidin   136

Rather puzzling was the low pK_a (<8) value of dibromophakellin (**137**) compared to other guanidine derivatives. The X-ray analysis, however, demonstrated that the aminoimidazole ring is somewhat twisted and cannot become planar on protonation. The mesomerism within the cation must therefore be considerably reduced.

It has been suggested that oroidin (**135**) and dibromophakellin (**137**) might be biogenetically related through a dihydrooroidin-type (**138**) intermediate, via an overall derivation of both sea sponge products from proline and histidine (*181*).

In 1982 Foley and Büchi (*182a*) described a biomimetic synthesis of racemic dibromophakellin (**137**) via an oxidative cyclization of dihydrooroidin (**138**). The ethyl ester of (+)-citrulline (**139**) was reduced with sodium amalgam and the crude aldehyde obtained was condensed with cyanamid and cyclized with hydrochloric acid, respectively, to give **140**. Compound **140** was hydrolyzed to the amine **141**. Acylation with 2,3-dibromo-5-trichloroacetylpyrrole yielded dihydrooroidin (**138**). Exposure of the hydrochloride of **138** to bromine in acetic acid and addition of methanol, followed by treatment with potassium *tert*-butanolate gave racemic **137**, identical to

139

140 R = CONH₂
141 R = H

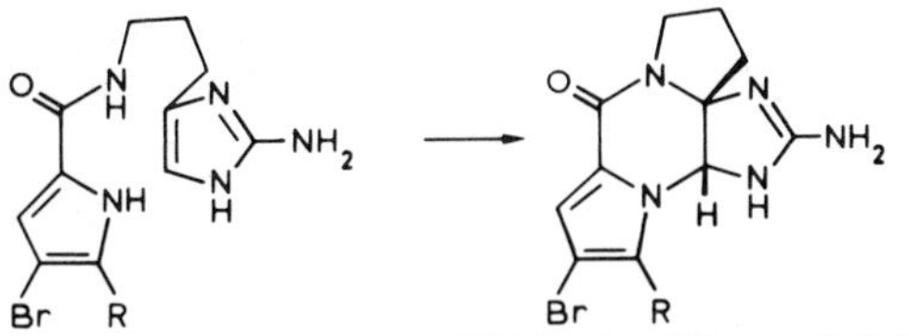

138 R = Br

137 R = Br, Dibromophakellin
142 R = H, 4-Bromophakellin

dibromophakellin. The sponge *P. Flabellata* also produces the closely related alkaloid 4-bromophakellin (**142**).

Recently a new bromo compound was isolated from the Red Sea sponge *Acanthella auzantiaca*. X-ray analysis established the structure: 4-(2-amino- 4-oxo-2- imidazolin-5-ylidene)-2-bromo- 4,5,6,7-tetrahydropyrrolo [2,3-*c*]azepin-8-one which shows structural relationship with (mono-bromo)oroidin (*182b*).

Both dibromo- and monobromphakellin exhibit very mild antibacterial activity against *B. subtilis* and *E. coli* (*181*).

K. OXALINE AND ROQUEFORTINE

Oxaline (**143**) was isolated in 1974 by Nagel *et al.* (*183,184*) from cultures of the toxicogenic fungus *Penicillium oxalicum*. The compound may be classified as an indole alkaloid, but it is one of the three indole alkaloids known at present that also contains an imidazole substituent; the other indole alkaloids being roquefortine (**144**) and neoxaline (**145**). The structure of oxaline was deduced from physicochemical data and confirmed by single-crystal X-ray analysis. It has been suggested that in the biosynthesis of oxaline, nature makes use of the amino acids tryptophan and histidine (*184*).

Neoxaline (**145**) has been isolated from *Aspergillus japonicus* (*185*). It has

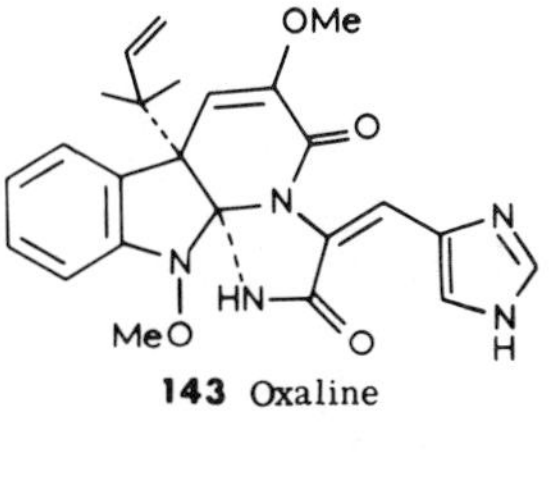

143 Oxaline

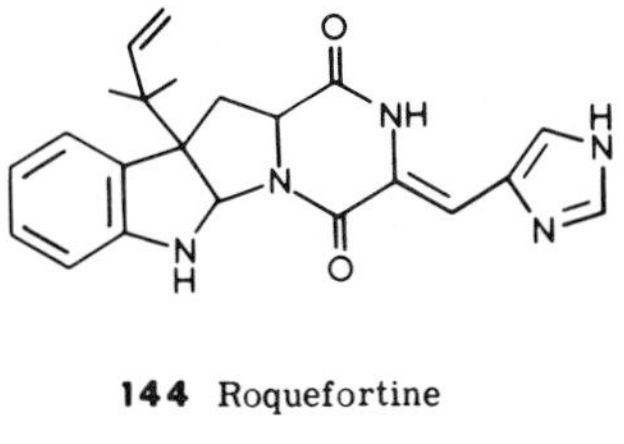

144 Roquefortine

a structural framework similar to that of oxaline, which was also isolated from another *Penicillium* species. For structure elucidation, neoxaline was correlated with oxaline through compound **146**, prepared as shown in Scheme 10.

In 1976 Scott *et al.* (*186*) isolated the neurotoxin roquefortine (**144**) from the fungus *Penicillium roqueforti*. Structure **144** was deduced on the basis of spectroscopic data and degradative products. The alkaloid appeared to be the same as roquefortine C, isolated by Ohmomo *et al.* in 1975 (*187*). The latter authors confirmed the structure by spectroscopic evidence (*188*); they also isolated roquefortine D. Reduction of roquefortine D with zinc in acetic acid yielded two dihydro derivatives. The properties of roquefortine C and of one of the isomers correlated very well. Thus, roquefortine D is dihydro-roquefortine C (*189*). The stereochemistry still remains to be solved.

The biosynthesis of roquefortine was investigated by feeding labeled mevalonic acid lactone, tryptophan, and histidine to *Penicillium roqueforti;* these compounds were incorporated (*190,191*). A biogenetic pathway for roquefortine obtained from Stilton cheese has been suggested (*191*).

The stereochemistry of the dehydrogenation of L-histidine in the biosynthesis of oxaline (**143**) and roquefortine (**144**) was deduced by feeding experiments (*192*). The dehydrogenation step in the biosynthesis of **143** and **144** involved syn elimination of H-2 and the pro-*S*-hydrogen from C-3 of L-histidine. The *E* configuration for the dehydrohistidine unit in both roquefortine and oxaline was confirmed by ^{13}C-NMR data. The acetate and mevalonic acid lactone incorporation in roquefortine was also investigated using high-field ^{13}C-NMR spectroscopy (*193*).

Biosynthetic investigations of roquefortine and indole metabolites have been reported with cultures of *P. commune* (*194*), *P. cyclopicum* (*195*), and *P. farinosum* (*196*).

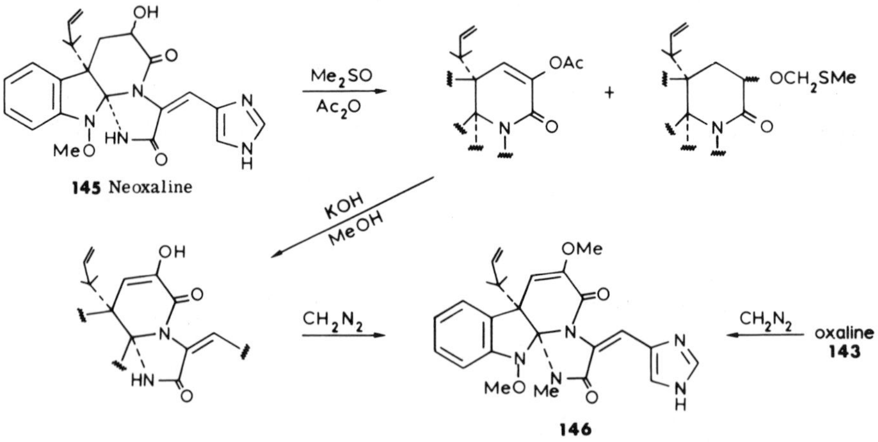

SCHEME 10. Correlation of neoxaline (**145**) with oxaline (**143**) via compound **146** (*185*).

L. ALKALOIDS OF *Parazoanthus* AND *Epizoanthus* SPECIES

The diimidazole alkaloid zoanthoxanthin (**147**) was isolated in 1973 by Cariello *et al.* (*197,198*), as a most important fluorescing pigment from the Mediterranean *Parazoanthus axinellae* (Zoanthidae), marine animals related to the sea anemones. The structure was determined with the aid of degradative reactions, spectroscopic methods, and single-crystal X-ray analysis, and proved to be 2-amino-3,4-dimethyl-6-dimethylamino-3*H*-1,3,5,7-tetrazacyclopent[*f*]azulene. The molecule appeared to be practically planar, with a slight boat conformation for the seven-membered ring and both imidazole rings bent in opposite directions (*198*). One may note the presence, once again, of the aminohistidine unit in zoanthoxanthin (**147**), which can formally be dissected and recombined in terms of two histidine-derived units as shown in Scheme 11 (*17*, p. 843).

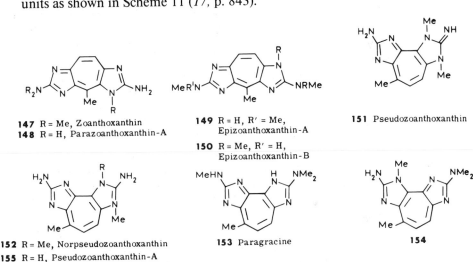

147 R = Me, Zoanthoxanthin
148 R = H, Parazoanthoxanthin-A

149 R = H, R′ = Me, Epizoanthoxanthin-A
150 R = Me, R′ = H, Epizoanthoxanthin-B

151 Pseudozoanthoxanthin

152 R = Me, Norpseudozoanthoxanthin
155 R = H, Pseudozoanthoxanthin-A

153 Paragracine

154

Other zoanthoxanthins have been found that differ in the number and position of the methyl group attached to the nitrogens (*199,200*). Parazoanthoxanthin A (**148**) is the least-substituted metabolite of the series.

From *Epizoanthus arenaceus,* another zoanthid closely related to *P. axinellae,* four new fluorescent pigments were isolated (*201*). Two possess the same skeleton as the previously described zoanthoxanthins. They are epizoanthoxanthin A (**149**) and B (**150**). The remaining two were characterized by an isomeric tetrazacyclopentazulene skeleton. The structures for the two compounds, pseudozoanthanoxanthin (**151**) and norpseudozoanthoxanthin (**152**), were proposed on the basis of chemical and spectroscopic evidence.

A diimidazole alkaloid with the same basic skeleton as **151** and **152** has been found as a biologically active and strongly fluorescent base in *Para-*

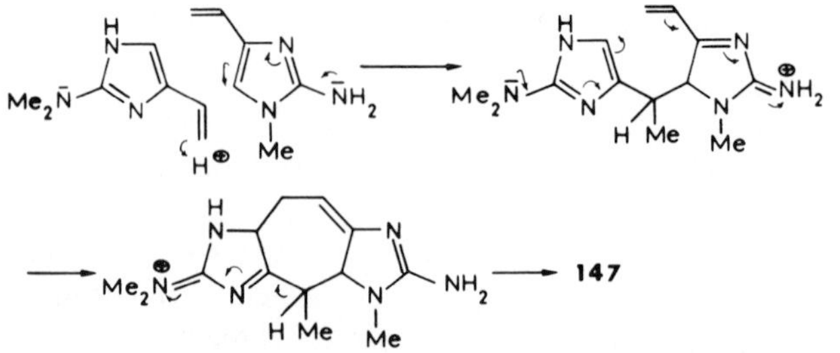

SCHEME 11. Biogenesis proposed for zoanthoxanthins (*17*, p. 843).

zoanthus gracilis Lwowsky (*202*). The structure of this compound, named paragracine (**153**), was determined to be 2-dimethylamino-6-methyl-8-methylamino-1*H*-1,3,7,9-tetrazacyclo[*e*]azulene. X-ray crystallographic analysis confirmed the structure (*203*). Several chemical reactions were carried out to investigate its reactivity (*203*). Recently, a similar metabolite was isolated from a *Parazonthus* species; its structure was found to be **154** from X-ray crystallographic analysis (*204*). It differs from the other compounds in the number and position of the *N*-methyl groups in the same manner as with the isomeric series. A survey on the distribution of zoanthoxanthins in some invertebrates has been given (*205*). Parazoanthoxanthin A (**148**) and pseudozoanthoxanthin A (**155**), the least-substituted zoanthoxanthins, have been synthesized by acid-catalyzed oxidative coupling of 2-amino-4(5)-(2-hydroxyethyl)imidazole, its dibenzoate, 2-benzoylamino-4(5)-vinylimidazole, and 2-amino-4(5)-(1-hydroxyethyl)imidazole. The ratio of **148** and **155** in these dimerizations depended mainly on the starting material used (*206*). The skeleton of **148** without the 4-methyl substituent has also been synthesized (*207*).

Schwartz *et al.* (*204*) remark that the trivial nomenclature of these pigments has become unnecessarily complex. They propose calling the linear system (as in **147–150**) zoanthoxanthins and the angular system (e.g. **151–155**) pseudozoanthoxanthins, with a prefix for the position of the saturated ring nitrogen.

The interaction of natural tetrazacyclopentazulenes with DNA, and their effects on the DNA and RNA polymerase reactions have been investigated (*208*). Paragracine (**153**) shows papaverine-like activity (*203*) and it selectively blocks sodium channels of squid axon membranes (*209*).

M. SPINACINE AND SPINACEAMINE

Spinacine (**156**) was isolated in 1936 by Ackermann and Mohr (*210*) from the liver of the shark *Acanthias vulgaris,* and in 1962 by Ackermann (*211*) from the crab-fish *Crango vulgaris.* The structure (**156**) was proved by a synthesis from histidine and formaldehyde (*211,212*).

156 Spinacine   **157** R = H, Spinaceamine
                       158 R = Me, 6-Methylspinaceamine

Spinaceamine (**157**) and 6-methylspinaceamine (**158**) were isolated in 1963 by Erspamer *et al.* (*22*) from the skin of an amphibian *Leptodactylus pentadactylus.* They also indicated a potential biogenesis of **156, 157,** and **158.** The structures proposed for spinaceamine and 6-methylspinaceamine were confirmed by NMR studies (*213*) and by single-crystal X-ray analysis (*214*). From X-ray diffraction studies, it was concluded that spinacine in the crystalline state occurs in the tautomeric form in which the imidazole nitrogen located farthest from the piperidine nitrogen is protonated. The absolute configuration at the C-5 position of spinacine has yet to be determined. However, because in the synthesis of spinacine successful use had been made of L-histidine as the starting material, it is assumed that **156** possesses the *S* configuration (*215*). Several preparations of spinaceamine and its derivatives have been described (*216–218*). Spinaceamine and 6-methylspinaceamine show antimicrobial activity (*219*).

From *Leptodactylus pentadactylus,* in addition to spinaceamine and 6-methylspinaceamine, the histidine-related bases histamine, N^{α}-methylhistamine, and N^{α},N^{α}-dimethylhistamine (*22*) were also isolated (Section II).

The acid–base properties of spinaceamine and spinacine and their complexing capacity with bivalent metals have been investigated (*220*).

N. Zooanemonin and Norzooanemonin

Zooanemonin (**159**) was isolated in 1933 by Kutscher and Ackermann (*221*) from the mussel *Arca noae.* Some 20 years later, Ackermann isolated zooanemonin from the sea anemone *Anemonia sulcata* (*Anthozoe*) (*222–224*). During that period, **159** was still called anemonin, for which structure **160** was assumed. It then transpired that the name anemonin had already been given to another compound and the name was changed to zooanemonin (*225*). In 1960 structure **160** was modified to **159,** following suggestions by Woodward who also reported to have synthesized zooanemonin (*226*).

Norzooanemonin (**161**) was isolated in 1973 by Weinheimer *et al.* (*227*), from the Caribbean bark coral *Pseudopterogorgia americana.* These authors confirmed structure **161** by synthesis. Methylation of imidazole-4-carboxylic acid with dimethyl sulfate yielded 85% norzooanemonin, identical to the natural product.

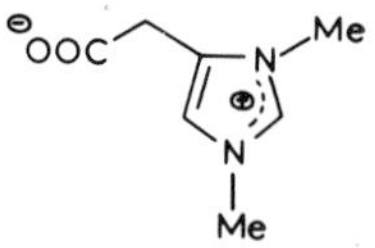

159 Zooanemonin 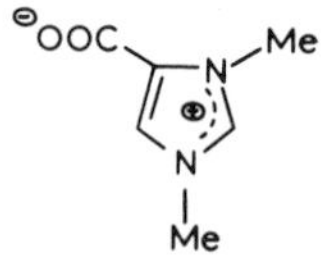   **161** Norzooanemonin 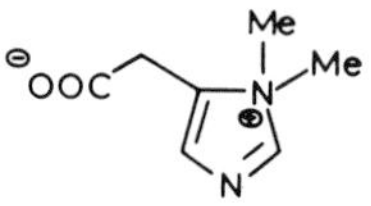   **160**

The hydrochloride of 1,3-dimethylimidazole-4-carboxylic acid (nor-zooanemonin) has also been isolated from the hydroid *Tubularia larynx*. Its structure was determined by X-ray analysis (*228*).

O. MISCELLANEOUS

Murexine (**162**) was isolated from *Murex brandaris, M. trunculus,* and *Tritonalia erinacea* (*229*). It has since been found in many widely distributed marine gastropod mollusks (*230*). N-Methylmurexine was reported to occur in the marine gastropod mollusk *Nucella emarginata* and was tentatively assigned structure **163** (*231*). A recent unambiguous synthesis of the two possible N-methyl derivatives of murexine (**163** and **164**) indicates that the natural compound is neither of these (*232*); therefore, further investigation is needed.

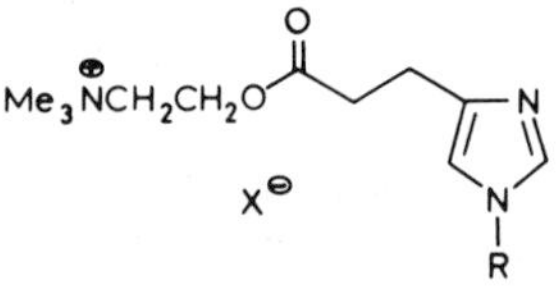

162 R = H, Murexine   **164**
163 R = Me

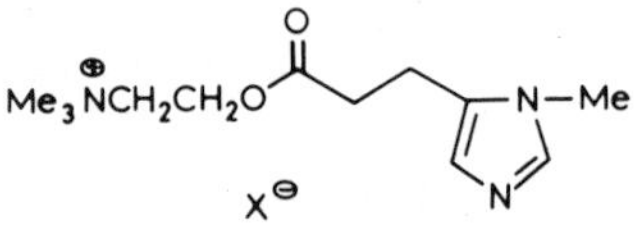

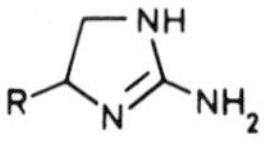

165 R = CH₂CHNH₂COOH, Enduracididine   **167** Monospermin
166 R = CH₂COOH

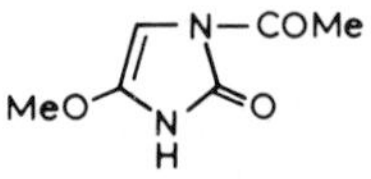

2-Aminoimidazole has been isolated from the seeds of *Mundulea sericea* (Willd.) A. Chev., a leguminous shrub from tropical Africa and Asia (*233*). The alanyl derivative of 2-aminoimidazole, enduracididine (**165**), and 2-[2-amino-2-imidazolin-4-yl]acetic acid have been obtained from the seeds of the legume *Lonchocarpus sericeus* (Poir) H.B. and K., native to the West Indies, tropical America, and West Africa (*234*).

From the seeds of *Butea monosperma* (Leguminaceae), the novel alkaloid monospermin (**167**) has been obtained (*235*). The structure of **167** was determined by spectroscopic methods.

TABLE I

IMIDAZOLE ALKALOIDS

Section	Alkaloid	Formula	mp (°C)	$[\alpha]_D$	Solvent	Ref.
II	Clithioneine	$C_{13}H_{22}N_4O_5S$	—	+44.2°	H_2O	27
	5-Mercapto-1-methyl-L-histidine (**4**)	$C_7H_{11}N_3O_2S$	—	—	—	28
	Disulfide of **4** (**5**)	$C_{14}H_{20}N_6O_4S_2$	202–205	+76°	0.1 N HCl	28
	N^α-trans-Cinnamoylhistamine (**6**)	$C_{14}H_{15}N_3O$	178–179	—	—	30
IV, B	Pilosine (**51**)	$C_{16}H_{18}N_2O_3$	171–172	+136.5°	EtOH	109
	Isopilosine (**50**)	$C_{16}H_{18}N_2O_3$	182–183	+37.6°	EtOH	109
	Epiisopilosine (**52**)	$C_{16}H_{18}N_2O_3$	179–180	−44.0°	EtOH	109
	(+)-Pilosinine (**32**)	$C_9H_{12}N_2O_2$	79–79.5	+13.4°	EtOH	69
	(±)-Pilosinine (**32**)	$C_9H_{12}N_2O_2$	69.5–70.5	—	—	69
IV, C	Epiisopiloturine (**57**)	$C_{16}H_{18}N_2O_3$	218–219	−11.0°	EtOH	117
	Epiisopiloturine · MeI	$C_{17}H_{21}IN_2O_3$	176–178	—	—	117
	Neopilosinine	$C_9H_{12}N_2O_2$	oil	—	—	117
	Neopilosinine picrate	$C_{15}H_{15}N_5O_9$	152–153	—	—	117
V, A	Alchornine (**73**)	$C_{11}H_{17}N_3O$	134–135	+74°	$CHCl_3$	127
	Alchornidine (**74a–b**)	$C_{16}H_{23}N_3O_2$	96–97	−18°	$CHCl_3$	127
	Isoalchornine (**75**)	$C_{11}H_{17}N_3O$	137–138	−84°	$CHCl_3$	127
	Alchorneine (**76**)	$C_{12}H_{19}N_3O$	43	−105°	$CHCl_3$	128
	Alchorneine · MeI	$C_{13}H_{22}IN_3O$	159 (dec.)	−28°	EtOH	128
	Alchorneine · MeBr	$C_{13}H_{22}BrN_3O$	193 (dec.)	−60°	EtOH	128
	Isoalchorneine (**78**)	$C_{13}H_{19}N_3O$	liq.	0°	$CHCl_3$	132
	Alchorneinone (**79**)	$C_{12}H_{20}N_2O_3$	liq.	−66°	$CHCl_3$	132
V, B	Casimiroedine (**81**)	$C_{17}H_{24}N_2O_5$	224–225	−36.5°	HCl	133
	Zapotidine (**85**)	$C_7H_9N_3S$	96–98	0°		139

(Continued)

TABLE I (*Continued*)

Section	Alkaloid	Formula	mp (°C)	[α]$_D$	Solvent	Ref.
V, C	Chaksine chloride (91)	C$_{11}$H$_{20}$ClN$_3$O$_2$	175	—	—	149
	Chaksine picrate	C$_{17}$H$_{22}$N$_6$O$_9$	236	—	—	149
	Isochaksine chloride	C$_{11}$H$_{20}$ClN$_3$O$_2$	248	—	—	149
	Isochaksine picrate	C$_{17}$H$_{22}$N$_6$O$_9$	182–184	—	—	149
V, D	Anantine (94)	C$_{15}$H$_{15}$N$_3$O	204	−549°		152
	Cynometrine (95)	C$_{16}$H$_{19}$N$_3$O$_2$	213	−30°		152
	Cynodine (96)	C$_{23}$H$_{23}$N$_3$O$_3$	155	+15°		152
	Isoanantine (98)	C$_{15}$H$_{15}$N$_3$O	190	−347°		155
	Isocynometrine (99)	C$_{16}$H$_{19}$N$_3$O$_2$	181	−66°		155
	Isocynodine (100)	C$_{23}$H$_{23}$N$_3$O$_3$	167	+139°		155
	Noranantine (101)	C$_{14}$H$_{13}$N$_3$O	205	−431°		155
	Hydroxyanantine (102)	C$_{15}$H$_{15}$N$_3$O$_2$	170	−433°		155
	Cynolujine (103)	C$_{16}$H$_{19}$N$_3$O	108	−6°		155
	Demethylcynometrine (104)	C$_{15}$H$_{17}$N$_3$O$_2$	204	−56.2°	CHCl$_3$	156
	Demethylcynodine (105)	C$_{22}$H$_{21}$N$_3$O$_3$	amorphous gum	—	—	156
V, E	Cypholophine (106)	C$_{18}$H$_{26}$N$_2$O$_3$	126–127	0°	CHCl$_3$	157
	O-Acetylcypholophine (107)	C$_{20}$H$_{28}$N$_2$O$_4$	gum	0°	CHCl$_3$	157
V, F	Octahydrodipyrido[1,2-*a*:1',2'-*c*]imidazol-10-ium bromide (110)	C$_{11}$H$_{17}$BrN$_2$	164–165	0°	CHCl$_3$	158
V, G	Dolichotheline (111)	C$_{10}$H$_{17}$N$_3$O	130–131	—	—	160
			131–132	—	—	167
	Dolichotheline acetate	C$_{12}$H$_{19}$N$_3$O$_2$	76–78	—	—	160
	Dolichotheline picrate	C$_{16}$H$_{20}$N$_6$O$_8$	150–152	—	—	160
			150–151	—	—	167
V, H	Glochidine (113)	C$_{15}$H$_{22}$N$_3$O	65–67	0°	CHCl$_3$	170
	Glochidine picrate	C$_{21}$H$_{26}$N$_6$O$_8$	143–144	—	—	171
	Glochidicine (114)	C$_{15}$H$_{23}$N$_3$O·½H$_2$O	103–105	0°	CHCl$_3$	170
			102–103	0°	CHCl$_3$	171

	Name	Formula	mp	[α]	Solvent	Ref.
	N^{α}-4'-Oxodecanoylhistamine (115)	$C_{15}H_{23}N_3O_2$	117–118	—	—	170
			115–117	—	—	171
V, I	Macrorine (116)	$C_{13}H_{11}N_3$	160	0°	$CHCl_3$	172
	Macrorine picrate	$C_{19}H_{14}N_6O_7$	180 (dec.)	—	—	172
	Macrorine perchlorate	$C_{13}H_{12}ClN_3O_4$	222	—	—	172
	Isomacrorine (117)	$C_{13}H_{11}N_3$	110	0°	$CHCl_3$	172
			105–107	—	—	173
	Isomacrorine perchlorate	$C_{13}H_{12}ClN_3O_4$	204–205.5	—	—	173
	Isomacrorine diperchlorate	$C_{13}H_{13}Cl_2N_3O_8$	290 (dec.)	—	—	172
			294–307	—	—	173
	Macrorungine (118)	$C_{13}H_{11}N_3O$	267–270 (dec.)	—	—	172
	Macrorungine picrate	$C_{19}H_{14}N_6O_8$	>250 (dec.)	—	—	172
	Macrorungine · HCl	$C_{13}H_{12}ClN_3O$	>290 (dec.)	—	—	172
	Macrorungine perchlorate	$C_{13}H_{12}ClN_3O_5 \cdot \frac{1}{2}H_2O$	>290 (dec.)	—	—	172
	Normacrorine (119)	$C_{12}H_9N_3$	156–157	—	—	174
	Normacrorine picrate	$C_{18}H_{12}N_6O_7$	211–212 (dec.)	—	—	174
	Longistrobine (120)	$C_{17}H_{19}N_3O_3$	145–148	—	—	175
	Isolongistrobine (121)	$C_{17}H_{19}N_3O_3$	132–136	—	—	175
			134–139	—	—	173, 177
	Dehydroisolongistrobine (122)	$C_{17}H_{17}N_3O_3$	131	—	—	175
			130.5–131.5	—	—	173, 176
	Anhydrolongistrobine	$C_{17}H_{17}N_3O_2$	218–222	—	—	175
V, J	Oroidin (135)	$C_{11}H_{11}Br_2N_5O$	—	—	—	178
	Dihydrooroidin (136)	$C_{11}H_{13}Br_2N_5O$	118–121	—	—	182
	Dibromophakellin (137)	$C_{11}H_{11}Br_2N_5O$	237–245 (dec.)	−203°	MeOH	180
	Dibromophakellin · HCl	$C_{11}H_{12}Br_2ClN_5O$	220–221	−205°	MeOH	181
	Dibromophakellin · MeOH	$C_{11}H_{11}Br_2N_5O\,MeOH$	237–245 (dec.)	—	—	181
	Monoacetyldibromophakellin	$C_{13}H_{13}Br_2N_5O_2$	245 (dec.)	−221°	MeOH	181
	4-Bromophakellin (142)	$C_{11}H_{12}BrN_5O$	170–180 (dec.)	—	—	180
	Monobromophakellin	$C_{11}H_{12}BrN_5O$	260–270 (dec.)	—	—	181
	Monobromophakellin · HCl	$C_{11}H_{13}BrClN_5O$	215–220	−123°	MeOH	181

(Continued)

TABLE I (Continued)

Section	Alkaloid	Formula	mp (°C)	[α]$_D$	Solvent	Ref.
V, K	Oxaline (143)	C$_{24}$H$_{25}$N$_5$O$_4$	220–221	−45°	MeOH	184
			230–232			183
	Roquefortine · MeOH (144)	C$_{22}$H$_{23}$N$_5$O$_2$,MeOH	195–200 (dec.)	−703°	MeOH	186
	Roquefortine C (144)	C$_{22}$H$_{23}$N$_5$O$_2$	225–228	−764°	pyridine	188
	Roquefortine D	C$_{22}$H$_{25}$N$_5$O$_2$	153–154	−370°	pyridine	189
	Neoxaline (145)	C$_{23}$H$_{25}$N$_5$O$_4$	202 (dec.)	−16.3°	CHCl$_3$	183
V, L	Zoanthoxanthin (147)	C$_{12}$H$_{16}$N$_6$	275–276 (dec.)	—	—	200
	Parazoanthoxanthin A (148)	C$_{10}$H$_{10}$N$_6$	>310	—	—	199
	Parazoanthoxanthin D	C$_{12}$H$_{14}$N$_6$	303–304 (dec.)	—	—	200
	Epizoanthoxanthin A (149)	C$_{13}$H$_{16}$N$_6$	191–192	—	—	201
	Epizoanthoxanthin B (150)	C$_{14}$H$_{18}$N$_6$	amorphous powder	—	—	201
	Pseudozoanthoxanthin (151)	C$_{12}$H$_{14}$N$_6$	>310	—	—	201
	3-Norpseudozoanthoxantin (152)	C$_{11}$H$_{12}$N$_6$	>230	—	—	201
	Paragracine (153)	C$_{13}$H$_{16}$N$_6$	258–262	—	—	203
	Paragracine dihydrobromide trihydrate	C$_{13}$H$_{24}$Br$_2$N$_6$O$_3$	280–282 (dec.)	—	—	203
	Pseudozoanthoxanthin A (155)	C$_{10}$H$_{10}$N$_6$	>310	—	—	206
V, M	Spinacine (156)	C$_7$H$_9$N$_3$O$_2$	265	—	—	236
			264	—	—	211
			263–265	—	—	213
	Spinaceamine · 2 HCl (157)	C$_6$H$_{11}$Cl$_2$N$_3$	277–279	—	—	213
	Spinaceamine dipicrate	C$_{18}$H$_{15}$N$_9$O$_{14}$	224–225 (dec.)	—	—	213
	6-Methylspinaceamine · 2 HCl (158)	C$_7$H$_{13}$Cl$_2$N$_3$	272–274	—	—	213
	6-Methylspinaceamine dipicrate	C$_{19}$H$_{17}$N$_9$O$_{14}$	230–231	—	—	213
V, N	Zooanemonin (159)	C$_7$H$_{10}$N$_2$O$_2$	178	—	—	223
	Zooanemonin · HCl	C$_7$H$_{11}$ClN$_2$O$_2$	180–184	—	—	223
	Norzooanemonin (161)	C$_6$H$_8$N$_2$O$_2$	260–263	—	—	227
	Norzooanemonin · HCl	C$_6$H$_9$ClN$_2$O$_2$	213–217	—	—	227
			216–219	—	—	228
V, O	Monospermin (167)	C$_6$H$_8$N$_2$O$_3$	161–163	0°	—	235

Acknowledgment

We greatly appreciate the help of Mr. T. S. Lie in checking the manuscript and references. We are indebted to Dr. A. Noordam and Mr. A. Steenks for preliminary overviews of imidazole alkaloids.

REFERENCES

1. A. R. Battersby and H. T. Openshaw, *Alkaloids* (*N.Y.*) **3**, 201–246 (1953).
2. V. A. Snieckus, *Alkaloids* (*London*) **1**, 456 (1971).
3. V. A. Snieckus, *Alkaloids* (*London*) **2**, 271–273 (1972).
4. J. Staunton, *Alkaloids* (*London*) **2**, 32 (1972).
5. V. A. Snieckus, *Alkaloids* (*London*) **3**, 301–302 (1973).
6. V. A. Snieckus, *Alkaloids* (*London*) **4**, 396–398 (1974).
7. V. A. Snieckus, *Alkaloids* (*London*) **5**, 265–269 (1975).
8. J. N. Reed and V. A. Snieckus, *Alkaloids* (*London*) **7**, 299–301 (1977).
9. J. R. Lewis, *Alkaloids* (*London*) **9**, 251–252 (1979).
10. J. R. Lewis, *Alkaloids* (*London*) **10**, 241–243, (1981).
11a. J. R. Lewis, *Alkaloids* (*London*) **11**, 238 (1981).
11b. J. R. Lewis, *Alkaloids* (*London*) **12**, 292 (1982).
12. G. A. Swan, "An Introduction to the Alkaloids," pp. 188–196. Blackwell, Oxford, 1967.
13. M. Luckner, *in* "Biosynthese der Alkaloide" (K. Mothes and H. R. Schütte, eds.), pp. 591–600. VEB Dtsch. Verlag Wiss., Berlin, 1969.
14. R. K. Hill, *in* "Chemistry of the Alkaloids" (S. W. Pelletier, ed.), pp. 424–429. Van Nostrand-Reinhold, Princeton, New Jersey, 1970.
15. W. Döpke, "Ergebnisse der Alkaloid-chemie," Vol. 1, 1960–1968, Part 2, pp. 1311–1318. Akademie-Verlag, Berlin, 1976.
16. M. Hesse, "Alkaloidchemie," p. 31. Thieme, Stuttgart, 1978.
17. G. A. Cordell, "Introduction to Alkaloids," pp. 833–845. Wiley, New York, 1981.
18. S. Jurisson, *Tartu Riikliku Ulik. Toim.* **270**, 71 (1971); *CA* **76**, 23018v (1972).
19. J. A. D. Jeffreys, *J. Chem. Soc. C.* 1091 (1970).
20. J. C. Madgwick, B. J. Ralph, J. S. Shannon, and J. J. Simes, *Arch. Biochem. Biophys.* **141**, 766 (1970).
21. M. Roseghini, R. Endean, and A. Temperilli, *Z. Naturforsch. C: Biosci.* **31**, 118 (1976).
22. V. Erspamer, T. Vitali, M. Roseghini, and J. M. Cei, *Experientia* **19**, 346 (1963).
23. R. T. Major and F. Dürsch, *J. Org. Chem.* **23**, 1564 (1958).
24. D. Ackermann, F. Holz, and H. Reinwein, *Z. Biol.* (*Munich*) **82**, 278 (1928).
25. V. F. German, *J. Pharm. Sci.* **60**, 495 (1971).
26a. H. Wagner and J. Grevel, *Planta Med.* **45**, 95 (1982).
26b. K. Konno, H. Shirahama, and T. Matsumoto, *Tetrahedron Lett.* **22**, 1617 (1981).
27. C. F. Huebner, *J. Am. Chem. Soc.* **73**, 4667 (1951).
28. A. Palumbo, M. d'Ischia, G. Misuraca, and G. Prota, *Tetrahedron Lett.* **23**, 3207 (1982).
29. H. C. Beyerman, L. Maat, and A. van Zon, *Recl. Trav. Chim. Pays-Bas* **91**, 246 (1972).
30. S. R. Johns, J. A. Lamberton, J. W. Loder, A. H. Redcliffe, and A. A. Sioumis, *Aust. J. Chem.* **22**, 1309 (1969).
31. C. Poupat and T. Sévenet, *Phytochemistry* **14**, 1881 (1975).
32. A. W. Gerrard, *Pharm. J.* **5**, 865 (1875).
33. B. Holmstedt, S. H. Wassén, and R. E. Schultes, *J. Ethnopharmacol.* **1**, 3 (1979).
34. F. Chemnitius, *J. Prakt. Chem.* **118**, 20 (1928).

328   L. MAAT AND H. C. BEYERMAN

35. V. Massa, P. Susplugas, and C. Taillade, *J. Pharm. Belg.* **28**, 69 (1973); *CA* **78**, 156626e (1973).
36. E. S. Vysotskaya, Yu. V. Shostenko, and S. Kh. Mushinskaya, *Tr. Voronezh. Gos. Univ.* **72**, 220 (1969); *CA* **77**, 156309h (1972).
37. E. Harnack and H. Meyer, *Justus Liebigs Ann. Chem.* **204**, 67 (1880).
38. H. A. D. Jowett, *J. Chem. Soc.* **77**, 473, 851 (1900).
39. A. Pinner and R. Schwarz, *Ber. Dtsch. Chem. Ges.* **35**, 192, 2441 (1902).
40. W. Langenbeck, *Ber. Dtsch. Chem. Ges.* **57**, 2072 (1924).
41. W. Langenbeck, *Ber. Dtsch. Chem. Ges.* **65**, 842 (1932).
42. N. A. Preobrazhenski, A. M. Poljakowa, and W. A. Preobrazhenski, *Ber. Dtsch. Chem. Ges.* **69**, 1835 (1936).
43. S. I. Zavyalov, *Dokl. Akad. Nauk SSSR* **82**, 257 (1952).
44. K. Nagarajan, C. Weissmann, H. Schmid, and P. Karrer, *Helv. Chim. Acta* **46**, 1212 (1963).
45. C. Bokhoven, J. C. Schoone, and J. M. Bijvoet, *Acta Crystallogr.* **4**, 275 (1951).
46. A. F. Peerdeman, *Acta Crystallogr.* **9**, 824 (1956).
47. R. K. Hill and S. Barcza, *Tetrahedron* **22**, 2889 (1966).
48. S. Fregerslev and S. E. Rasmussen, *Acta Chem. Scand.* **22**, 2541 (1968).
49. J. Haase and E. Kussäther, *Z. Naturforsch. B: Anorg. Chem., Org. Chem., Biochem., Biophys., Biol.* **27B**, 212 (1972).
50. A. V. Chumachenko, M. E. Maurit, A. D. Treboganov, G. V. Smirnova, R. B. Teplinskaya, L. V. Vokova, E. N. Zvonkova, and N. A. Preobrazhenskii, *Dokl. Akad. Nauk SSSR* **178**, 1352 (1968).
51. A. V. Chumachenko, E. N. Zvonkova, and N. A. Preobrazhenskii, *J. Org. Chem. USSR (Engl. Transl.)* **5**, 571 (1969).
52. T. D. Inch and G. J. Lewis, *Carbohydr. Res.* **22**, 91 (1972).
53. S. Kang, *Int. J. Quantum Chem., Quantum Biol. Symp.* **1**, 109 (1974).
54. W. Döpke and G. d'Heureuse, *Tetrahedron Lett.* 1807 (1968).
55. M. A. Nunes, Ph.D. Dissertation, University of California, San Francisco, California 1974; *Diss. Abstr. Int. B* **35**, 748 (1974).
56. M. A. Nunes and E. Brochmann-Hanssen, *J. Pharm. Sci.* **63**, 716 (1974).
57. J. A. Ryan, *Anal. Chim. Acta* **85**, 89 (1976).
58. J. D. Weber, *J. Assoc. Off. Anal. Chem.* **59**, 1409 (1976).
59. G. A. Neville, F. B. Hasan, and I. C. P. Smith, *Can. J. Chem.* **54**, 2094 (1976).
60. K. Baeschlin, J. C. Etter, and H. Moll, *Pharm. Acta Helv.* **44**, 301 (1969).
61. P.-H. Chung, T.-F. Chin, and J. L. Lach, *J. Pharm. Sci.* **59**, 1300 (1970).
62. R. Tulus and G. Iskender, *Istanbul Univ. Eczacilik Fak. Mecm.* **5**, 130, (1969); *CA* **73**, 69887u (1970).
63. A. S. C. Wan, *J. Chromatogr.* **60**, 371 (1971).
64. S. Ebel, W. D. Mikulla, and K. H. Weisel, *Dtsch. Apoth.-Ztg.* **111**, 931 (1971); *CA* **75**, 80317v (1971).
65. J. Kalman, K. Toth, and D. Kuttel, *Acta Pharm. Hung.* **41**, 267 (1971); *CA* **76**, 50003t (1972).
66. V. Massa, F. Gal, P. Susplugas, and G. Maestre, *Trav. Soc. Pharm. Montpellier* **30**, 267 (1970); *CA* **75**, 25455p (1971).
67. W. F. Bayne, L.-C. Chu, and F. T. Tao, *J. Pharm. Sci.* **65**, 1724 (1976).
68. S. W. Dziedzic, S. E. Gitlow, and D. L. Krohn, *J. Pharm. Sci.* **65**, 1262 (1976).
69. H. Link and K. Bernauer, *Helv. Chim. Acta* **55**, 1053 (1972).
70. M. S. Karawya and M. G. Ghourab, *J. Assoc. Off. Anal. Chem.* **55**, 1180 (1972).
71. E. Ermer, *Pharm. Ztg.* **120**, 1771 (1975).

72. G. A. Neville, F. B. Hasan, and I. C. P. Smith, *J. Pharm. Sci.* **66**, 638 (1977).
73. B. S. Scott, D. L. Dunn, and E. D. Dorsey, *J. Pharm. Sci.* **70**, 1046 (1981).
74. S. El-Masry and R. Soliman, *J. Assoc. Off. Anal. Chem.* **63**, 689 (1980).
75. E. Smith, *J. Assoc. Off. Anal. Chem.* **55**, 248 (1972).
76. T. Urbányi, A. Piedmont, E. Willis, and G. Manning, *J. Pharm. Sci.* **65**, 257 (1976).
77. J. I. DeGraw, J. S. Engström, and E. Willis, *J. Pharm. Sci.* **64**, 1700 (1975).
78. A. Noordam, K. Waliszewski, C. Olieman, L. Maat, and H. C. Beyerman, *J. Chromatogr.* **153**, 271 (1978).
79. A. Noordam, L. Maat, and H. C. Beyerman, *J. Pharm. Sci.* **70**, 96 (1981).
80. J. M. Kennedy and P. E. McNamara, *J. Chromatogr.* **212**, 331 (1981).
81. J. J. O'Donnell, R. Sandman, and M. V. Drake, *J. Pharm. Sci.* **69**, 1096 (1980).
82. H. Y. Aboul-Enein and R. F. Borne, *Chem. Biomed. Environ. Instrum.* **10**, 231 (1980).
83. G. C. F. Clark, G. J. Moody, and J. D. R. Thomas, *Anal. Chim. Acta* **98**, 215 (1978).
84. G. Canti, A. Scozzafava, G. Ciciani, and G. Renzi, *J. Pharm. Sci.* **69**, 1220 (1980).
85. J. I. DeGraw, *Tetrahedron* **28**, 967 (1972).
86. A. V. Chumachenko, E. N. Zvonkova, and R. P. Evstigneeva, *J. Org. Chem. USSR (Engl. Transl.)* **8**, 1112 (1972).
87. N. A. Preobrazhenski, A. F. Wompe, W. A. Preobrazhenski, and M. N. Schutschukina, *Ber. Dtsch. Chem. Ges.* **66**, 1536 (1933).
88. A. Noordam, Ph.D. Dissertation, Delft University of Technology, Delft (1979).
89. A. Noordam, L. Maat, and H. C. Beyerman, *Recl. Trav. Chim. Pays-Bas* **98**, 467 (1979).
90. A. Noordam, L. Maat, and H. C. Beyerman, *Recl. Trav. Chim. Pays-Bas* **100**, 441 (1981).
91. A. Noordam, L. Maat, and H. C. Beyerman, *Recl. Trav. Chim. Pays-Bas* **97**, 293 (1978).
92. H. A. D. Jowett, *J. Chem. Soc.* **87**, 794 (1905).
93. H. C. Beyerman, A. W. Buijen van Weelderen, L. Maat, and A. Noordam, *Recl. Trav. Chim. Pays-Bas* **96**, 191 (1977).
94. H. C. Beyerman, L. Maat, A. Noordam, and A. van Zon, *Recl. Trav. Chim. Pays-Bas* **96**, 222 (1977).
95. L. Maat, H. C. Beyerman, and A. Noordam, *Tetrahedron* **35**, 273 (1979).
96. F. R. Stermitz and H. Rapoport, *J. Am. Chem. Soc.* **83**, 4045 (1961); M. L. Louden and E. Leete, *ibid.* **84**, 4507 (1962).
97. H.-G. Boit, "Ergebnisse der Alkaloid-Chemie bis 1960," p. 753. Akademie-Verlag, Berlin, 1961.
98. E. Leete, *in* "Biogenesis of Natural Compounds" (P. Bernfeld, ed.), p. 791. Pergamon, Oxford, 1963.
99. E. Leete, H. Gregory, and E. G. Gros, *J. Am. Chem. Soc.* **87**, 3475 (1965).
100. E. Brochmann-Hanssen, M. A. Nunes, and C. K. Olah, *Planta Med.* **28**, 1 (1975).
101. J. M. van Rossum, M. J. W. J. Cornelissen, C. T. P. de Groot, and J. A. T. M. Hurkmans, *Experientia* **16**, 373 (1960).
102. H. Y. Aboul-Enein and A. A. Al-Badr, *Methods Find. Exp. Clin. Pharmacol.* **4**, 321 (1982).
103. P. Taylor, *in* "The Pharmacological Basis of Therapeutics" (A. G. Gilman, L. S. Goodman, and A. Gilman, eds.), 6th ed., p. 96, Macmillan, New York, 1980.
104. H. O. Schild, *J. Physiol. (London)* **153**, 26 (1960).
105. I. Hanin, D. J. Jenden, and A. K. Cho, *Mol. Pharmacol.* **2**, 325 (1966).
106. J. M. van Rossum, *Arch. Int. Pharmacodyn. Ther.* **140**, 592 (1967).
107. "British National Formulary 1981," No. 1, p. 265. British Medical Association and The Pharmaceutical Society of Great Britain, Pharmaceutical Press, London, 1981.
108. H.-W. Voigtländer and W. Rosenberg, *Arch. Pharm. (Weinheim, Ger.)* **292**, 579 (1959).
109. W. Löwe and K.-H. Pook, *Justus Liebigs Ann. Chem.* 1476 (1973).

110. W. E. Oberhänsli, *Cryst. Struct. Commun.* **1**, 203 (1972).
111. J. K. Mehrotra and A. N. Dey, *J. Indian Chem. Soc.* **38**, 971 (1961).
112. E. Tedeschi, J. Kamionsky, D. Zeider, S. Fackler, S. Sarel, V. Usieli, and J. Deutsch, *J. Org. Chem.* **39**, 1864 (1974).
113. H. Link, K. Bernauer, and W. E. Oberhänsli, *Helv. Chim. Acta* **57**, 2199 (1974).
114. S. Sarel, V. Usieli, and E. Tedeschi, *Tetrahedron Lett.* 97 (1975).
115. E. Tedeschi, J. Kamionsky, S. Fackler, and S. Sarel, *Isr. J. Chem.* **11**, 731 (1973).
116. P. Pfeffer (Plantex Ltd.), *S. Afr. Pat.* **67/03,807** (1967); *CA* **70**, 31682y (1969).
117. H.-W. Voigtländer, G. Balsam, M. Engelhardt, and L. Pohl, *Arch. Pharm. (Weinheim, Ger.)* **311**, 927 (1978).
118. A. Pinner, *Ber. Dtsch. Chem. Ges.* **38**, 2560 (1905).
119. M. Polonovski and M. Polonovski, *Bull. Soc. Chim. Fr.* [4] **31**, 1204 (1922).
120. H. Y. Aboul-Enein, *Acta Pharm. Suec.* **11**, 387 (1974).
121. W. Döpke and U. Mücke, *Z. Chem.* **13**, 177 (1973).
122. R. T. Koda, F. J. Dea, K. Fung, C. Elison, and J. A. Biles, *J. Pharm. Sci.* **62**, 2021 (1973).
123. R. F. Borne, H. Y. Aboul-Enein, I. W. Waters, and J. Hicks, *J. Med. Chem.* **16**, 245 (1973).
124. H. Y. Aboul-Enein, A. A. Al-Badr, S. E. Ibrahim, and M. Ismail, *Pharm. Acta Helv.* **55**, 278 (1980).
125. R. F. Borne and H. Y. Aboul-Enein, *J. Heterocycl. Chem.* **17**, 1609 (1980).
126. N. K. Hart, S. R. Johns, and J. A. Lamberton, *J. Chem. Soc., Chem. Commun.* 1484 (1969).
127. N. K. Hart, S. R. Johns, J. A. Lamberton, and R. I. Willing, *Aust. J. Chem.* **23**, 1679 (1970).
128. F. Khuong-Huu, J. P. Leforestier, G. Maillard, and R. Goutarel, *C. R. Hebd. Seances Acad. Sci., Ser. C* **270**, 2070 (1970).
129. M. Cesario and J. Guilhem, *C. R. Hebd. Seances Acad. Sci., Ser. C* **271**, 1552 (1970).
130. M. Cesario and J. Guilhem, *Acta Crystallogr., Sect. B* **B28**, 151 (1972).
131. R. Goutarel and F. Khuong-Huu-Laine (Agence Nationale de Valorisation de la Recherche, Anvar), *Fr. Demande* **2,087,982** (1972); *CA* **77**, 88757a (1972).
132. F. Khuong-Huu, J.-P. le Forestier, and R. Goutarel, *Tetrahedron* **28**, 5207 (1972).
133. F. B. Power and T. Callan, *J. Chem. Soc.* **99**, 1993 (1911).
134. C. Djerassi, J. Herrán, H. N. Khastgir, B. Riniker, and J. Romo, *J. Org. Chem.* **21**, 1510 (1956).
135. C. Djerassi, C. Bankiewicz, A. L. Kapoor, and B. Riniker, *Tetrahedron* **2**, 168a (1958).
136. S. Raman, J. Reddy, W. N. Lipscomb, A. L. Kapoor, and C. Djerassi, *Tetrahedron Lett.* 357 (1962).
137. R. P. Panzica and L. B. Townsend, *J. Am. Chem. Soc.* **95**, 8737 (1973).
138. F. A. Kincl, J. Romo, G. Rosenkranz, and F. Sondheimer, *J. Chem. Soc.* 4163 (1956).
139. R. Mechoulam, F. Sondheimer, A. Melera, and F. A. Kincl, *J. Am. Chem. Soc.* **83**, 2022 (1961).
140. R. Mechoulam and A. Hirshfeld, *Tetrahedron* **23**, 239 (1967).
141. D. Ackermann and G. Hoppe-Seyler, *Hoppe-Seyler's Z. Physiol. Chem.* **336**, 283 (1964).
142. A. Kjaer, R. Gmelin, and R. B. Jensen, *Acta Chem. Scand.* **10**, 432 (1956).
143. S. Siddiqui and Z. Ahmad, *Proc. — Indian Acad. Sci. Sect. A* **2A**, 421 (1935).
144. K. Wiesner, Z. Valenta, B. S. Hurlbert, F. Bickelhaupt, and L. R. Fowler, *J. Am. Chem. Soc.* **80**, 1521 (1958).
145. G. Singh, G. V. Nair, K. P. Aggarwal, and S. S. Saksena, *J. Sci. Ind. Res., Sect. B* **17**, 332 (1958).
146. L. R. Fowler, Z. Valenta, and K. Wiesner, *Chem. Ind. (London)* 95 (1962).

147. S. C. Biswas and S. K. Talapatra, *Indian J. Phys.* **40**, 492 (1966).
148. S. Siddiqui and M. Hasan, *Pak. J. Sci. Ind. Res.* **8**, 73 (1965).
149. R. V. Krishna Rao, J. V. L. N. Seshagiri Rao, and M. Vimaladevi, *J. Nat. Prod.* **42**, 299 (1979).
150. A. Qayum, K. Khanum, M. Ahmad, and S. Babar, *Pak. J. Sci. Ind. Res.* **12**, 378 (1970).
151. A. Qayum, K. Khanum, and G. A. Miana, *Pak. Med. Forum* **6**, 35 (1971); CA **77**, 148526m (1972).
152. F. Khuong-Huu, X. Monseur, G. Ratle, G. Lukacs, and R. Goutarel, *Tetrahedron Lett.* 1757 (1973).
153. L. Tchissambou, M. Bénéchie, and F. Khuong-Huu, *Tetrahedron Lett.* 1801 (1978).
154. A. Chiaroni, C. Riche, L. Tchissambou, and F. Khuong-Huu, *J. Chem. Res., Synop.* 182 (1981).
155. L. Tchissambou, M. Bénéchie, and F. Khuong-Huu, *Tetrahedron* **38**, 2687 (1982).
156. P. G. Waterman and D. F. Faulkner, *Phytochemistry* **20**, 2765 (1981).
157. N. K. Hart, S. R. Johns, J. A. Lamberton, J. W. Loder, and R. H. Nearn, *J. Chem. Soc., Chem. Commun.* 441 (1970); *Aust. J. Chem.* **24**, 857 (1971).
158. K. Leander and B. Lüning, *Tetrahedron Lett.* 905 (1968).
159. E. Söderberg and P. Kierkegaard, *Acta Chem. Scand.* **24**, 397 (1970).
160. H. Rosenberg and A. G. Paul, *Tetrahedron Lett.* 1039 (1969).
161. J. J. Dingerdissen and J. L. McLaughlin, *J. Pharm. Sci.* **62**, 1663 (1973).
162. H. Rosenberg and A. G. Paul, *Phytochemistry* **9**, 655 (1970).
163. H. Rosenberg and A. G. Paul, *Lloydia* **34**, 372 (1971).
164. H. Rosenberg and S. J. Stohs, *Lloydia* **37**, 313 (1974).
165. H. Rosenberg and S. J. Stohs, *Phytochemistry* **15**, 501 (1976).
166. D. G. O'Donovan and T. J. Forde, *Tetrahedron Lett.* 3637 (1970).
167. H. Horan and D. G. O'Donovan, *J. Chem. Soc. C* 2083 (1971).
168. H. Rosenberg and A. G. Paul, *J. Pharm. Sci.* **62**, 403 (1973).
169. H. Rosenberg, S. J. Stohs, and A. G. Paul, *Phytochemistry* **13**, 823 (1974).
170. S. R. Johns and J. A. Lamberton, *J. Chem. Soc., Chem. Commun.* 312 (1966).
171. S. R. Johns and J. A. Lamberton, *Aust. J. Chem.* **20**, 555 (1967).
172. R. R. Arndt, A. Jordaan, and V. P. Joynt, *J. Chem. Soc.* 5969 (1964).
173. M. A. Wuonola and R. B. Woodward, *Tetrahedron* **32**, 1085 (1976).
174. A. Jordaan, V. P. Joynt, and R. R. Arndt, *J. Chem. Soc.* 3001 (1965).
175. R. R. Arndt, S. H. Eggers, and A. Jordaan, *Tetrahedron* **25**, 2767 (1969).
176. M. A. Wuonola and R. B. Woodward, *J. Am. Chem. Soc.* **95**, 284 (1973).
177. M. A. Wuonola and R. B. Woodward, *J. Am. Chem. Soc.* **95**, 5098 (1973).
178. S. Forenza, L. Minale, R. Riccio, and E. Fattorusso, *J. Chem. Soc., Chem. Commun.* 1129 (1971).
179. E. E. Garcia, L. E. Benjamin, and R. I. Fryer, *J. Chem. Soc., Chem. Commun.* 78 (1973).
180. G. M. Sharma and P. R. Burkholder, *J. Chem. Soc., Chem. Commun.* 151 (1971).
181. G. Sharma and B. Magdoff-Fairchild, *J. Org. Chem.* **42**, 4118 (1977).
182a. L. H. Foley and G. Büchi, *J. Am. Chem. Soc.* **104**, 1776 (1982).
182b. C. A. Mattia, L. Mazzarella, and R. Puliti, *Acta Crystallogr., Sect. B* **B38**, 2513 (1982).
183. D. W. Nagel, K. G. R. Pachler, P. S. Steyn, P. L. Wessels, G. Gafner, and G. J. Kruger, *J. Chem. Soc., Chem. Commun.* 1021 (1974).
184. D. W. Nagel, K. G. R. Pachler, P. S. Steyn, R. Vleggaar, and P. L. Wessels, *Tetrahedron* **32**, 2625 (1976).
185. Y. Konda, M. Onda, A. Hirano, and S. Ōmura, *Chem. Pharm. Bull.* **28**, 2987 (1980).
186. P. M. Scott, M. K. A. Merrien, and J. Polonsky, *Experientia* **32**, 140 (1976).
187. S. Ohmomo, T. Sato, T. Utagawa, and M. Abe, *Agric. Biol. Chem.* **39**, 1333 (1975).

188. S. Ohmomo, T. Utagawa, and M. Abe, *Agric. Biol. Chem.* **41**, 2097 (1977).
189. S. Ohmomo, K. Oguma, T. Ohashi, and M. Abe, *Agric. Biol. Chem.* **42**, 2387 (1978).
190. S. Ohmomo, T. Ōhashi, and M. Abe, *Agric. Biol. Chem.* **43**, 2035 (1979).
191. K. D. Barrow, P. W. Colley, and D. E. Tribe, *J. Chem. Soc., Chem. Commun.* 225 (1979).
192. R. Vleggaar and P. L. Wessels, *J. Chem. Soc., Chem. Commun.* 160 (1980).
193. C. P. Gorstallman, P. S. Steyn, and R. Vleggaar, *J. Chem. Soc., Chem. Commun.* 652 (1982).
194. R. E. Wagener, N. D. Davis, and U. L. Diener, *Appl. Environ. Microbiol.* **39**, 882 (1980).
195. R. F. Vesonder, L. Tjarks, W. Rohwedder, and D. O. Kieswetter, *Experientia* **36**, 1308 (1980).
196. A. G. Kozlovsky, T. F. Solovieva, T. A. Reshetilova, and G. K. Skryabin, *Experientia* **37**, 472 (1981).
197. L. Cariello, S. Crescenzi, G. Prota, F. Giordano, and L. Mazzarella, *J. Chem. Soc., Chem. Commun.* 99 (1973).
198. L. Cariello, S. Crescenzi, G. Prota, S. Capasso, F. Giordano, and L. Mazzarella, *Tetrahedron* **30**, 3281 (1974).
199. L. Cariello, S. Crescenzi, G. Prota, and L. Zanetti, *Experientia* **30**, 849 (1974).
200. L. Cariello, S. Crescenzi, G. Prota, and L. Zanetti, *Tetrahedron* **30**, 3611 (1974).
201. L. Cariello, S. Crescenzi, G. Prota, and L. Zanetti, *Tetrahedron* **30**, 4191 (1974).
202. Y. Komoda, S. Kaneto, M. Yamamoto, M. Ishikawa, A. Itai, and Y. Itata, *Chem. Pharm. Bull.* **23**, 2464 (1975).
203. Y. Komoda, M. Shimizu, S. Kaneko, M. Yamamoto, and M. Ishikawa, *Chem. Pharm. Bull.* **30**, 502 (1982).
204. R. E. Schwartz, M. B. Yunker, P. J. Schreuer, and T. Ottersen, *Tetrahedron Lett.* 2235 (1978).
205. L. Cariello, S. Crescenzi, L. Zanetti, and G. Prota, *Comp. Biochem. Physiol. B* **63**, 77 (1979).
206. M. Braun and G. Büchi, *J. Am. Chem. Soc.* **98**, 3049 (1976).
207. M. Yasunami, Y. Sasagawa, and K. Takase, *Chem. Lett.* 205 (1980).
208. F. Quadrifoglio, V. Crescenzi, G. Prota, L. Cariello, A. di Marco, and F. Zunino, *Chem.-Biol. Interact.* **11**, 91 (1975).
209. I. Seyama, C. H. Wu, and T. Narahashi, *Biophys. J.* **29**, 531 (1980).
210. D. Ackermann and M. Mohr, *Z. Biol.* (*Munich*) **98**, 37 (1937).
211. D. Ackermann, *Hoppe-Seyler's Z. Physiol. Chem.* **328**, 275 (1962).
212. O. G. Eilazyan and Y. M. Yutilov, *Khim. Prom-st., Ser.: Reakt. Osobo Chist. Veshchestva* 30 (1981); *CA* **96**, 85467n (1982).
213. T. Vitali and G. Bertaccini, *Gazz. Chim. Ital.* **94**, 296 (1964).
214. M. Nardelli, T. Vitali, and F. Mossini, *Ric. Sci., Parte 2: Sez. A* **7**, 718 (1964).
215. G. D. Andreetti, L. Cavalca, and P. Sgarabotto, *Gazz. Chim. Ital.* **101**, 625 (1971).
216. T. Vitali, F. Mossini, and G. Bertaccini, *Farmaco, Ed. Sci.* **22**, 821 (1967); *CA* **68**, 87234e (1968).
217. G. G. Habermehl and W. Ecsy, *Heterocycles* **5**, 127 (1976).
218. Y. M. Yutilov and O. G. Eilazyan, *Khim. Geterotsikl. Soedin.* 992 (1981); *CA* **95**, 169075a (1981).
219. H. J. Preusser, G. Habermehl, M. Sablofski, and D. Schmann-Haury, *Toxicon* **13**, 285 (1975); *CA* **83**, 189060x (1975).
220. A. Braibanti, F. Dallavalle, E. Leporati, and G. Mori, *J. Chem. Soc., Dalton Trans.* 323 (1973).
221. F. Kutscher and D. Ackermann, *Hoppe-Seyler's Z. Physiol. Chem.* **221**, 38 (1933).
222. D. Ackermann, *Hoppe-Seyler's Z. Physiol. Chem.* **294**, 1 (1953); cited in Ref. 223, but probably equal to Ref. 224.

223. D. Ackermann and R. Janka, *Hoppe-Zeyler's Z. Physiol. Chem.* **294,** 93 (1953).
224. D. Ackermann, *Hoppe-Seyler's Z. Physiol. Chem.* **295,** 1 (1953).
225. D. Ackermann, *Hoppe-Seyler's Z. Physiol. Chem.* **296,** 286 (1959).
226. D. Ackermann and P. H. List, *Hoppe-Seyler's Z. Physiol. Chem.* **318,** 281 (1960).
227. A. J. Weinheimer, E. K. Metzner, and M. L. Mole, Jr., *Tetrahedron* **29,** 3135 (1973).
228. K. C. Gupta, R. L. Miller, J. R. Williams, and J. F. Blount, *Experientia* **33,** 1556 (1977).
229. V. Erspamer and F. Dordoni, *Arch. Int. Pharmacodyn. Ther.* **74,** 263 (1947); *CA* **43,** 1491b (1949).
230. J. T. Baker and V. Murphy, "CRC Handbook of Marine Science, Compounds from Marine Organisms," Vol. 1, p. 96. CRC Press, Cleveland, Ohio, 1976.
231. J. A. Bender *et al., Comp. Gen. Pharmacol.* **5,** 191 (1974); cited in Ref. 232.
232. C. C. Duke, J. V. Eichholzer, and J. K. MacLeod, *Tetrahedron Lett.* 5047 (1978).
233. L. E. Fellows, E. A. Bell, and G. S. King, *Phytochemistry* **16,** 1399 (1977).
234. L. E. Fellows, R. C. Hider, and E. A. Bell, *Phytochemistry* **16,** 1957 (1977).
235. B. Mehta and M. M. Bokadia, *Chem. Ind.* (*London*) 98 (1981).
236. D. A. Ackermann and S. Kraup, *Hoppe Seyler's Z. Physiol. Chem.* **284,** 129 (1949).

INDEX

A

Acanthias vulgaris alkaloids, 320, 321
Acarnidines, 130–132, 174
Acarnus erithacus, acarnidines, 130
O-Acetylcypholophine, 309, 310, 324
N^1-Acetyl-N^1-deoxymayfoline, 122, 125, 159
2-Acyl-1-benzylideneisoquinolines,
    photocyclization, 258–260
N-Acylenamines, *see* Enamides
Aerothionin, isolation and structure, 96, 97,
    151
Agelas oroides alkaloids, 316, 317
Agmatine, 93, 94
    derivatives, 93–95
Agrobacterium tumefaciens, spermidine
    alkaloids, 97
Agrobactine, 97, 99, 160
Alamarine
    isolation, 21
    occurrence, 5
    structure, 4, 22
    synthesis, 209, 233, 240
Alangamide
    occurrence, 5
    structure, 2
Alangicine
    occurrence, 5
    structure, 2, 15, 16
    synthesis, 15, 20
Alangimarckine
    dehydrogenation, 21
    occurrence, 5
    structure, 3, 20, 21
    synthesis, 20
Alangimaridine
    isolation, 21, 240
    occurrence, 5
    structure, 4
Alangimarine
    isolation, 21
    occurrence, 5

structure, 4
synthesis, 240
Alangiside
    isolation, 21
    occurrence, 5
    structure, 3
Alangium lamarckii alkaloids, 1–50
Alangium vitiense, alkaloid $C_{28}H_{35}N_3O_3$, 5, 32
Alchornea alkaloids, 302–304
Alchorneine, 303, 323
Alchorneinone, 303, 304, 323
Alchornidine, 302, 303, 323
Alchornine, 302, 303, 323
Alkaloid LBX, 160
Alkaloid LBY, *see* Lunarinol II
Alkaloid LBZ, 161
Alloyohimban, synthesis, 242–244
Alloyohimbone, synthesis, 242–244
Amaryllidaceae alkaloids, synthesis, 210–214
Amines, naturally occurring, 88, 89
2-[2-Amino-2-imidazolin-4-yl]acetic acid, 322
Anabasine, 4, 5
Anantine, 307–309, 324
Ancanthella auzantiaca alkaloid, 317
Anemonia sulcata alkaloids, 321
Angustidine
    isolation, 243, 247
    synthesis, 249
Angustidine thia analog, synthesis, 249
Angustine, isolation, 247
Angustoline
    isolation, 243, 247
    synthesis, 249
Angustoline thia analog, synthesis, 249
Anhydrocannabisativine, 113, 161
Anhydrolongistrobine, 325
Anhydrolycorine, synthesis, 210–212
Ankorine
    occurrence, 5
    stereochemistry, 8, 9
    structure, 3, 7
    synthesis, 7–9

DATE DUE			
Chemistry Dept			

Manske   195938